# NANO, QUANTUM AND MOLECULAR COMPUTING

# Nano, Quantum and Molecular Computing

## Implications to High Level Design and Validation

Edited by

**Sandeep K. Shukla**
*Virginia Polytechnic and State University, Blacksburg, U.S.A.*

and

**R. Iris Bahar**
*Brown University, Providence, U.S.A.*

**KLUWER ACADEMIC PUBLISHERS**
NEW YORK, BOSTON, DORDRECHT, LONDON, MOSCOW

eBook ISBN: 1-4020-8068-9
Print ISBN: 1-4020-8067-0



**Shukla & Bahar: Nano, Quantum and Molecular Computing**



First Indian Reprint 2010
Second Indian Reprint 2014

ISBN 978-81-8489-371-7



This edition is published by Springer (India) Private Limited, (a part of Springer Science+Business Media), Registered Office: 7th Floor, Vijaya Building, 17 Barakhamba Road, New Delhi - 110 001, India

Printed in India by Chennai Micro Print Pvt. Ltd., Chennai

***Sandeep dedicates this book to his mother Atashi, grandmother Nirupama, and brother Rajiv. Iris dedicates this book to her husband Andrew and daughters Jasmine and Maya, who provide endless support even with her crazy work schedule.***

# Contents

# Preface

> *...no one expects conventional silicon-based micro-electronics to continue following Moore's Law forever. At some point, chip-fabrication specialists will find it economically infeasible to continue scaling down microelectronics. As they pack more transistors onto a chip, phenomena such as stray signals on the chip, the need to dissipate the heat from many closely packed devices, and the difficulty of creating the devices in the first place will halt or severely slow progress.*

The above quotation is taken from an essay titled "Computing with Molecules" written by Mark Reed and James Tour in 2002. The quote clearly shows that as computer engineers we are at a technological and scientific inflection point. However, the advent of nanotechnology might be the recourse for continuing improvement of our computing power. Computer engineers and scientists face new challenges as nano-technological innovations grow in different fields of science and technology. As industry experts argue on the scaling of Moore's law beyond 2015, one fact is certain about the future of electronics: *The certainty is in the uncertainty germane in the size, nature, and physics of electronic devices on which we will build our future computing and communications infrastructures*.

One of the grand challenges in the nano-scopic computing era is guarantees of robustness. Robust computing system design is confronted with quantum physical, probabilistic, and even biological phenomena, and guaranteeing high reliability is much more difficult than ever before. Scaling devices down to the level of single electron operation will bring forth new challenges due to probabilistic effects and uncertainty in guaranteeing 'zero-one' based computing. Minuscule devices imply billions of devices on a single chip, which may help mitigate the challenge of uncertainty by replication and redundancy. However, such device densities will create a design and validation nightmare with the shear scale.

Much of the nanotechnology research taking place today is confined in the domain of material science, electrical engineering, quantum and device physics, chemistry, and even biology. However, computer engineers and scientists will be forced to confront the effects that we described above as nanostructured material with unreliable and defective substrates begin to enter into the main stream of computer design.

According to estimates made by nanotechnology experts, we should see such substrates in the upcoming years (see Table 0.1), and hence the associated problems of guaranteeing reliable computing and scaling in design automation tools. This table, published in Wired Magazine about 10 years ago, might not reflect the current predictions. Nonetheless, a point to remember about this table is that breakthroughs in nanotechnology are happening every day and hence predictions of this nature are being updated accordingly. As an example, last April at a conference in Washington DC, a company announced that they have been able to capture images at $10^{-8}$ meter scale, which is a breakthrough in the capability of Scanning Tunneling Microscopes. Such inventions often lead to a quick progression unprecedented and heretofore unthinkable.

*Table 0.1.* Prediction on Arrival of Nano Computing by Nano Technology Experts, Wired Magazine in 1995

| Expert Name | Birge | Brenner | Drexler | Hall | Smalley |
|---|---|---|---|---|---|
| Arrival Prediction | 2005 | 2025 | 2015 | 2010 | 2100 |

The questions that confront computer engineers regarding the current status of nanocomputing material and the reliability of systems built from such miniscule devices, are difficult to articulate and answer. We have found a lack of resources in the confines of a single volume that at least partially attempts to answer these questions.

In November 2003, during the International Conference on Computer-Aided Design (ICCAD) we started a discussion with Mark De Jhong of Kluwer on the idea of putting together a book that would serve as a single source for addressing many of the questions researchers would confront when designing systems based on nano-scale or quantum effect devices. Fortunately, that year, following ICCAD in San Jose, the IEEE High Level Design and Validation Workshop (HLDVT) was held in San Francisco. At the workshop, we organized a special session entitled "Science of The Small Coming the Bigway: Are We Ready for the Design and Validation Challenges?" The speakers included Seth C. Goldstein, Forrest Brewer, and Sankar Basu. Seth Goldstein spoke about his work on defect-tolerant, dynamically reconfigurable architectures that are based on future molecular devices, which drew much interest from the audience. Forrest Brewer talked about coherence effects in today's computing, and how the future computing paradigms need to preserve some of these coherence effects. Finally, Sankar Basu from the National Science Foundation spoke about important issues in nanocomputing and the NSF's interest in nanotechnology. A very interesting panel discussion followed, and from the discussions and queries from the audience, we were more convinced that we need a single source put together for bringing these issues to the forefront. This resulted in our effort to collect some of the most relevant work that deals with the issue of design and validation of architectures on top of nano-scale devices.

As we discussed the possibility of this edited book with the experts in the field, they were immediately ready to help out by contributing various chapters. Their immediate response and enthusiam has made this very informative volume possible, within a span of six months. Based on the emphasis of the chapters, we have divided the book into four major parts. The first part is meant to introduce the readers to the physical realities of implementing nano-computing, using various technologies such as carbon nano tubes (CNT), quantum dots, and molecular switches. This part also ties in physical layer design issues such as variational effects. The sole chapter in this part is written by Arijit RoyChowdhury and Kaushik Roy from Purdue University. In this opening chapter they introduce the readers to the intricacies as well as the promises of the nano-scopic technologies for implementing computing.

Part two is a relatively larger part of this volume, and mainly focuses on *defect-tolerance*. As we have hinted earlier, defects will be a feature of nano-technology by the estimates by experts. One can no longer assume that pieces of silicon that pass the post-silicon tests are the only one used for system design. As a result, designers and system architects have to consider defect-tolerance as a first-order parameter when making design decisions. These defects are not only due to manufacturing imperfection, but also due to signal noise, quantum effects, deformation, and aging. Techniques from traditional fault-tolerant literatures need to be borrowed and enhanced, reconfigurability needs to be designed in, and reliability figures of merits need to be computed. The six chapters in part II are dedicated to this important aspect of nano-computing. This part starts with an article by Paul Graham and Maya Gokhale of Los Alamos Labs, where various defect-tolerant techniques are surveyed. Given this introduction, the next chapter by Mahim Mishra and Seth Goldstein discusses reconfiguration centric defect-tolerance. The next chapter written by Lisa Durbeck and Nicholas Macias of Cell Matrix, introduces more abstract reconfigurable fabric implementation in the form of cells, which will allow dynamic reconfiguration for localization of defects, and thereby guarantee fault-containment. After the readers have thorougly familiarized themselves with reconfiguration based defect tolerance in the first three chapters, the following chapter by Iris Bahar, Jie Chen and Joseph Mundy of Brown University presents a novel model of computation to capture the coherence effects in nanocomputing devices, and how Markov Random Fields can be used to capture the computing with such effects. This chapter is a great introduction to the next chapter by Debayan Bhaduri and Sandeep Shukla of Virginia Polytechnic on automated tools for reliability/redundancy trade-offs for defect-tolerant architectures. The final paper in this part is by André DeHon from Caltech, titled "Law of Large Numbers for System Design" which brings perspective to the uncertainty at the device level in the nanoscopic substrate and where the abstraction boundaries are, and at what levels we assume determinism approximating the uncertainties.

One very important technology for nano-scale computing is projected to be Quantum Dot Cellular automata and variants of quantum computing devices. Given that such devices work with single electron quantum dots, and the computation often blurs the demarkation between 1 and 0s, implications to high level design when using these technologies may be quite broad. Part three of this book is organized in three chapters dealing with quantum dot devices and their design parameters and rules. The first chapter is a great introduction to the challenges of Quantum Computation by Diana Franklin of California Polytechnic State University, and Fred Chong of University of California at Davis. This opening chapter is followed by two chapters dealing with design rules of QCA devices. The first one by Michael Niemier of Georgia Tech and Peter Kogge of the University of Notre Dame provides a detailed look at the Quantum Dots, Quantum Wires, and clocking issues. The second one by Sung Kyu Lim and Michael Niemier from Georgia Tech follows on this to discuss clock zoning in such systems and details the algorithms for partitioning systems into clock zones. These papers will be of great importance to design automation tool designers for future QCA based systems. Part four complements the rest of the chapters with a single chapter on a look at formal verification technology for large scale designs and for designs with unknowns or uncertainties. This last chapter is written by Michael Hsiao and Rajat Arora of Virginia Tech, Shuo Sheng of Mentor Graphics, Ankur Jain from University of California at Berkeley, and Vamsi Boppana of Zenasis. This part introduces to the readers possibilities of verifying large scale designs with redundancy and replications as we have seen common in defect-tolerant design.

In summary, we believe that this volume contains a large amount of research material as well as new ideas that will be very useful for some one starting research in the arena of nanocomputing, not at the device level, but addressing the problems one would face at system level design and validation when nanoscopic physicality will be present at the device level.

SANDEEP K. SHUKLA AND R. IRIS BAHAR

# Foreword

The manufacturing approaches that have been developed and evolved for modern computational hardware represent what is undoubtedly the highest level of technological achievement that the world has ever witnessed. For some 30+ years this technology has advanced at a pace and through a number of generations that is unsurpassed by any other. However, it has also become increasingly apparent that over the next decade, this technology paradigm will mature. Issues such as power consumption, lithographic patterning limitations, and others are beginning to come up as red flags.

Nevertheless, the physics of computation is still young. A number of emerging approaches, each one of which represents, to some degree, a paradigm shift, are beginning to gain at least scientific credibility. These approaches include quantum computation, spintronics, quantum cellular automata (QCA), molecular electronics and neural networks, to name just the major ones. All of these approaches have so-called 'killer applications'. For quantum computing, it is the reduced scaling of various classes of NP-hard problems. For QCA it is an energy efficient computational approach that should get better as the components are reduced in size. For spintronics, it is a memory density that scales exponentially with the number of coupled spin transistors. For molecular electronics, it is an improved energy efficiency per bit operation as well as the potential for continued device scaling to true molecular dimensions. True neural networks possess a greatly increased connectivity and therefore the potential for a greatly increased rate of information flow through a circuit.

All of these alternative computing approaches face tremendous challenges that must be overcome before they can transition into technologies. These include, for example, issues related to materials and/or molecular components that are completely foreign to modern electronics manufacturing. Furthermore, most of them are nanotechnologies — that is, they will require a near atomic level control over the manufacturing steps, and manufacturing 'noise' will translate directly into defective components. Finally, each of them will require new thinking and significant breakthroughs with respect to architecture. This last point is a critical one. Just as it is foolish to build a house without appropriate design guidelines, it is foolish to build a new computational technology without co-designing a framework for that construction. This emerging arena of 'alternative computer architectures' constitutes a new branch of computer engineering that is both fraught with challenges and rich with opportunity.

This book will serve a unique purpose. Several times over the past few years I have had students ask me to point them toward literature references on defect tolerant architectures for nano-electronics, or nanoelectronics design concepts, etc. For quantum computing and neural nets, well established research communities exist and such references are easy to point to. For some of these other paradigms, papers often appear in journals that biologists, chemists, materials scientists, physicists, etc., rarely encounter — or, worse yet — students must turn to patent literature! This book brings together, for the first time, many of these modern architectural concepts into a single text, with chapters written by a terrific group of experts. It is sure to become a mainstay in my group, and I expect that it will be a valuable resource for many years to come.

Jim Heath
Elizabeth W. Gilloon Professor
California Institute of Technology

## Acknowledgments

The editors would like to thank the following people and agencies for helping make this book become a reality. First, we would like to thank Sankar Basu from the National Science Foundation for his initial encouragement and support behind this book. In addition, we would like to acknowledge the NSF for providing support for our own nanotechnology research under grant numbers CCR-0340740 and CCR-0304284. We would also like to thank Mark De Jongh from Kluwer Academic Publishers for working with us and helping to ensure a fast turnaround time for the book. We also thank Cindy Zitter from Kluwer, for all her timely help with administrative issues.

We are grateful to Jim Heath, the Elizabeth W. Gilloon Professor of Chemistry at California Institute of Technology, for encouraging the project with his foreword for the book.

We would also thank the authors of the different chapters in this book for their patience and timely production of the chapters. We thank Forrest Brewer for his encouragement at the inception of this project. Lisa Durbeck has also helped us in planning some of the topic areas.

The students of FERMAT Lab at Virginia Tech, especially Hiren D. Patel and Debayan Bhaduri have helped with a lot of LaTex issues, and with reviewing some of the chapters. We thank them for their help and dedication.

Finally, we would like to thank Nikil Mehta for the countless hours he devoted to helping us format and compile this book. His great attitude and resourcefulness made him a pleasure to work with. He's more of a LaTex wizard than he gives himself credit for.

I

# NANO-COMPUTING AT THE PHYSICAL LAYER

# Preface

Feymann said in one of his famous lectures *"There is plenty of room at the bottom,"* in which he referred to the possibility that computing advances may not have to stop with the gradually diminishing stature of Moore's Law for silicon technologies. New possibilities are abundant when we can harness the computing powers inherent in miniscule particles, atoms, molecules, and their Coulombic, van der Waals, and quantum interactions, which so far have gone unexploited.

In a way, silicon technology is already at the threshold of nano-technological wonders. At the time of publishing this book, most semiconductor companies have been manufacturing devices at the 90nm scale, some going to 65 nanometer. Companies have even figured out how to scale below these numbers. However, with device scaling, comes a plethora of problems related to reliability in the face of cosmic particles, quantum physical interactions, and other physical phenomena which did not play such an important role in the recent past. As we scale down in size, we also scale up in number. This throws a challenge as well for the designers of tools that need to simulate, validate and compute various performance measures for systems built on nano-scale technologies.

The International Technology Roadmap has projected that silicon technology could easily continue scaling down at least until the middle of the next decade. To continue beyond this point, we have to think of alternative technologies. Already we have been seeing some such alternatives in the form of Carbon Nanotube devices, molecular switches, and Quantum Dot cellular automata to name a few. These not only provide novel challenges to the technologists who physically try to make them but also to designers of systems who have so far assumed perfect or near perfect non-linear devices as basic components of their systems. These new technologies not only bring forth new physical challenges, but also imply system level challenges, where the probabilistic nature of devices become a reality, and a first class parameter for high level design.

Parts II, III, and IV of this book discuss issues pertaining to system level design based on nano-scopic technologies. However, in this first part, in order to ground ourselves to physical reality, we present a chapter that discusses both the scaling issues in silicon based technologies in the nano-era, as well as carbon



*S.K. Shukla and R.I. Bahar (eds.), Nano, Quantum and Molecular Computing*, 3-4.

nanotube (CNT) based technology issues at the physical layer. We believe this chapter will provide the readers with the appropriate background to appreciate the later chapters.

Chapter 1

# NANOMETER SCALE TECHNOLOGIES: DEVICE CONSIDERATIONS

Arijit Raychowdhury
*Department of Electrical and Computer Engineering*
*Purdue University*
*IN, USA*
araycho@ecn.purdue.edu

Kaushik Roy
*Department of Electrical and Computer Engineering*
*Purdue University*
*IN, USA*
kaushik@ecn.purdue.edu

**Abstract** This chapter discusses the problems and challenges in scaling Silicon transistors in the nanotechnology era. The principle bottle necks to the scaling of Silicon devices have been discussed. In the latter half of this chapter, novel devices, particularly carbon nanotubes, have been introduced as possible alternatives to Silicon. The material properties, principal device characteristics and circuit issues relating to these revolutionary devices have been discussed.

**Keywords:** Keywords: Scaling of Silicon, Leakage current, Process variation, Molecular transistors, Carbon Nanotubes.

## 1.1 Introduction

For the last three decades the semiconductor industry has witnessed an exponential growth in accordance with Moore's Law. Integration density has attained incredible heights and on-chip functionality has advanced from simple adders to systems-on-chip in the same time frame. Challenges in device design, circuit engineering, and fabrication have been met and overcome. As we advance to an era of nanotechnology, the promise is enormous. Wearable



*S.K. Shukla and R.I. Bahar (eds.), Nano, Quantum and Molecular Computing*, 5-33.

computers, bio sensors, adaptive control systems are all set to have a large impact on life. In this new era, the semiconductor devices will be scaled down to their physical limits. In the process, the circuit and the system engineer is faced with the challenges of scaling. The Silicon MOSFET is no longer a perfect switch and the ratio of the on current to the off current is decreasing [14, 43, 57, 21]. Further, process variation has led to variation of the critical transistor parameters like length, width and threshold voltage ($V_{th}$) thereby reducing the production yield. Research has started in the earnest to gauge the possibility of newer device structures to mitigate these problems. The novel devices include modifications of bulk silicon into FINFETs, trigate structures, and double gate MOSFETs [55]. For example, these modified MOSFETs have better short channel immunity and better subthreshold slopes. Device designers are also looking at revolutionary devices like carbon nanotube transistors, molecular diodes and nano electromechanical systems. These devices with characteristics different from Silicon could potentially have better scalability and increase the on current to off current ratio.

In this chapter, a brief overview will be presented on the issues associated with super-scaled bulk silicon devices and an introduction to non-silicon alternatives will be put forward. Carbon nanotubes have emerged as the most promising alternative device in the nanotechnology era and there has been considerable interest in the design and understanding of carbon nanotube field-effect transistors. In the second part of this chapter, the nature and properties of carbon nanotubes and carbon nanotube field-effect transistors will be discussed from a circuit designer's point of view.

## 1.2 Silicon Nanoelectronics

To achieve higher density and performance at lower power consumption, MOS devices have been scaled for more than 30 years [48], [6], [1]. Transistor delay times have decreased by more than 30% per technology generation resulting in doubling of microprocessor performance every two years. Supply voltage (VDD) has been scaling down at the rate of 30% per technology generation in order to keep power consumption under control. Hence, the transistor threshold voltage ($V_{th}$) has to be commensurately scaled to maintain high drive current and achieve performance improvement of at least 30% per technology generation. The semiconductor industry has enjoyed the fruits of scaling; but with shorter and shorter devices the problems of scaling are becoming more and more predominant. In the first section of this chapter we would visit some of the scaling issues of bulk silicon transistors.

## Short Channel Effects

Short channel effect in scaled MOSFET devices is the lowering of the threshold voltage $V_{th}$ with decreasing channel length [55]. In long-channel devices, the source and drain are separated far enough that their depletion regions have no effect on the potential or field pattern in most part of the device, and hence, the threshold voltage is virtually independent of the channel length and drain bias. In a short-channel device, however, the source and drain depletion width in the vertical direction is comparable to the effective channel length. This causes the depletion regions from the source and the drain to interact with each other. The obvious consequence of this is lowering of the potential barrier between the source and the channel. This causes lowering of the threshold voltage of the MOSFET with decreasing channel length, a phenomenon referred to as short channel effect [55, 40, 51]. Figure 1.1 illustrates the effect of channel length scaling on the surface potential of the device along the length.

Apart from the channel length, the drain voltage also has a significant effect on the potential barrier for short channel devices. Under off conditions, this potential barrier between the source and the channel prevents electrons from flowing to the drain. For a long-channel device, the barrier height is mainly controlled by the gate voltage and is not sensitive to $V_{ds}$. However when a high drain voltage is applied to a short-channel device, barrier height is lowered resulting in further decrease of the threshold voltage. The source then injects carriers into the channel surface without the gate playing a role. This is known as drain induced barrier lowering (DIBL). DIBL is enhanced at higher drain voltage and shorter effective lengths Surface DIBL typically happens before deep bulk punch through. Ideally, DIBL does not change the subthreshold slope, S, but does lower $V_{th}$. The DIBL effect increases as $V_{ds}$ increases. Higher surface and channel doping and shallow source/drain junction depths reduce the DIBL effect [55, 40, 51].

In scaled Silicon MOSFETs several other effects like Gate induced drain leakage (GIDL), hot electron effect and punch-through are important for different biasing regions [55].

## Leakage Current in Scaled Devices

It has already been established that one of the primary concerns in Silicon MOSFETs is the increasing drain control over the channel. Figure 1.2 illustrates how leakage power is fast catching up with active power in scaled CMOS and Figure 1.2 shows the principle leakage current components in scaled Silicon devices. As DIBL increases, the Vth of the device gets significantly lowered resulting in higher subthreshold leakage current. Subthreshold or weak inversion current is the drain current of the MOSFET when the gate is biased voltage less than $V_{th}$. The minority carrier in the weak inversion region is small but

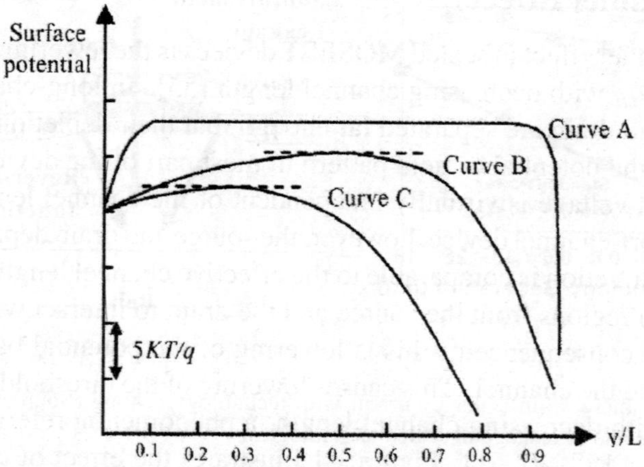


*Figure 1.1.* Surface potential versus lateral distance (normalized to the channel length L) from the source to the drain for (a) long-channel MOSFET ($L = 6.25\mu m,\ V_{ds} = 0.5V$), (b) a short-channel MOSFET ($L = 1.25\mu m,\ V_{ds} = 0.5V$), (c) a short channel MOSFET at high drain bias ($L = 1.25\mu m,\ V_{ds} = 5V$). The gate voltage is same for all three cases [55].

not zero. This results in a diffusion current from the drain to the source of the device even when the gate to source voltage ($V_{gs}$) is at zero potential. The current increases exponentially as the threshold voltage is lowered [55, 51].

To mitigate the problem of lowered $V_{th}$ in MOSFETs, the channel doping is increased in a region below the drain and the source (retrograde well) and near the source-bulk and drain-bulk junctions (halo implants) [15] (refer to Figure 1.3). These higher doping regions serve to increase the threshold voltage of the device and lower the subthreshold current. However, for scaled devices ($l_{eff}$ < 50nm) the increased halo doping creates a high electric field across the reverse biased drain-bulk junction. This causes a junction band-to-band tunneling (BTBT) current to flow from the drain to the source of an n-MOS device. A similar current flows across the source-body junction too depending on the biasing conditions. Thus the halo doping decreases the subthreshold current at the cost of higher BTBT leakage in scaled devices. Figure 1.3 illustrates how the subthreshold and the BTBT currents vary with increasing halo doping.

For scaled devices the oxide thickness is scaled commensurately to increase the gate control over the channel [51]. This has resulted in yet another significant leakage current called gate tunneling leakage [20, 9, 29, 60]. Gate tunneling current is the current due to tunneling of electrons from the conduction band of bulk silicon and the source/ drain overlap regions through the potential barrier of the oxide into the gate of the device. This tunneling current becomes more significant as the oxide thickness is scaled and for sub-100nm devices where the oxide thickness is about or below 20°A, gate tunneling forms an important

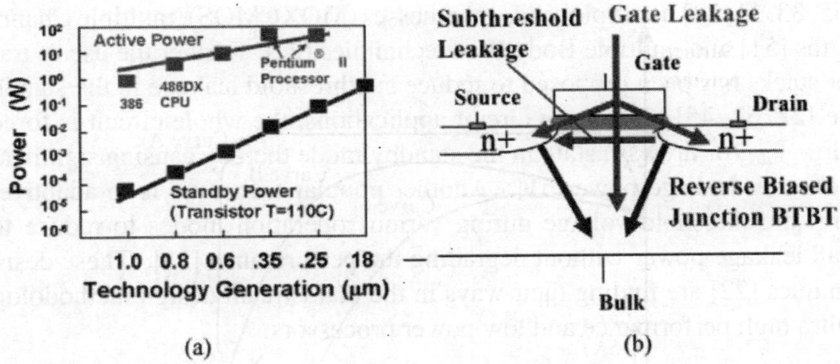


*Figure 1.2.* (a) Increase in leakage current with technology scaling (Source: Intel) (b) The different components of leakage in scaled technologies.

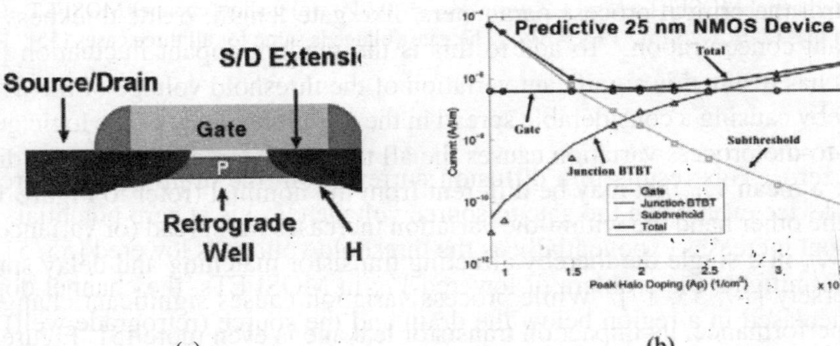


*Figure 1.3.* Typical channel profile showing the retrograde well and the halo regions. (b) The variation of the different leakage components with increasing halo doping.

leakage component. Simulation results from a super scaled device ($l_{eff} \sim$ 25nm), (Figure 1.3) show the contribution of different leakage components.

In scaled Silicon devices, the leakage current increases almost exponentially with scaling. This reduces the on current to off current ratio of transistors and also consume considerable amount of power even in the standby mode. In order to alleviate the leakage power consumption in densely integrated logic and memory designs, circuit engineers and researchers have proposed different schemes. One of the popular methods to reduce leakage power is to use multiple $V_{th}$ transistors in the design, where the performance critical transistors are allowed to have a lower $V_{th}$ whereas the $V_{th}$ of the off-critical transistors are set at a higher value. This can be implemented using multiple channel dopings [56,

58, 55, 33, 11, 24], multiple oxide thicknesses (MOXCMOS), multiple Channel Lengths [51] and multiple Body Bias techniques [22]. Further the use of transistor stacks has been proposed to reduce subthreshold leakage in the standby mode [22, 52, 45]. In certain circuit applications, the whole circuit is forced to a low $V_{dd}$ (or drowsy) state in the standby mode thereby causing significant reduction in leakage power[51]. Another popular technique is to adaptively control the threshold voltage during various operation modes to reduce the overall leakage power without degrading the performance [11]. These design techniques [22] are finding their ways in the mainstream design methodology for ultra high performance and low power processors.

## Process Variation

Process parameter variation has also been identified as one of the principle bottlenecks in scaling of Silicon MOSFETs beyond 100nm [46, 5, 19]. As the device dimensions continue to shrink, it is becoming increasingly difficult to control the critical process parameters, like gate length, oxide thickness and dopant concentration. To add to this is the random dopant fluctuation [55]. This has resulted in significant variation of the threshold voltage of the device thereby causing a considerable spread in the switching delay of the logic gates. Die-to-die process variation causes the all the transistors in a particular die to have a mean $V_{th}$ that may be different from the nominal (refer to Figure 1.4). On the other hand the within-die variation increases the spread (or variance) of the $V_{th}$ in a single die thereby affecting transistor matching and delay spread adversely [19, 53, 17]. While process variation causes significant change in the performance, its impact on transistor leakage is even more [5]. Figure 1.4 illustrates, a 2X variation in the 'ON' current of the device corresponds to 100X variation in the leakage current. The increase in leakage current and process variation in scaled Silicon devices may prove to be showstoppers. Even in circuits with low activity (for example, level 2 caches) the increase in leakage power has led to high power consumption thereby reducing battery lifetime. With process variation production yield has gone down drastically and new design methodologies like statistical timing analysis, statistical sizing for yield are becoming popular. In the regime where the gate lengths are scaled below 50nm, predictable circuit design with tolerable power budgets may become uneconomical for production [54]. Hence, leading researchers are investigating modified device structures like FINFETs, partially depleted SOI and double gate MOSFETS [55]. A discussion on these modified device structures is beyond the scope of this chapter. If Silicon electronics reach the limits of scalability, it will only be prudent to look for other materials to replace Silicon. Carbon nanotube has emerged as a promising replacement to Silicon in future nanoelectronic designs. In the following sections we will give an overview of the non-Silicon

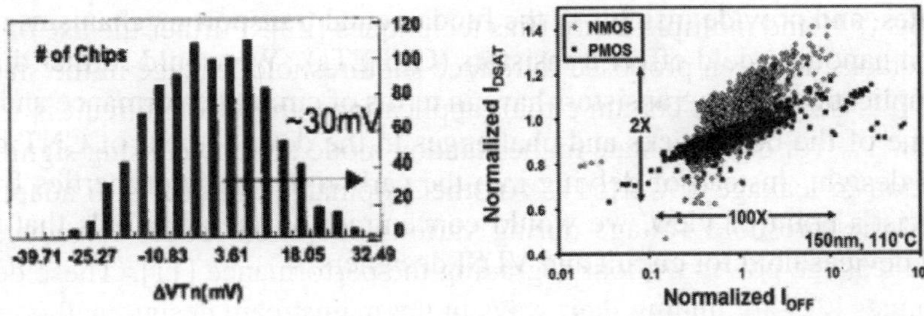


(a) (b)

*Figure 1.4.* (a) Variation of Vth for scaled technologies [14]. (b) The variation of the ON and OFF currents with process variation.

based molecular devices such as carbon nanotubes and molecular diodes which show promise for the future. Carbon nanotube transistors are predicted to have about ten times the current density of silicon MOSFETs and maintain an on-current ot off-current ratio of more than $10^3$. Further, carbon nanotubes will allow successful integration of high $\kappa$ gate dielectrics because there is no dangling bond in carbon nanotubes. This would result in the possible use of thicker gate dielectrics thereby reducing gate leakage at no performance penalty. Several other molecular diodes that are being currently investigated show enormous promise as ultra-scaled switches for future technologies.

## 1.3 Carbon Nanotube Electronics

In a pursuit for novel materials in a post Silicon electronics era, scientists and engineers worldwide have already started active research in carbon nanotube electronics[16, 4, 32]. Although carbon filaments of nanoscaled diameters ($\sim$10nm ) were extensively grown in the '70s and the '80s, it was only after the pioneering work of Iijima in 1991 [31] that the potential of carbon nanotube as a possible device material has been recognized and extensively studied. Owing to their excellent electrical, mechanical and thermal properties, researchers have identified an array of potential applications for carbon nanotubes. Even in a the short span since their inception, field effect transistors, diodes, optical and cathode ray emitters, bio-sensors and energy storage elements have been demonstrated [16].

Carbon nanotube (CNT) electronics is still in its infancy and the transport, contacts, interfaces and electrostatics of these devices require detailed understanding. Experimentalists and theoreticians have been working closely to unravel the unique properties of semiconducting as well as metallic carbon nanotubes. In the next few sections, we will provide an introduction to carbon nanotube electronics. We will discuss the bandstructure and the density of states, and provide insights to the fundamental transport mechanisms of the carbon nanotube field effect transistors (CNFETs). We would further discuss the implications these transistors have in terms of circuit performance and look at some of the bottlenecks and challenges in the development of CNT based VLSI design. Instead of delving into the carbon nanotube properties from a physicist's point of view, we would concentrate on the potentials that these novel devices hold for circuit and VLSI designers.

## Bandstructure and Density of States

To understand the properties of a device structure it is essential to have a clear understanding of its energy-band structure. Before calculating the E-k (energy vs. wave vector) relation of a carbon nanotube, it is only prudent to understand the bandstructure of graphene [13]. Carbon nanotubes are sheets of graphene and the simplest way to calculate their bandstructure would be to quantize the graphene E-k along the circumference.

Let us consider a sheet of graphene where the carbon atoms are packed in a regular hexagonal structure with a C-C bond length $a_{cc}$ (~0.142nm) and a C-C bond energy $t_0$ (~3eV). It can be noted that all the carbon atoms in the graphene lattice do not see an identical environment. Hence two atoms can be lumped together to form a unit cell, as has been illustrated in Figure 1.5. The entire graphene sheet can thus be constructed by translating the unit cell along the linear combinations of the basis vectors, $\mathbf{a_1}$ and $\mathbf{a_2}$. Let us define the translation vector as $\mathbf{T} = \mathbf{m}\mathbf{a_1} + \mathbf{n}\mathbf{a_2}$ where m and n are integers. It can be noted from Figure 1.5 that the basis vectors can be expressed in terms of the geometrical parameters as,

$$\mathbf{a_1} = a_0\left(\frac{\sqrt{3}}{2}\bar{x} + \frac{1}{2}\bar{y}\right) \quad (1.1)$$

$$\mathbf{a_2} = a_0\left(\frac{\sqrt{3}}{2}\bar{x} - \frac{1}{2}\bar{y}\right)$$

where, $a_0 = \sqrt{3}a_{cc}$. In graphene, the $2p_z$ orbitals gives rise to $\pi$ bonds, which determine the electronic properties of graphene. We may also assume that the electrons are tightly bound to the respective atoms and only the wave functions of neighboring atoms overlap to a small extent. This is known as the tight binding approximation. This tight binding approximation applied to the $2p_z$ orbitals of carbon yields the following E-k relationship of graphene.

$$E(\mathbf{k}) = \pm|t_0|\sqrt{3 + 2\cos(\mathbf{k.a_1}) + 2\cos(\mathbf{k.a_2}) + 2\cos(\mathbf{k.a_3})} \quad (1.2)$$

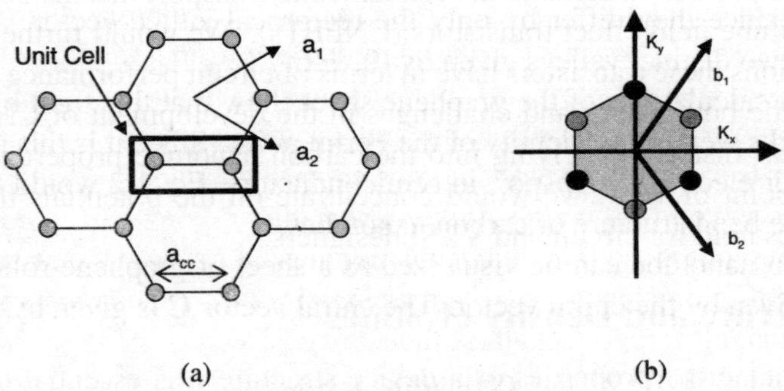


*Figure 1.5.* (a) Real Space lattice of Graphene (b) The reciprocal space representation of graphene showing the first Brillouin zone.

where $\mathbf{a_3} = \mathbf{a_1} - \mathbf{a_2}$. The positive sign in Eq. 1.2 corresponds to the conduction band and the negative sign corresponds to the valence band. It should be noted that the conduction and the valence bands are exact mirror images of each other. For transport properties, this implies identical hole and electron mobilities. Having obtained the real space lattice, it will be beneficial to obtain the lattice vectors in reciprocal space. Let the basis vectors be $\mathbf{b_1}$ and $\mathbf{b_2}$, and they satisfy the relations,

$$\mathbf{a_i}.\mathbf{b_j} = 2\pi\delta_{ij} \quad \text{for } i,j = 1,2 \tag{1.3}$$

where, $\delta_{ij}$ is the Kronecker delta function, being 1 when i=j and 0 otherwise. The electronic properties can be obtained by translating the real space to the **k** (wave-vector) space also called the reciprocal space. Using the real space vectors we obtain the reciprocal space vectors for graphene,

$$\begin{aligned} \mathbf{b_1} &= \left(\frac{\pi}{a}\bar{x} + \frac{\pi}{b}\bar{y}\right) \\ \mathbf{b_2} &= \left(\frac{\pi}{a}\bar{x} - \frac{\pi}{b}\bar{y}\right) \end{aligned} \tag{1.4}$$

where $a = \sqrt{3}a_0/2$ and $b = a_0/2$. Figure 1.4 shows the reciprocal lattice space. It can be shown that the energy at the three corners of the Brillouin zone (the smallest volume in the reciprocal vector space such that the total reciprocal space is a periodic repetition of this volume) is zero. These corners can be represented in the $k_x - k_y$ plane as,

$$(k_x a, k_y b) = \begin{bmatrix} \left(0, -\frac{2\pi}{3}\right), & \left(-\pi, \frac{\pi}{3}\right), & \left(\pi, \frac{\pi}{3}\right), \\ \left(0, -\frac{2\pi}{3}\right), & \left(-\pi, \frac{-\pi}{3}\right), & \left(\pi, \frac{-\pi}{3}\right) \end{bmatrix} \tag{1.5}$$

It should be noted that the first three are equivalent and the last three are equivalent since they differ by only the reciprocal lattice vector. Hence we have only two distinct valleys given by (0, $\pm 2\pi/3$).

Detailed calculations of the graphene sheet show that this tight binding approach works well in the vicinity of the Fermi point. Since it is this region that accounts for electronic transport in semiconductors, Eq. 1.2 would be used in deriving the bandstructure of carbon nanotubes.

A carbon nanotube can be visualized as a sheet of graphene rolled up in a direction given by the chiral vector. The chiral vector **C** is given by,

$$\mathbf{C} = m\hat{\mathbf{a}}_\mathbf{1} + n\hat{\mathbf{a}}_\mathbf{2} \tag{1.6}$$

where, $\hat{\mathbf{a_1}}$ and $\hat{\mathbf{a_2}}$ are the unit vectors along $\mathbf{a_1}$a1 and $\mathbf{a_1}$ respectively. This chiral vector (m,n) specifies most of the electronic properties of the nanotube as would be evident subsequently. The diameter of the nanotube is given by,

$$d = a_0\sqrt{n^2 + m^2 + mn} \tag{1.7}$$

and the translation vector is given by,

$$\mathbf{T} = t_1\hat{\mathbf{a}}_\mathbf{1} + t_2\hat{\mathbf{a}}_\mathbf{2} \tag{1.8}$$

where,

$$t_1 = \frac{2m+n}{d_R} \tag{1.9}$$

$$t_2 = -\frac{2n+m}{d_R}$$

and $d_R$ is the highest common divisor of (2m+n) and (2n+m). The number of hexagons in a unit cell is given by,

$$N = \frac{2(m^2 + n^2 + mn)}{d_R} \tag{1.10}$$

Owing to the periodic boundary condition of the carbon nanotube, the following needs to be satisfied,

$$\mathbf{k.C} \equiv k_x a(m+n) + k_y b(m-n) = 2\pi u \tag{1.11}$$

where, u is an integer. It can be observed that Eq. 1.11 represents a series of lines in the $k_x - k_y$ plane and if (m-n)/3 is an integer then one of these lines pass through (0, $\pm 2\pi/3$). Hence such nanotubes are metallic. This leads to an important conclusion that without accurate control over chirality (as in the present state of the art) one-third of the nanotubes are metallic and not usable for transistor applications.

Eq. 1.2 gives the E-k relationship in the entire Brillouin zone of graphene. Using a Taylor series expansion of Eq. 1.2 near a Fermi point, $k_F$ and neglecting higher order terms, the E-k relation near the Fermi point is

$$E(k) = \frac{\sqrt{3}a_{cc}\,|t_0|}{2}\,|k - k_F| \tag{1.12}$$

Based on this linear approximation the E-k relation in graphene and imposing the periodic boundary condition of carbon nanotubes, the band structure of carbon nanotube is obtained as

$$E(k) = \frac{\sqrt{3}a_{cc}\,|t_0|}{2}\sqrt{k_c^2 - k_t^2} \tag{1.13}$$

where, $k_c$ and $k_t$ are the circumferential and tangential components of the k vectors. For metallic nanotubes the minimum value of $k_c$ is zero and hence we obtain a linear dispersion relation

$$E(k) = \frac{\sqrt{3}a_{cc}\,|t_0|}{2}k_t \tag{1.14}$$

whereas, for semiconducting carbon nanotubes the minimum value of $k_c$ is 2/3d thereby giving

$$E(k) = \frac{\sqrt{3}a_{cc}\,|t_0|}{2}\sqrt{(2/3d)^2 + k_c^2} \tag{1.15}$$

Since the conduction and the valence band are mirror images for the carbon nanotube, the bandgap is

$$E(k) = \frac{2a_{cc}\,|t_0|}{d} \approx \frac{0.8eV}{d} \tag{1.16}$$

Figure 1.6 shows typical E-k diagrams of one semiconducting and one metallic nanotube.

## Carbon Nanotube Field-effect Transistors

The principle of carrier transport in carbon nanotubes is of active research in the device community [18, 3, 59, 37, 26, 27, 30, 36, 38, 23, 28]. Both metallic as well as semiconducting nanotubes are being studied experimentally and theoretically. Although the principle transport mechanisms have not yet been established without reasonable doubt, some of the pioneering works in this field have given us insights into the behavior of carbon nanotube based transistors. In the following few sub-sections we would review a few of these results and discuss their implications.

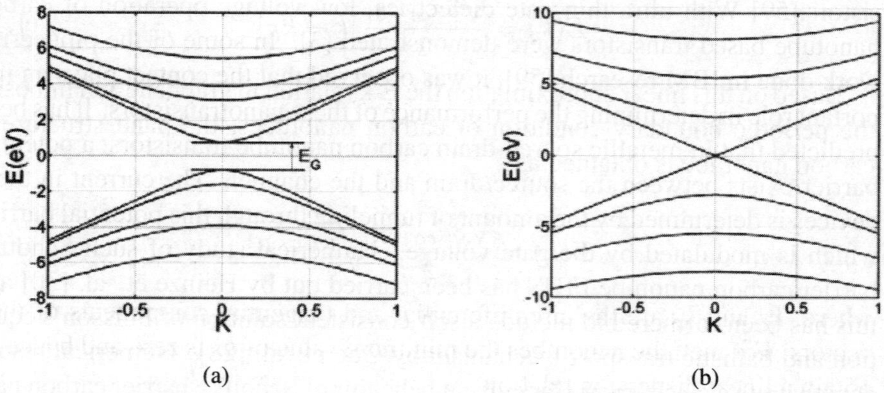


*Figure 1.6.* (a) E-k diagram of a semiconducting carbon nanotube (b) E-k diagram of a metallic carbon nanotube. Note: k has been normalized with respect to the maximum value of k ($k_{max}$).

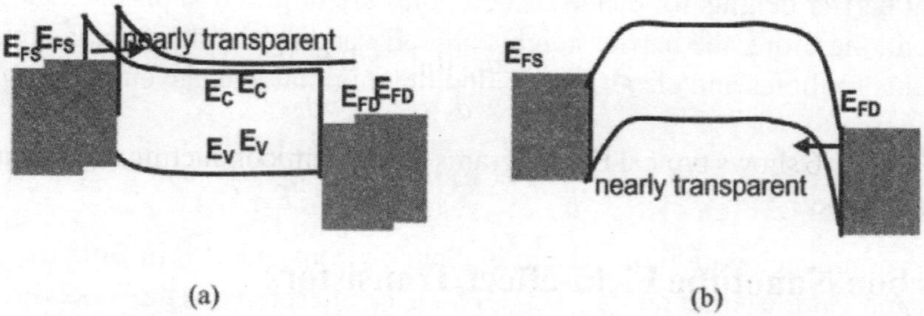


*Figure 1.7.* (a) Band diagram of a Schottky barrier carbon nanotube with $V_G$>0. Note that the barrier seen by electrons is thin and nearly transparent (b) The band diagram of the same Schottky barrier carbon nanotube with $V_G$<0. The bands bend upwards and the holes see a thin barrier.

**Schottky Barrier Carbon Nanotube FETs.** Since it was experimentally shown that semiconducting carbon nanotubes (CNTs) can work as channel material of field-effect transistors (FET) [30], significant progress has been made in understanding and modeling the principle transport properties of these transistors.[59] With ultra thin gate dielectrics, low voltage operation of carbon nanotube based transistors were demonstrated [3]. In some of the pioneering work done in IBM research [59], it was observed that the contact plays an important role in determining the performance of these nanotransistors. It has been predicted that in metallic source-drain carbon nanotube transistors, a potential barrier exists between the source/drain and the channel. The current in these devices is determined by the amount of tunneling through this potential barrier, which is modulated by the gate voltage. Numerical study of such Schottky barrier carbon nanotube FETs has been carried out by Heinze et. al. [30] and this has been furthered to include a self consistent solution of Poisson's equation and ballistic transport mechanism by Guo et. al. [28]. Numerical results demonstrate a clear ambipolar current behavior of Schottky barrier carbon nanotube transistors. The band diagram of the Schottky barrier transistor has been illustrated in Figure 1.7 for positive and negative gate voltages and the physical diagram is illustrated in Figure 1.8a. It can be noted that for positive gate voltages there is electron current and for negative drain voltages a strong hole current is established. The minimal leakage current in Figure 1.8b can be estimated by noticing that it occurs when the electron and hole currents are equal. The tunneling barrier for holes at the drain end is nearly transparent when the gate oxide is thin, thus the off-current for holes is limited by thermionic emission over the barrier, , in the bulk body. The total current of these devices is minimum when the electron current and the hole current are equal to each other. Equal barrier heights for electrons and holes are required to produce the same current, therefore, the barrier heights are . By adding the thermionic emission currents for holes and electrons, we find the minimal leakage current as

$$I \sim \frac{8ek_BT}{h} \times exp(-\frac{E_g - eV_D}{2k_BT}) \tag{1.17}$$

in the non-degenerate limit. (Here is Planck's constant. $k_B$ in Boltzman constant and T is the temperature.) Eq. 1.17 can be interpreted in the following way. At equilibrium, the largest barrier height that limits electron and hole current is one half of the band gap, and it decreases by an amount of after the drain voltage is applied. For equal barrier heights of electrons and holes (mid-gap source and drain materials) the minimum drain current occurs when the gate voltage is equal to half the drain voltage, as illustrated in Figure 1.8.

Two important aspects of these nanotube transistors are worth mentioning. First, the energy barrier at the Schottky barrier severely limits the transconductance of the nanotube transistors in the 'ON' state and reduces the current

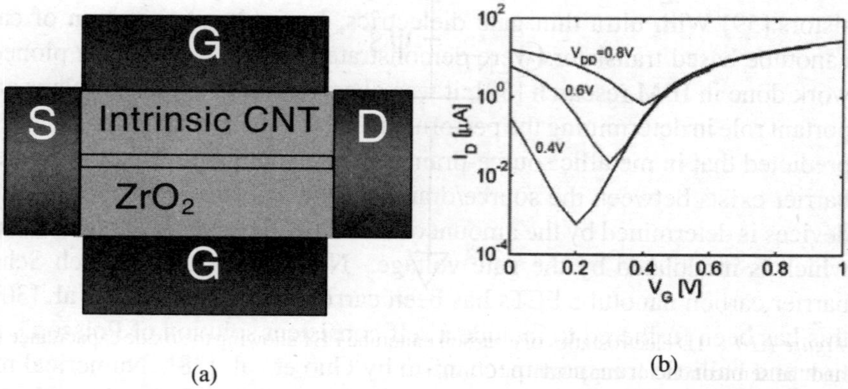


*Figure 1.8.* (a) A Schottky barrier carbon nanotube FET. Note that the Source and Drain are metallic and a high K dielectric has been used (b) The $I_D - V_G$ characteristics of the Schottky barrier FET showing ambipolar conduction.

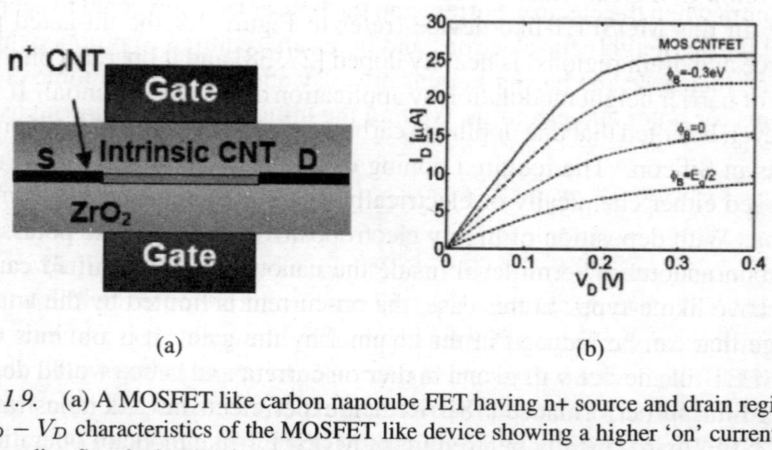


*Figure 1.9.* (a) A MOSFET like carbon nanotube FET having n+ source and drain regions (b) The $I_D - V_D$ characteristics of the MOSFET like device showing a higher 'on' current than a corresponding Scottky barrier device (for different source drain metal work-functions).

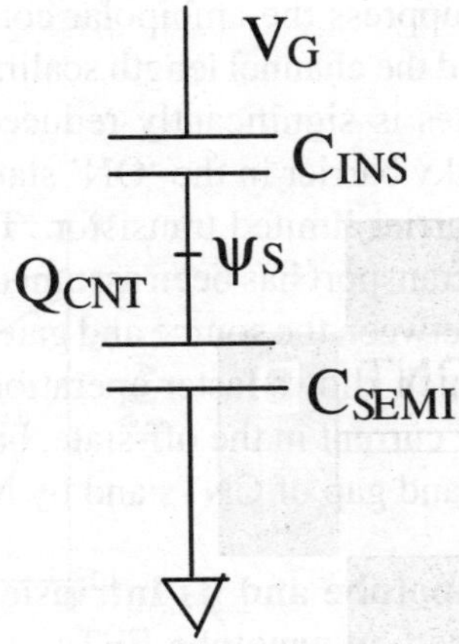


*Figure 1.10.* 1D electrostatics of a carbon nanotube FET showing the oxide capacitance ($C_{OX}$) and the semiconductor capacitance ($C_{semi}$)

delivery capability- a key metric to transistor performance. Second, Schottky barrier CNFETs exhibit strong ambipolar characteristics and this constraints the use of these transistors in conventional CMOS logic families. Some of the design issues with these transistors would be subsequently visited.

**Ideal MOSFET-like Carbon Nanotube FETs.** To overcome these handicaps associated with the Schottky barrier CNFETs, there have been attempts to develop CNFETs, which would behave like normal MOSFETs (Figure 1.9) [18]. In this MOSFET-like device (refer to Figure 1.9 the un-gated portion (source and drain regions) is heavily doped [27, 38] and it operates on the principle of barrier height modulation by application of the gate potential. It should, however, be noted that that doping in carbon nanotubes is not substitutional doping as in Silicon. The required doping of the source/drain extension may be achieved either chemically or electrically. Carbon nanotubes are intrinsically p-type. With deposition of highly electropositive materials like potassium on a carbon nanotube, Fermi level inside the nanotube can be shifted causing it to behave like n-type. In this case, the on-current is limited by the amount of charge that can be induced in the channel by the gate. It is obvious that the MOSFET-like device will give a higher on current and hence would define the upper limit of performance [18]. Recent experiments have demonstrated that the CNFET can typically be used in the MOSFET-like mode of operation with near ballistic transport [36]. Although the feasibility of MOSFET-like carbon nanotube transistors is questionable, some of the numerical studies have caught the fancy of circuit designers. Guo et al. [18] were the first to predict performance parameters of ideal MOSFET-like carbon nanotube transistors. These

MOSFET-like CNFETs suppress the ambipolar conduction that occurs in SB CNFETs. They also extend the channel length scaling limit because the density of metal-induced-gap-states is significantly reduced. A MOSFET like CN-FET has a negative Schottky barrier in the 'ON' state and hence delivers more current than a Schottky barrier limited transistor. This has been illustrated in Figure 1.9 where ballistic transport has been assumed in both the devices. Also, the parasitic capacitance between the source and gate electrode is reduced in the MOSFET-like devices, which allows faster operation. MOSFET-like CNFETs will also display a leakage current in the off-state, but that leakage current will be controlled by the full band gap of CNTs and by band to band tunneling.

**Charge on a Carbon Nanotube and its Intrinsic Capacitance.** So far in our discussion on the carbon nanotube FETs, we have investigated the I-V characteristics of the device in the Schottky barrier and the MOSFET-like modes of operation. For circuit design, the extrinsic as well as the intrinsic capacitances need detailed understanding [26]. Figure 1.10 demonstrates the intrinsic capacitances (with 1D electrostatics) of the device, where $C_{ox}$ is the oxide capacitance, $C_{semi}$ is the semiconducting capacitance. It is a well known theory that for Silicon MOSFETs in the inversion region of operation, the inversion layer is limited to a very narrow width and the semiconducting capacitance is way larger than the oxide capacitance. In carbon nanotubes however, this assumption does not hold and even under inversion, the surface potential, $\psi_s$ s is strongly modulated by the gate voltage. This results in a semiconducting capacitance that is comparable or even less than the oxide capacitance and the charge (on the channel) is strongly modulated by the gate potential.

**Principle Scattering Mechanisms.** For nanoelectronic applications of carbon nanotube transistors, it is essential to identify the nature of transport and the principle scattering mechanisms. Scattering determines not only the current through the device but also the locations of power dissipation in the nanotube transistor [16]. A single nanotube should provide an ideal conductance of $4e^2/h$ but in practice a much lower value is typically observed. The anomaly can be attributed to the scattering effects and the imperfections in the contacts. Yao et al. [61] have demonstrated a current handling capacity of $25\mu A$ at high biases for metallic carbon nanotubes due to electron backscattering from optical or zone boundary phonons. In more recent experimental data [39] electron-phonon interactions in single walled CNTs have been studied both in the high bias as well as the low bias regimes. It has been observed that for low biases ($<\sim$0.1V), channel lengths less than 200nm has a bias independent conductance, one of the hallmarks of ballistic transport. However for longer channel lengths at low biases, a length dependant conductance has been observed. This has been illustrated in Figure 1.11. In low bias regime, the principle scattering mechanism is due to electronic interactions with acoustic phonons and the corresponding mean free path (mfp) of acoustic phonons has been predicted to be $1.6\mu m$. However at high bias (>0.16V) optical or zone boundary phonons (mfp $\sim$ 10nm) become relevant and channel lengths less than 100nm shows significant scattering effects. Several other experimental results have been reported in recent publications and with high quality chemical vapor deposition (CVD) materials and ohmic contact strategies, several groups have observed ballistic electron transport in metallic nanotubes [41] and more recently, in semiconducting SWNTs by Javey et al.[35]. Numerical studies of metallic nanotubes have been extensively carried out [34] and the mfp for acoustic phonon scattering has been estimated to be $l_{ap} \sim 300nm$, and that for optical phonon scattering is lop 15 nm. Transport through these short macromolecular ($\sim$ 10 nm) nanotubes has been shown to be free of significant acoustic and optical phonon scattering and thus essentially ballistic at both high and low voltage limits.

From a circuit designer's point of view, ballistic transport through a semiconducting carbon nanotube FET would imply greater 'ON' current and faster speed of operation. The current would be governed by the well-known transport equation for 1D carriers,

$$I_D = \frac{4ek_BT}{h}\left[ln(1 + exp(\xi_S)) - ln(1 + exp(\xi_D))\right] \tag{1.18}$$

where, $\xi_S$ and $\xi_D$ are the source-channel and drain-channel potential barriers. Several circuit simulations in this ideal performance limit incorporating the intrinsic capacitance of the CNFETs have been presented in [8, 7, 49]. These results show several terahertz of performance of these nanotransistors. However, the ballistic nature of transport is under close scrutiny and with increasing perfection of the fabrication process, ballistic transport for low voltage applications seem plausible.

## Circuit Design and Circuit Compatible Modeling

Along with extensive experimental results that are being published and corresponding numerical simulations of these carbon nanotube devices, circuit designers and VLSI experts have started to gauge the performance of carbon nanotubes in digital and analog circuits [8, 7, 49]. Three stage ring oscillators with an oscillation frequency of 220Hz have been demonstrated in [38] (refer to Figure 1.12). Although 220 Hz is indeed a low frequency of operation compared to the terahertz performance predictions, this was one of the first attempts to reveal ac measurements of CNFETs. Simple logic gates and inverters have been experimentally demonstrated by the leading experimentalists [3] and has been illustrated in Figure 1.13. More recent RF measurements [23] have shown performances upto 250MHz thereby confirming the high frequency nature of the

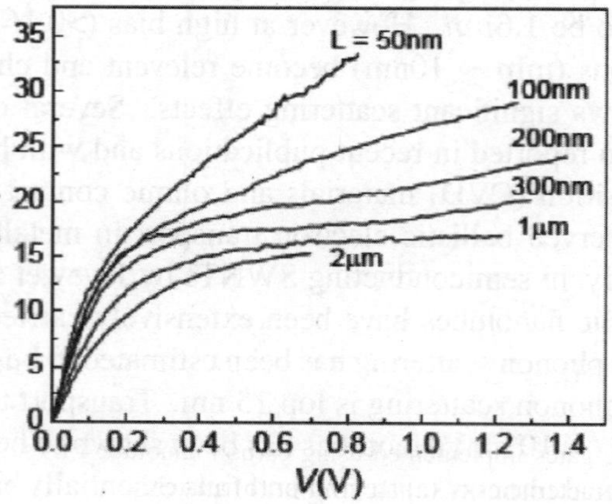


*Figure 1.11.* The experimentally measured I-V characteristics of a 1.8nm diameter metallic carbon nanotube [39]. Note that for high biases the current saturates showing clear evidence of increased scattering events at high biases.

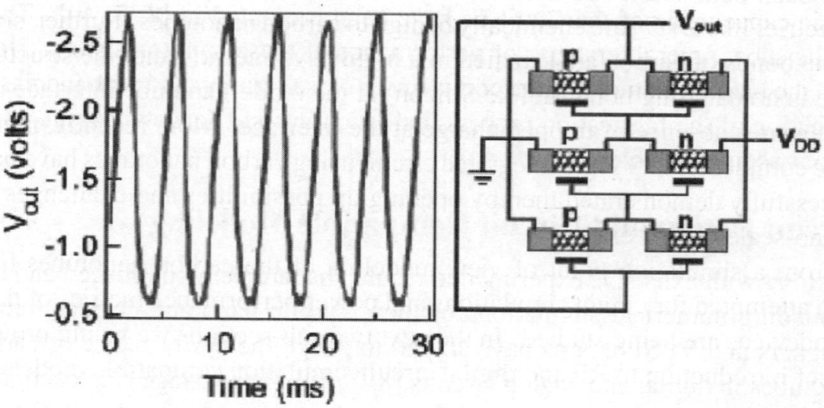


*Figure 1.12.* A three stage ring oscillator made out of carbon nanotube FETs. The frequency of operation is about 220Hz [38] and this is limited by the parasitic capacitances of the device geometry and the measuring instruments.

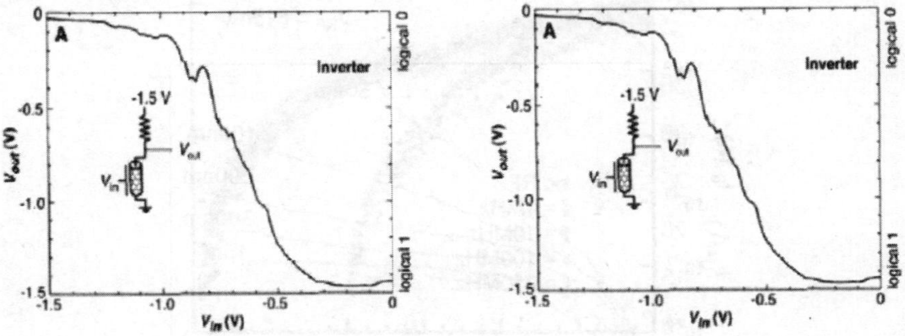


(a) (b)

*Figure 1.13.* Logic gates implemented using carbon nanotube FETs. Experimental data [3] showing transfer characteristics of (a) an inverter (b) a latch.

CNTs (refer to Figure 1.14). Till now, all the high frequency measurements are limited by the parasitic capacitances and the low current drive of the individual CNTs. The integration of materials having a high dielectric constant (high-$\kappa$) into carbon-nanotube transistors further promises to push the performance limit for molecular electronics [36]. The p-type transistors with subthreshold swings of S $\sim$ 70 mV per decade and n type transistors exhibiting S $\sim$ 90 mV per decade have been experimentally demonstrated. High voltage gains of up to 60 have been obtained for complementary nanotube based inverters. The high $\kappa$ dielectrics like $ZrO_2$ are chemically benign to carbon nanotubes. Further, since all the bonds of carbon are satisfied in a highly symmetric nanotube structure, there is no dangling bond (unlike Silicon) at the oxide-nanotube interface and this ensures absence of trapped charge at the interface. More recently, monolithic complementary logic integrated circuit using carbon nanotubes have been successfully demonstrated thereby opening up possibilities and challenges for a nano-scaled VLSI era.

From a simulation point of view modeling of the carbon nanotubes have been attempted for circuit simulations and power/performance metrics of these nanodevices are being studied. In the next two sub-sections we would provide a brief introduction to RF and digital circuit simulation compatible models for CNTs.

**An RF Circuit Model of Metallic Carbon Nanotubes.** A carbon nanotube, because of its band structure has two propagating channels [8]. To add to this, there is spin up and spin down that results in four channels in the Landauer-Büttiker formalism [7]. Based on the Lüttinger liquid theory for a 1D electron gas [7] spin charge separation can be considered for each of the modes of

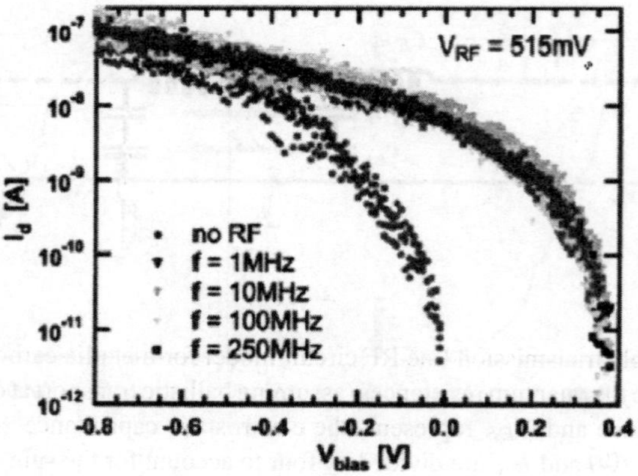


*Figure 1.14.* One of the pioneering efforts in measuring RF characteristics of carbon nanotube FETs. Note that the limit 250MHz comes from the parasistic capacitors and not the carbon nanotube itself [23]. Researchers are working on improved methodologies to measure GHz performance of carbon nanotube FETs. Physics poses no restriction on high frequency operation of these nano-devices.

propagation. Consequently these modes can be decoupled and each of them can be modeled as a transmission line (refer to Figure 1.15). The principle circuit elements of the transmission line model incorporate an accurate model of the capacitance and the inductance. For a CNT of diameter d placed on a dielectric of thickness h (backplane connected to ground), the electrostatic capacitance comes out to be

$$C_E = \frac{2\pi\epsilon}{cosh^{-1}(2h/d)} \tag{1.19}$$

which for typical experimental setups is in the order of tens of $aF/\mu m$. To add to this, the quantum capacitance in carbon nanotubes is of considerable significance. The quantum capacitance can be expressed as [8]

$$C_Q = \frac{2e}{v_F} \tag{1.20}$$

where, $v_F$ is the Fermi velocity of electrons in the CNT. Numerically $C_Q$ comes out to be $100aF/\mu m$ proving thereby that both electrostatic as well as the quantum capacitances play pivotal role in the CNTs.

The inductance of a CNT comprises of a series connection of of the kinetic inductance and the magnetic inductance. The kinetic inductance can be expressed as,

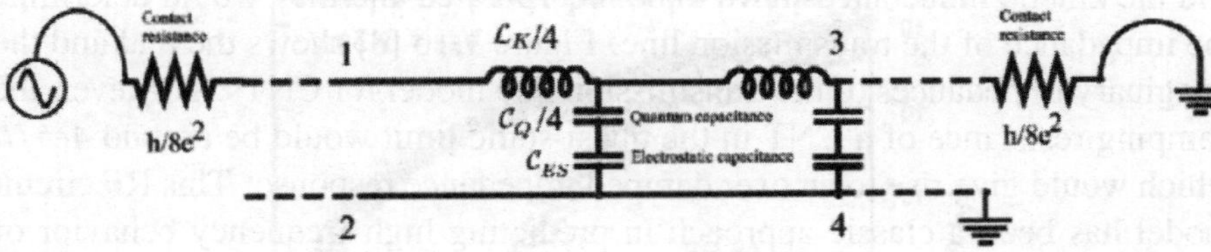


*Figure 1.15.* A simple transmission line RF circuit model for metallic carbon nanotubes. The contact resistances are the quantum resistances (assuming ballistic transport at dc). $C_Q$ represents the quantum capacitance and $C_{ES}$ represents the electrostatic capacitance. $L_K$ represents the magnetic inductance. $C_Q$ and $L_K$ are divided by four to account for the spin-charge separation. The magnetic inductance (which is much smaller than the kinetic inductance) has been neglected.

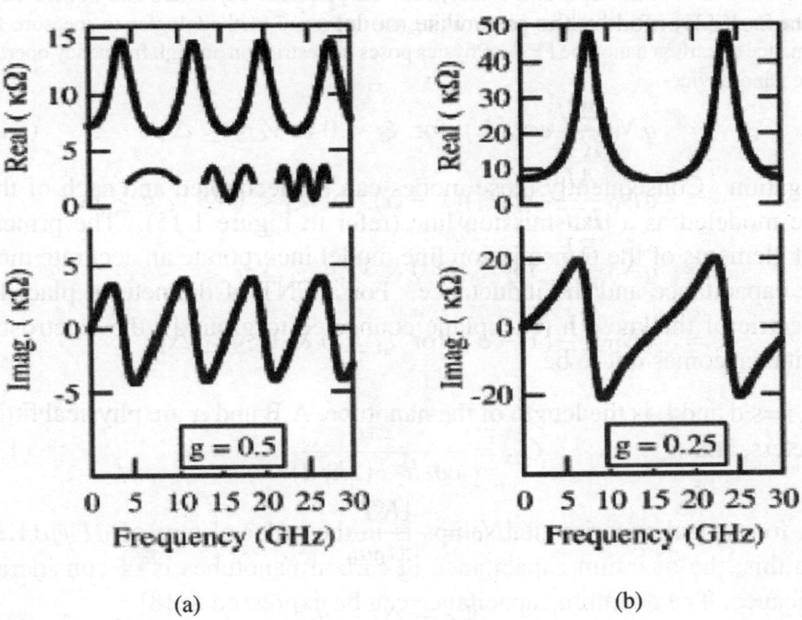


*Figure 1.16.* The real and imaginary parts of impedance for two different values of g. 'g' is the ratio of the semiconducting to the quantum capacitances.

$$L_K = \frac{h}{2e^2 v_F} \tag{1.21}$$

and numerically it turns out to be $16nH/\mu m$. This is larger than the magnetic inductance of CNTs, the latter being numerically in the order of $pF/\mu m$. Thus in nanoscaled transmission line model of the CNTs, the quantum capacitance and the kinetic inductances have to be incorporated and they would determine the impedance of the transmission line. Figure 1.16 [8] shows the real and the imaginary impedances of the transmission line model for CNTs. However, the damping resistance of a CNT in the quasi-static limit would be around $4e^2/h$ which would give rise to an over damped impedance response. This RF circuit model has been a classic approach in predicting high frequency behavior of CNTs.

**Spice Compatible Model of CNFETs in the Ballistic Performance Limit.** Attempts have been made to incorporate CNFETs in a circuit simulation (SPICE) environment by proper modeling of the current-voltage and the capacitance-voltage relations. One such methodology to model ballistic CNFETs has been discussed extensively in [49]. The circuit compatible model of the CNFET has been illustrated in Figure 1.17. The current source $I_{DS}$ is a non-linear current source and the capacitances $C_{GS}$ and $C_{GD}$ are non linear capacitances governed by the piecewise model as,

$$\begin{aligned} C_{Gi} &= qN_0\frac{AL}{KT}exp(\xi_i) \text{ for } \xi_i < 0 \,\&\, V_{GS} \leq \Delta_1 \qquad (1.22)\\ &= qN_0\frac{AL}{KT}exp(\xi_i)(1-\alpha) \text{ for } \xi_i < 0 \,\&\, V_{GS} \geq \Delta_1 \\ &= qN_0\frac{BL}{KT} \text{ for } \xi_i \geq 0 \,\&\, V_{GS} \leq \Delta_1 \\ &= qN_0\frac{BL}{KT}(1-\alpha) \text{ for } \xi_i \geq 0 \,\&\, V_{GS} \geq \Delta_1 \end{aligned}$$

where, i =s,d and L is the length of the nanotube, A,B and $\alpha$ are physical fitting parameters, and

$$N_0 = \frac{4KT}{3\pi t_0 a_0} \qquad (1.23)$$

$$\xi_i = \frac{\Psi_S - \Delta_1 - \mu_i}{KT} \qquad (1.24)$$

$t_0$ is the carbon-carbon (C-C) bonding energy ($\approx$ 3eV), $a_0$ the C-C bonding distance ($\approx$0.142nm), $\Delta_1$ is half the band-gap of the CNFET, $\Psi_S$ is the surface potential and $\mu_S$ and $\mu_d$ are the source and the drain Fermi-levels respectively. The current $I_{ds}$ is given by the Eq. 1.18. Figure 1.17 illustrates the match between this circuit compatible model and a detailed atmistic numerical simulation. Simulations have also been carried out with ideal CNFET based logic gates and ripple carry adders [49] (where, all extrinsic parasitic capacitances have been neglected).

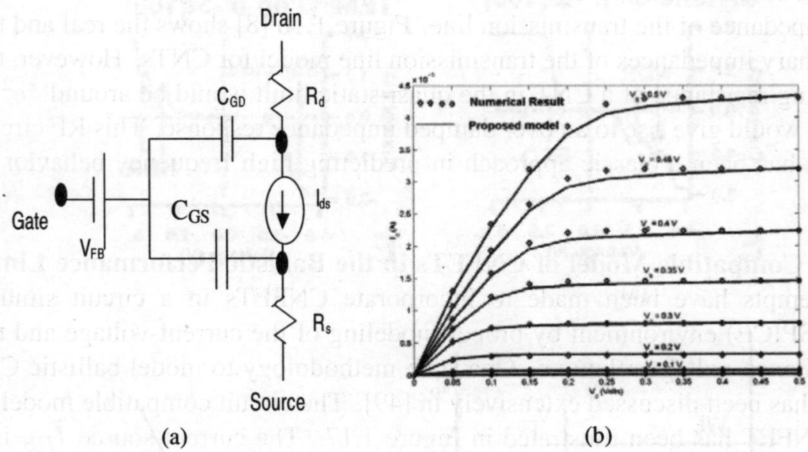


*Figure 1.17.* **(a) The proposed compact model for ballistic CNFETs (b) The $I_D - V_G$ characteristics of a CNFET showing a close match between numerical simulations and the proposed model.**

Carbon nanotubes have thus attracted the fancy of physicists, device engineers and circuit designers. Research has begun to harness the potential of these nano devices and use carbon nanotube based transistors in integrated circuit design for the future generations. Although our understanding of these devices needs to be furthered and a considerable portion of the theoretical work has not yet been demonstrated in experiments, the promise is enormous. Like any other device that is in its premature state, reliable production of these devices is definitely an issue and an enormous amount of research is necessary to build CNFETs with performances matrices comparable to the modern day Silicon MOSFETs. However, with their super-scaled dimensions, reliable and high current carrying capabilities and strong mechanical properties, CNTs have emerged as champions among the different revolutionary non-silicon devices that are being explored worldwide.

## 1.4 Molecular Diodes and Switches

Although carbon nanotubes have been the centre of exploratory research as an alternative to Silicon devices, other molecular devices have also gained popularity [50, 2, 44, 12, 10, 47, 42]. The simplest molecular electronic component: one molecule between two metal electrodes: has recently been demonstrated by several groups, leading to diverse charge transport behavior including gaps in conduction [50], Coulomb blockade [2], current rectification [44], bistable switching [12], negative differential resistance (NDR) [25], and Kondo reso-

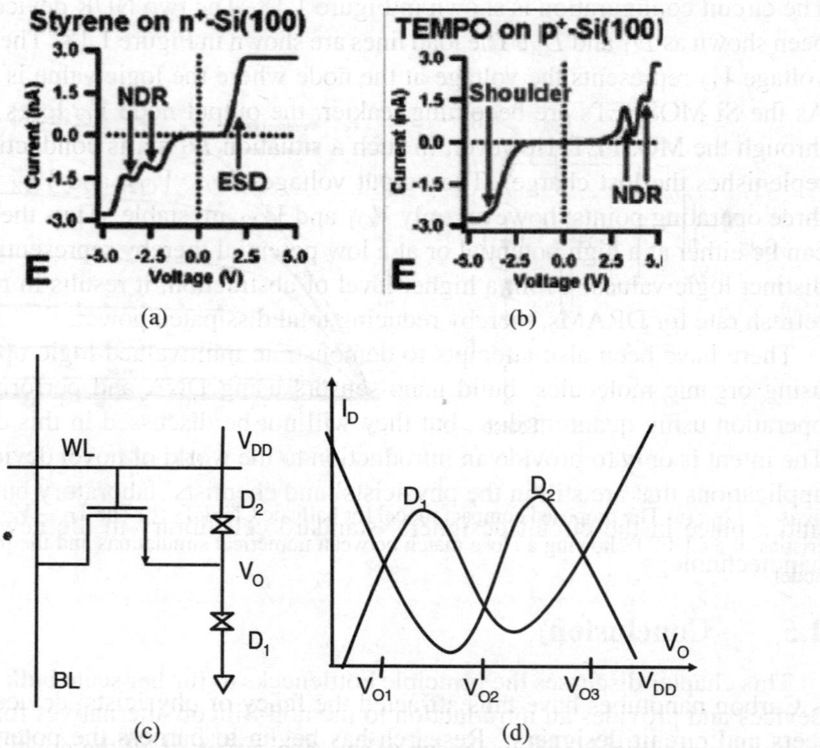


*Figure 1.18.* NDR effects in styrene (a) and TEMPO (b) [25]. DRAM cell with molecular RTDs (c) the circuit schematic and (d) the I-V characteristics.

nance [10, 47]. Although the choice of molecule plays a critical role in the operation of metal-molecule-metal junctions, the interface between the molecule and the electrodes has come under increasing scrutiny due to limited agreement between experimental and theoretical analyses. The apparent importance of the contacts in molecular electronic devices suggests that alternative electrode materials may lead to unique charge transport phenomena. For example, recent results show that self-assembled organic monolayers on Silicon lead to current rectification. Recently, styrene and 2,2,6,6-tetramethyl-1-piperidinyloxy (TEMPO) have been shown to have Negative Differential Resistance (NDR) when connected between metal (Au) on one hand and Silicon on the other [25]. This provides concrete motivation for an integration of molecular devices with nanoscaled Silicon.

Figure 1.18 illustrate the NDR effect when styrene is in contact with n+ Silicon and TEMPO is in contact with p+ Silicon. At a bias voltage of around 2.5 V, both the molecules exhibit distinct NDR effects. This can be utilized in a number of ways, the most popular being to reduce the refresh rate of DRAMs.

The circuit configuration is shown in Figure 1.18. The two NDR devices have been shown as $D_1$ and $D_2$. The load lines are shown in Figure 1.18. The output voltage $V_O$ represents the voltage at the node where the logic value is stored. As the Si MOSFETs are becoming leakier, the output node $V_O$ loses charge through the MOSFET. However, in such a situation $D_1$ starts conducting and replenishes the lost charge. The output voltages $V_{O1}$, $V_{O2}$, and $V_{O3}$ are the three operating points; however only $V_{O1}$ and $V_{O2}$ are stable. Thus the output can be either at a high potential or at a low potential thereby representing two distinct logic values. From a higher level of abstraction, it results in reduced refresh rate for DRAMs, thereby reducing total dissipated power.

There have been also attempts to demonstrate multivalued logic operation using organic molecules, build nano-sensors using DNA and perform logic operation using quantum dots, but they will not be discussed in this chapter. The intent is only to provide an introduction to the world of novel devices and applications that are still in the physicists' and chemists' laboratory but might find a place in the circuit designers' standard cell library in the new era of nanotechnology.

## 1.5 Conclusion

This chapter discusses the principle bottlenecks to further scale bulk Silicon devices and provides an introduction to the non-Silicon alternatives for future technologies. It is, by no means, an exhaustive survey of all the novel devices that are being extensively studied, but is principally an attempt to make bring them within the circuit designers' horizon. The physical properties of carbon nanotubes and fundamentals of carbon nanotube circuit design have been discussed. With the need for higher and higher integration density and complex on-chip functionality, the laws of physics would be taken to their limits and circuit and system designers; would have an increasing important role to play. Understanding the principles of operation of such ultra-scaled devices and using them in the ICs of the future generation would be a hurdle that the device, circuit and the architecture communities have to overcome together.

## References


[1] 2001 international technology roadmap for semiconductors. http://public.itrs.net/.

[2] R.P. Andres, T. Bein, M. Dorogi, S. Feng, J.I. Henderson, C.P. Kubiak, W. Mahoney, R.G. Osifchin, and R. Reifenberger. *Science*, 272:1323–1325, 1996.

[3] A. Bachtold, P. Hadley, T. Nakanishi, and C. Dekker. Logic circuits with carbon nanotube transistors. *Science*, 294:1317–1320, 2001.

[4] R. H. Baughman, A. A. Zakhidov, and W. A. de Heer. Carbon nanotubes-the route toward applications. *Science*, 297:787, 2002.

[5] S. Borkar, T. Karnik, S. Narendra, J. Tschanz, A. Keshavarzi, and V. De. Parameter variations and impact on circuits and microarchitecture. In *Proceedings of Design Automation Conference*, pages 338–342, June 2003.

[6] J. Brews. *High Speed Semiconductor Devices*. John Wiley & Sons, New York, 1990.

[7] P.J. Burke. Luttinger liquid theory as a model of the gigahertz electrical properties of carbon nanotubes. *IEEE Transactions on Nanotechnology*, 1(3):129–144, September 2002.

[8] P.J. Burke. An RF circuit model for carbon nanotubes. *IEEE Trans Nanotechnology*, 2(1):55–58, March 2003.

[9] K. Cao, W.-C.Lee, W.Liu, X.Jin, P.Su, S. Fung, J. An, B.Yu, and C. Hu. Bsim4 gate leakage model including source drain partiotion. *IEDM Technical Digest. Electron Devices Meeting*, 2000.

[10] J. Chen, M.A. Reed, A.M. Rawlett, and J.M. Tour. *Science*, 286:1550–1552, 1999.

[11] T. Chen and S. Naffziger. Comparison of adaptive body bias (abb) and adaptive supply voltage (asv) for improving delay and leakage under the presence of process variation. *IEEE Transactions on Very Large Scale Integration (VLSI) Systems*, 11(5):888–899, October 2003.

[12] C.P. Collier, G. Mattersteig, E.W. Young, Y. Luo, K. Beverly, J. Sampaio, F.M. Raymo, J.F. Stoddart, and J.R. Heath. *Science*, 289:1172–1175, 2000.

[13] S. Datta. *Quantum transport: from Atom to Transistor*. Cambridge Unievrsity Press, 2003.

[14] V. De and S. Borkar. Technology and design challenges for low power and high performance. In *Proc. of Intl. Symp. on Low Power Electronics and Design*, pages 163–168, August 1999.

[15] R. H. Dennard, F. H. Gaensslen, H. N. Yu, V. L. Rideout, E. Bassous, and A. R. LeBlanc. Design of ion-implanted MOSFETs with very small physical dimensions. *IEEE J. Solid-State Circuits*, 1974.

[16] M. Dresselhaus, G. Dresselhaus, and P. Avouris. *Carbon Nanotubes: Synthesis, Structure, Properties, and Applications*. Springer-Verlag, New York, 2001.

[17] M. Dresselhaus, G. Dresselhaus, and P. Eklund. *Science of Fullerenes and Carbon Nanotubes*. Academic Press, New York, 1996.

[18] J. Guo et. al. Assessment of silicon mos and carbon nanotube fet performance limits using a general theory of ballistic transistors. *Technical Digest IEDM*, 2002.

[19] K. Bowman et. al. Impact of die-to-die and within die parameter fluctuations on the clock frequency distribution in gigascale integration. *Journal of Solid State Circuits*, 37:183–190, 2002.

[20] Lo et. al. Modeling and characterization of N/sup +/- and P/sup +/- polysilicon-gated ultra thin oxides (21-26 /spl aring/). *Symposium on VLSI Technology*, 1997.

[21] R. Dennard et al. Design of ion-implanted mosfet's with very small physical dimensions. *IEEE J. Solid State Ckt.*, page 256, October 1974.

[22] Vivek De et. al. *Techniques for leakage power reduction.* IEEE Press, Piscataway NJ, 2000.

[23] D.J. Frank and J. Appenzeller. High-frequency response in carbon nanotube field-effect transistors. *Electron Device Letters*, 25(1):34–36, January 2004.

[24] A.S. Grove. *Physics and Technology of Semiconductor Devices*. John Wiley & Sons, 1967.

[25] Nathan P. Guisinger, Mark E. Greene, Rajiv Basu, Andrew S. Baluch, and Mark C. Hersam. Room temperature negative differential resistance through individual organic molecules on silicon surfaces. *Nano Lett.*, 4(1):55–59, 2004.

[26] J. Guo, S. Goasguen, M. Lundstrom, and S. Datta. Metal-insulator-semiconductor electrostatics of carbon nanotubes. *Appl. Phys. Lett.*, 81:1486, 2002.

[27] J. Guo, M. Lundstrom, and S. Datta. Performance projections for ballistic carbon nanotube field-effect transistors. *Appl. Phys. Lett.*, 80:3192, 2002.

[28] S. Guo, S. Datta, and M. Lundstrom. A numerical study of scaling issues for schottky barrier carbon nanotube transistors. *IEEE Transactions on Electron Devices*, 51:172, February 2004.

[29] F. Hamzaoglu and M. Stan. Circuit-level techniques to control gate leakage for sub-100 nm cmos. *International Symposium on Low Power Design*, 2002.

[30] S. Heinze, J. Tersoff, R. Martel, V. Derycke, J. Appenzeller, and Ph. Avouris. Carbon nanotubes as schottky barrier transistors. *Phys. Rev. Lett.*, 89:106801, 2002.

[31] S. Iijima. Helical microtubules of graphitic carbon. *Nature*, 354:56, 1991.

[32] S. Iijima and T. Ichilashi. Single-shell carbon nanotube of 1nm diameter. *Nature*, 363:603–605, 1993.

[33] J. Jacobs and D. Antoniadis. Channel profile engineering for mosfet's with 100 nm channel lengths. *IEEE Transactions on Electron Devices*, 42:870–875, May 1995.

[34] A. Javey, J. Guo, M. Paulsson, Q. Wang, D. Mann, M. Lundstrom, and H. Dai. High-field, quasi-ballistic transport in short carbon nanotubes. *Physical Review Letters*, 92:106804, 2004.

[35] A. Javey, J. Guo, and Q. Wang. Nanotechnology: A barrier falls. *Nature*, 424:654, 2003.

[36] A. Javey, H. Kim, M. Brink, Q. Wang, A. Ural, J. Guo, P. McIntyre, P. McEuen, M. Lundstrom, and H. Dai. High dielectrics for advanced carbon nanotube transistors and logic. *Nature Materials*, 1:241, 2002.

[37] A. Javey, Q. Wang, A. Ural, Y. Li, and H. Dai. Carbon nanotube transistor arrays for multi-stage complementary logic and ring oscillators. *Nano Lett.*, 2:929–932, 2002.

[38] A. Javey, Q. Wang, A. Ural, Y. Li, and H. Dai. Carbon nanotube transistor arrays for multi-stage complementary logic and ring oscillators. *Nano Lett. 2*, pages 929–932, 2002.

[39] Ji-Yong, Sami Rosenblatt, Yuval Yaish, Vera Sazonova, Hande Üstünel, Stephan Braig, T. A. Arias, Piet W. Brouwer, and Paul L. McEuen. Electron-phonon scattering in metallic single-walled carbon nanotubes. *cond-mat/0309641*, September 2003.

[40] Keshavarzi, K. Roy, , and C. F. Hawkins. Intrinsic leakage in low power deep submicron cmos ics. *Int. Test Conf.*, pages 146–155, 1997.

[41] W. Liang, M. Bockrath, and D. Bozovic. *Nature*, 411:665, 2001.

[42] W. Liang, M.P. Shores, M. Bockrath, J.R. Long, and H. Park. *Nature*, 417:725–729, 2002.

[43] C. Mead. Scaling of MOS technology to submicrometer feature sizes. *Analog Integ. Ckt. and Signal Process.*, 6:9–25, 1994.

[44] R.M. Metzger and J. Mater. *Chem.*, 10:55–62, 2000.

[45] S. Mukhopadhyay, C. Neau, R.T. Cakici, A. Agarwal, C.H. Kim, and K. Roy. Gate leakage reduction for scaled devices using transistor stacking. *Very Large Scale Integration (VLSI) Systems, IEEE Transactions on*, 11(4), August 2003.

[46] S. Mukhopadhyay and K. Roy. Modeling and estimation of total leakage current in nano-scaled-CMOS devices considering the effect of parameter variation. In *Proceedings of the 2003 International Symposium on ISLPED '03.*, pages 172–175, Aug 2003.

[47] J. Park, A.N. Pasupathy, J.I. Goldsmith, C. Chang, Y. Yaish, J.R. Petta, M. Rinkoski, J.P. Sethna, H.D. Abruna, P.L. McEuen, and D.C. Ralph. *Nature*, 417:722–725, 2002.

[48] R. Pierret. *Semiconductor Device Fundamentals*. Addison-Wesley, MA, 1996.

[49] A. Raychowdhury, S. Makhopadhya, and K. Roy. Modeling of ballistic carbon nanotube field effect transistors for efficient circuit simulation. *International Conference on Computer Aided Design*, pages 487–490, November 2003.

[50] M.A. Reed, C. Zhou, J. Muller T.P. Burgin, and J.M. Tour. *Science*, 278:252–254, 1997.

[51] K. Roy, S. Mukhopadhyay, and H. Mahmoodi-Meimand. Leakage current mechanisms and leakage reduction techniques in deep-submicrometer CMOS circuits. In *Proceedings of the IEEE*, volume 91, Feb 2003.

[52] K. Roy and S. C. Prasad. Wiley Interscience Publications, New York, 2000.

[53] G. Sery. Life in CMOS: Why chase life after that. *Proceeding of Design Automation Conference*, pages 78–83, June 2002.

[54] Y. Taur. *CMOS Scaling and Issues in sub-0.25?m Systems*. IEEE Press, Piscataway NJ, 2000.

[55] Y. Taur and T. H. Ning. *Fundamentals of Modern VLSI Devices*. Cambridge University Press, New York, 1998.

[56] Thompson, P. Packan, , and M. Bohr. Linear versus saturated drive current: tradeoffs in super steep retrograde well engineering. *Symposium on VLSI Technology*, pages 154–155, 1996.

[57] S. Thompson, P. Packan, and M. Bohr. MOS scaling: Transistor challenges for the 21st century. *Intel Technology Journal*, 1998.

[58] S. Venkatesan, J.W. Lutze, C. Lage, and W.J. Taylor. Device drive current degradation observed with retrograde channel profiles. *International Electron Devices Meeting*, pages 419–422, 1995.

[59] S. Wind, J. Appenzeller, R. Martel, V. Derycke, and P. Avouris. Vertical scaling of carbon nanotube field-effect transistors using top gate electrodes. *Appl. Phys. Lett.*, 80:3817–3819, 2002.

[60] N. Yang, W. Henson, and J. Hauser. Modelling study of ultrathin gate oxides using tunneling current and capacitance-voltage measurement in mos devices. *IEEE Trans On Elec Dev*, 46(7), july 1999.

[61] Z. Yao, C. L. Kane, and C. Dekker. High-field electrical transport in single-wall carbon nanotubes. *Phys. Rev. Lett.*, 84:2941–2494, 2000.

II

# DEFECT TOLERANT NANO-COMPUTING

# Preface

In his famous book from 1956, "Probabilistic Logics and the Synthesis of Reliable Organisms from Unreliable Components", John von Neumann had talked about building reliable computing systems from fault-prone components. Historically at that point in time, it made perfect sense. The valves that were used as switching devices at that time used to burn out frequently, and hence the ability to maintain reliability in the presence of such possibilities was very important. However, with the advent of silicon transistors, and with the increasing perfection of semiconductor manufacturing, for decades, this work was of lesser importance. Of course, fault-tolerant computing has always been a major research area, especially for mission critical systems. Nevertheless, the idea of designing systems assuming non-zero defect probabilities in the substrate, has not been common until recently, as researchers consider the use of nanotechnology based devices as switching elements.

In current semiconductor manufacturing processes, immediately after a silicon wafer is fabricated, tests are carried out and defective parts are rejected. Manufacturing yield is determined by the percentage of parts that come out without defects. Unfortunately, with silicon-based devices scaled down to a few nanometers, or even newer technologies (such as molecular self-assembled substrates, quantum dot cellular automata type devices, or carbon nanotube based switches), the defect probability will be quite high. As a result, throwing away parts that contain any defect at all will not be an option. Instead, robust systems will have to be designed using these fault-prone substrates. As a result, the notion of defect-tolerance will be a prime force in design. Instead of computing manufacturing yield in the present manner, parts will likely be graded according to their reliability characterizations.

With this in mind, researchers have already started looking into defect-tolerant computing. There are two major schools of thoughts in defect-tolerance. One is based on reconfigurability, and often self-reconfigurability of systems. Such systems are designed to diagnose and map their own fault locations, and reconfigure the computing around the faulty regions, so as to provide reliable results. The other strain of research is based on probabilistic characterization of the faulty substrates. Redundancy based defect-tolerance is



*S.K. Shukla and R.I. Bahar (eds.), Nano, Quantum and Molecular Computing*, 37-38.

then designed into the system in order to guarantee a certain level of reliability given the particular characterization of the faults.

In this part of the book, we present six chapters. The first two focus mainly on fault detection and reconfigurability-based defect tolerance. The last four are based on defect-tolerance through probabilistic assumptions. Due to the very nature of the computation at the nano-scopic level, new models of computation — distinguishable from Boolean logic — may actually help in understanding the interaction between quantum phenomena and valid computation. In particular, Chapter 5 in this part presents one such model of computation that takes into account thermal energy and information theoretic bounds on energy expenditure in bit transformations. In addition, the final chapter in this section takes a statistical view of computing with quantum devices, and draws contrast against current silicon technologies.

We believe this part of the book will provide the readers with an interesting sampling of defect-tolerance work pertaining to nanotechnology. Also included in these chapters is discussion regarding the corresponding design automation problems and emerging tools and techniques needed to support these defect-tolerant systems.

Chapter 2

# NANOCOMPUTING IN THE PRESENCE OF DEFECTS AND FAULTS: A SURVEY

Paul Graham
*Los Alamos National Laboratory*
*Los Alamos, NM, USA*
grahamp@lanl.gov

Maya Gokhale
*Los Alamos National Laboratory*
*Los Alamos, NM, USA*
maya@lanl.gov

**Abstract** Computing systems implemented with nanotechnology will need to employ defect- and fault-tolerant measures to improve their reliability due to the large number of factors that may lead to imperfect device fabrication as well as the increased susceptibility to environmentally induced faults when using nanometer-scale devices. Researchers have approached this problem of reliability from many angles and this survey will discuss many promising examples, ranging from classical fault-tolerant techniques to approaches specific to nanocomputing. The research results summarized here also suggest that many useful, yet strikingly different solutions may exist for tolerating defects and faults within nanocomputing systems. Also included in the survey are a number of software tools useful for quantifying the reliability of nanocomputing systems in the presence of defects and faults.

**Keywords:** nanocomputing, redundancy, fault, defect, fault tolerance, defect tolerance

## Introduction

With the increasing fabrication costs of CMOS-based computing devices and the ever-approaching physical limits of their fabrication and use, a significant community of researchers are exploring a number of nanometer-scale



S.K. Shukla and R.I. Bahar (eds.), *Nano, Quantum and Molecular Computing*, 39-72.

alternatives to existing CMOS silicon technology. Devices and architectures of interest include carbon nanotubes[34], single-electron transistors (SETs)[29], quantum-dot cellular automata (QCA)[47], molecular devices[48], and quantum computing (see Chapter 8), to name just a few. With this demand to use nano-scale devices to create more powerful and complex computers, a significant concern is the reliability of these devices and the systems built with them. For instance, the chemical processes for building molecular devices will have significantly lower yields than those obtained through current fabrication practice, resulting in aggregates with high defect rates. Since they manipulate single electrons, devices such as QCA and SETs are susceptible to background charge fluctuations that can cause faults during operation. The decrease in the device scales used also means that radiation effects, electromagnetic interference, and power and temperature effects will all be more challenging to counteract. These effects are already being seen: upsets in 150-nm CMOS devices by atmospheric neutrons have already been observed [15].

The large number of factors that will affect the reliability of nanocomputing devices suggests that defect and fault tolerance will be an integral part of device and system design for nanocomputing. In this chapter, we will provide a brief survey of the many techniques and tools being researched to aid in constructing reliable nanocomputers from unreliable nano-scale devices. These techniques range from applications of classical defect- and fault-tolerant techniques to techniques that are unique to nano-scale devices. We will first define key terms and concepts that will be used throughout the paper and the remaining sections will cover classical error masking and reconfiguration techniques (Section 2.2), non-traditional computing models and architectures (Section 2.3), and tools for defect and fault tolerance (Section 2.4). We will conclude the chapter with some general summary comments.

## 2.1 Background

Before surveying current research into reliable nanocomputing, we will define several terms and concepts that will be used throughout the chapter. Additionally, we will provide a framework for discussing the specific ideas being pursued in the nanocomputing community for building reliable systems.

A *defect*, or more specifically, a *manufacturing defect* is a physical problem with a system that appears as a result of an imperfect fabrication process. By contrast, a *fault* is an incorrect state of the system due to manufacturing defects, component failures, environmental conditions, or even improper design[46]. Faults can be:

- permanent, as in the case of physical defects or permanent device failures during the lifetime of the system;

- intermittent, in which faults may periodically stop and start, but are potentially detectable and repairable; or

- transient, where faults are due to temporary environmental conditions.

With this context, *defect tolerance* is the ability of a system to operate correctly in the presence of manufacturing defects while *fault tolerance* is the ability of the system to operate correctly in the presence of permanent, intermittent, and transient faults. Clearly, fault tolerance encompasses defect tolerance but also implies the ability to withstand temporary faults as well. Generally, both defect and fault tolerance require redundancy to overcome problems within the system. This redundancy may be in terms of the replication of functions temporally or physically, or using techniques such as error-control coding, which uses a redundancy in the code space for the data to detect and correct faults. Often, the system must be able to reconfigure its resources to take advantage of redundant components.

In the following sections of this chapter, the techniques used fall into two broad categories:

- classical error detection, masking, and reconfiguration, and

- non-classical and/or nano-scale-specific defect- and fault-tolerant techniques.

Classical error-masking includes techniques such as N-modular redundancy and error-control coding. Reconfiguration involves using "spare" redundant hardware to replace defective hardware and can be performed at a large scale because of the massive amount of redundant hardware that is feasible to assemble at the nano-scale. This level of redundancy may well be necessary, considering the expected high hardware defect rates at the nanometer scale. The second category includes techniques that are not included in the above more classical defect- or fault-tolerant techniques, such as the use of artificial neural networks for fault tolerance [41], and includes some techniques that are very specific to nano-scale devices.

Additionally, tolerance to defects and faults can be accomplished at several different levels of abstraction: the physical device level, the architectural level, and the application level. The *physical device level* refers to specific features of nano-scale devices (such as QCA cells, SETs, CNTs, nanocells, etc.) that provide tolerance to defects or device failures. At the *architectural level*, defect and fault tolerance is achieved through techniques of assembling collections of these nano-scale devices. Finally, defect and fault tolerance at the *application level* involves features of the computing applications themselves that allow them to operate correctly despite defects and faults in the computing system on which they execute. Though nano-scale devices have some features that may make them tolerant to some defects and faults and though some applications themselves may have some inherent defect or fault tolerance, most of the techniques being investigated relate to the architectural level.

To maximize the reliability of a system based on nano-scale devices for a given or minimal cost, we expect that it may require a combination of techniques at various levels of abstraction, from the device to the application level. For instance, an extremely high degree of hardware reliability may not be necessary if the application itself can handle a certain degree of noise in its data. In such a case, it may not be cost effective to use extreme levels of redundancy when the application doesn't require it.

## 2.2 Error Detection, Masking, and Reconfiguration

To mitigate the effects of both faults and defects in designs, several traditional techniques have been suggested and evaluated for nanocomputing, assuming a high rate of defects and/or faults. Some of these more traditional approaches include triple-modular redundancy, N-modular redundancy, cascaded triple-modular redundancy, and NAND multiplexing [49]. Some of these techniques are based on the early work of John von Neumann on creating reliable systems using unreliable components [49]—a significant concern with early computing systems and a concern with nanocomputing as well. Other techniques which may be applicable include error-control coding and reconfiguration. We will briefly describe each of these techniques and summarize the results of applying them to nanocomputing.

### Triple- and N-Modular Redundancy

With triple-modular redundancy (TMR), three copies of the same hardware are executed with common inputs so that, ideally, they produce the same outputs if all modules are defect or fault free. Since it is assumed that, at most, any one module may either have a defect or fault during operation, the outputs of the three modules are then combined using majority vote circuitry, which selects the output which is most common, i.e., provided by two or three of the modules. TMR with a single voter is illustrated in Figure 2.1. This technique is fairly easy to apply to digital logic at the cost of increased circuit area and power and decreased circuit speed. One of the strengths of TMR is that it can tolerate multiple failures in a single module. Faults or defects in two or more of the three modules will cause the logic to fail. To improve the reliability of the whole system, three voters can be used instead of just one, as illustrated in Figure 2.2, removing the voter as the single point of failure.

As a generalization of TMR, N-modular redundancy (NMR) uses $N$ copies of the hardware, $N$ being an odd number so no "tie" votes are possible. NMR is illustrated in Figure 2.3. Again, the output of the hardware is determined

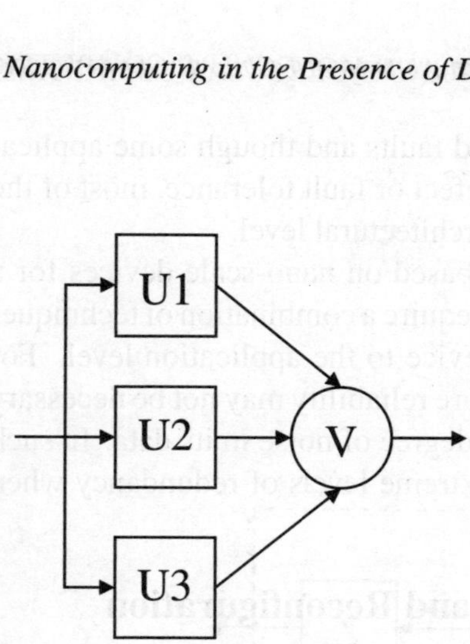


*Figure 2.1.* Triple-Modular Redundancy with a Single Voter

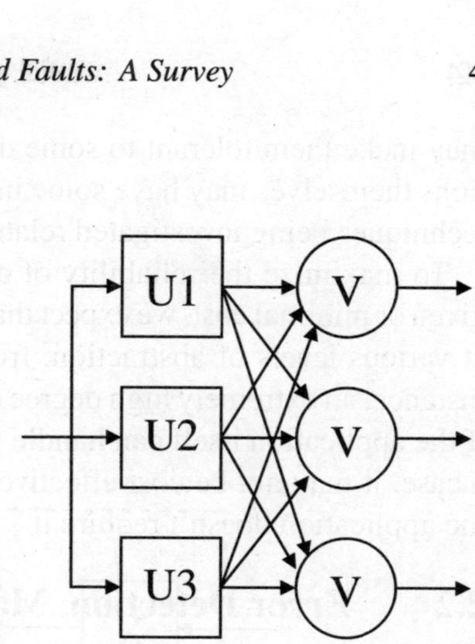


*Figure 2.2.* Triple-Modular Redundancy with Three Voters

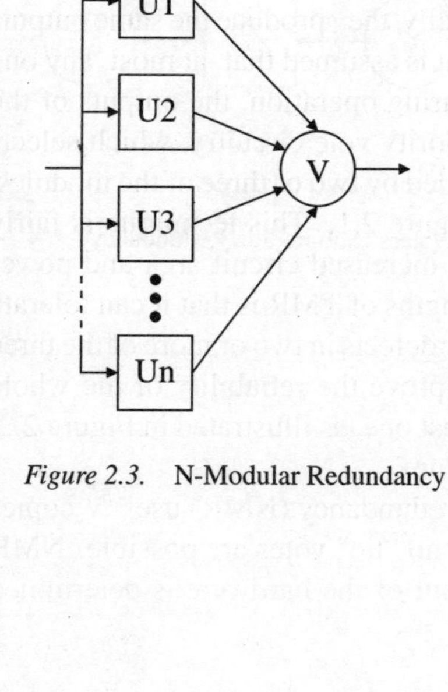


*Figure 2.3.* N-Modular Redundancy

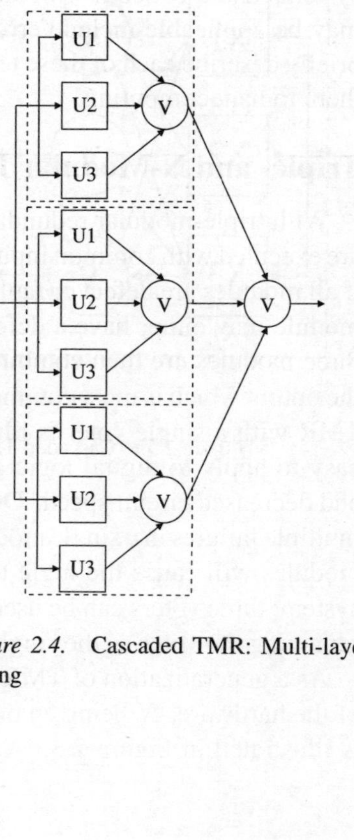


*Figure 2.4.* Cascaded TMR: Multi-layer Voting

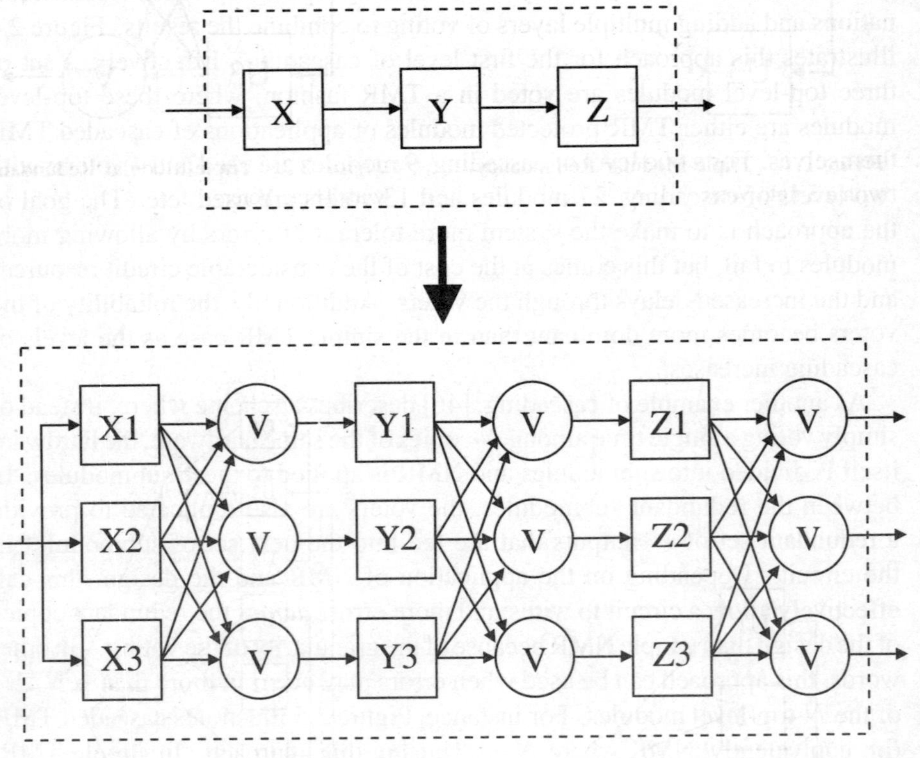


*Figure 2.5.* Cascaded TMR with Triple Voters: Submodule Redundancy

using a majority ($\lfloor (N/2) + 1 \rfloor$ or more outputs) voting scheme. With a cost of more than a factor of $N$ when considering the voting circuitry and modular redundancy, the benefit of NMR over TMR is that it will correctly compute the output with multiple defects or faults in $\lfloor (N/2) \rfloor$ of the modules. As with TMR, the NMR technique can take advantage of redundant voters to reduce the probability of system failure due to a single defective or faulty voter.

Another variation on these approaches is to use cascaded NMR. Many approaches to cascaded NMR exist. For instance, [37] describes a cascading scheme to increase the reliability of a module by replicating TMR-voter combinations and adding multiple layers of voting to combine the results. Figure 2.4 illustrates this approach for the first level of cascading. Effectively, a set of three top-level modules are voted in a TMR fashion, where these top-level modules are either TMR-protected modules or applications of cascaded TMR themselves. For a one level of cascading, 9 modules are used and 4 voters; with two levels of cascading, 27 modules and 13 voters are used; etc. The goal of the approach is to make the system more tolerant of errors by allowing more modules to fail, but this comes at the cost of the considerable circuit resources and the increased delays through the voters. Additionally, the reliability of the voters becomes more dominant than in the simple TMR case as the levels of cascading increases.

As another example of cascading, [46] describes a scheme where, instead of simply voting a single time among $N$ copies of the same hardware, the hardware itself is divided into submodules and NMR is applied to these submodules. In between the redundant submodules, the voters are also replicated to provide a redundant set of $N$ outputs that are fed into the next set of submodules in the circuit. Depending on the application of NMR and the design, this can effectively allow a circuit to withstand more errors *across* the redundant copies of the design than simple NMR because of the granularity of the voting—in other words, this approach can be used when errors may occur in more than $\lfloor (N/2) \rfloor$ of the $N$ top-level modules. For instance, Figure 2.5 illustrates cascaded TMR (or, equivalently, NMR where $N = 3$) using this approach. In simple TMR, only one of the three modules can have any faults. In the cascaded approach, errors could exist in submodules $X1$, $Y2$, and $Z3$ (i.e., any single submodule from each set of submodules) without affecting the performance of the design.

A good example of this second form of cascading can be found in work related to the reliability of SRAM-based field-programmable gate arrays (FPGAs) in the presence of radiation-induced soft errors (or single-event upsets—SEUs)[30, 8, 44, 51]. In addition to simply applying cascaded TMR to submodules, these papers also apply another variation of TMR that illustrates an important issue. Effectively, TMR is performed on the digital circuits, but voters are only placed in feedback loops within the circuit and at circuit outputs (as necessary). With this approach, the researchers assume that very few faults exist in the circuit due to SEUs at any one time—something they can assume due to constant repair of the FPGA's programming data, but an assumption that will probably not be applicable to nanocomputing—and that the faults occur with uniform probability across an entire FPGA. Consequently, there is a very low probability that a TMR-protected submodule will fail. By placing voters in the feedback loops, the submodule affected by the SEU will be able to recover from soft errors in a short amount of time and resynchronize with the other two submodules in the TMR-protected unit. The net result is that the system can withstand transient faults in multiple submodules of the TMR *over time*, making the system more robust. The variations of TMR mentioned before do not address this synchronization issue with regards to feedback, meaning that even a transient fault at different times in two of the modules may cause the system to fail because the modules are no longer synchronized. Also, notice that the more control structures, counters, and other forms of feedback that exist in the circuit, the more submodules there are for applying cascaded TMR.

In each of these cases the defect and fault tolerance of the voters themselves must also be considered and balanced with the reliability of the modules being voted. If voting is performed at too small of a granularity, the reliability of the voters will dominate the system reliability. If the voting is performed at too coarse of a level, the number of defects and faults tolerated will be non-optimal. Both [5] and [6] illustrate the need for balancing the reliability of voters with the reliability of the redundant logic.

## NAND Multiplexing

Von Neumann, the originator of TMR, also developed a theory that has been termed "NAND multiplexing," which can be used to produce the expected function in the presence of a high number of defects and faults in its components—up to a failure probability of about .0107 for each component. This theory was developed during an era when the reliability of individual components used in building computers was low, thus, requiring designers to consider both defect and fault tolerance in their designs.

The scheme involves replicating the function to be multiplexed $N$ times. $N$ wires are used to carry the signal of each input and each output. Processing is performed in two stages: an executive stage and a restorative stage—see Figure 2.6 for an example. As described in [49], the executive stage performs the function using the $N$ copies of the original function unit. For each bundle of $N$ wires, the bundle is considered "stimulated" (logic '1') if at least $(1 - \Delta) \cdot N$ of the wires are "stimulated", where $0 \leq \Delta < 0.5$; likewise, an input is considered "unstimulated" (logic '0') if no more than $\Delta \cdot N$ wires are "stimulated." Stimulation levels in between these two values are considered undecided and would indicate that the circuit has failed or malfunctioned. Based on the probability of function failure and the probability of correctness of the inputs to the executive stage, the percentage of wires in the output bundle that are in the correct stimulated or unstimulated states may be lower than the fractions of the input-wire bundles that were correct.

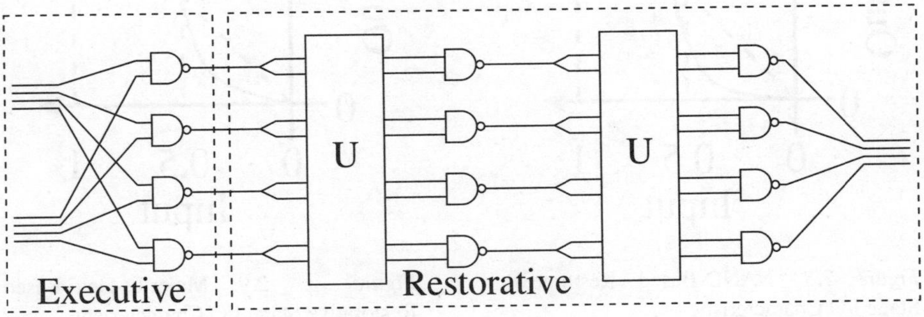


*Figure 2.6.* NAND Multiplexing with $N = 4$

As a result of this potential reduction in effective signal strength, von Neumann developed the restorative stage as a part of this multiplexing technique. The purpose of the restorative stage is to increase the number of wires in the output bundle that are in the majority state, whether the bundle is in the stimulated or unstimulated state. As suggested in [37], the restorative stage is, in effect, a non-linear amplifier. Figure 2.8 is the output characteristic of the restorative stage made up of NAND gates, as illustrated in Figure 2.6. Figures 2.7 and 2.9 illustrate the restorative stage made up of majority gates and its output characteristic, respectively. The restorative stage made from majority gates is more ideal, while the NAND-based version is made of simpler com-

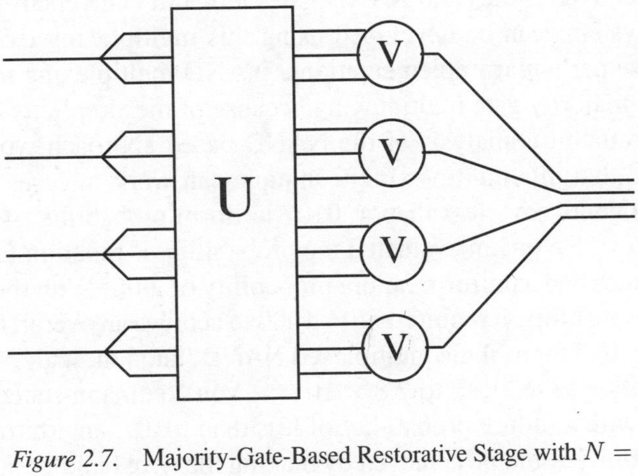


*Figure 2.7.* Majority-Gate-Based Restorative Stage with $N = 4$

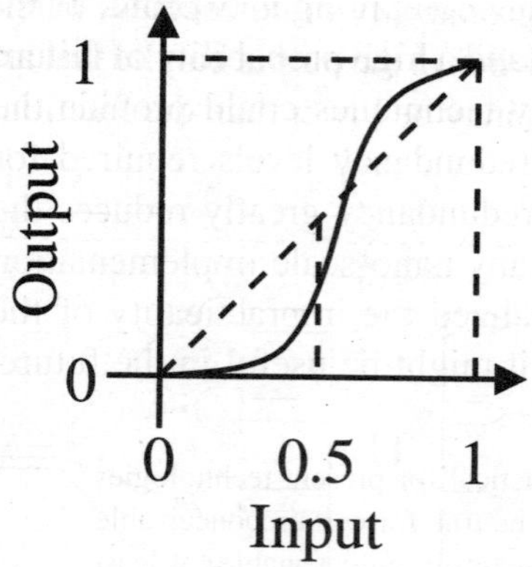


*Figure 2.8.* NAND-Based Restorative Stage I/O Characteristic

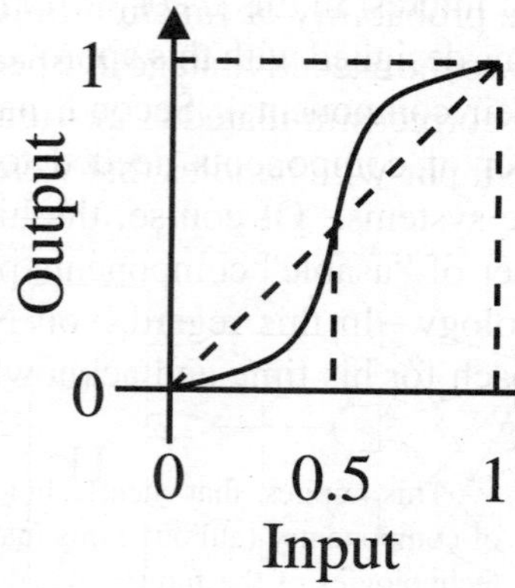


*Figure 2.9.* Majority-Gate-Based Restorative Stage I/O Characteristic

ponents. The boxes labeled "U" in the figures represent the multiplexing part of the technique. Effectively, they randomly permute their inputs to reduce the correlation of the error within the input bundles. The random permutation can also be performed in the executive stage, though, von Neumann's original work did not require it. Randomization of the error is made so that the restorative logic has a better chance of properly selecting the majority state for its output based on localized majority or majority-like logic decisions.

Several research groups in the nanotechnology community (and elsewhere) have described von Neumann's technique as "NAND multiplexing" because one of his analyses used NAND gates for both the executive and restorative stages. The technique is more general. For example, von Neumann initially presented the technique using majority gates for both stages. Other formulations of the technique are also possible, for example, some combination of NOR gates could be used. Since NOR gates and NAND gates are both "universal" gates, entire computing systems can be developed using this multiplexing technique with either of these particular implementations. NAND multiplexing may be more popular than majority gate multiplexing because of the simplicity of analysis.

Through reliability analysis of the NAND-based approach, von Neumann demonstrated that this multiplexing technique can work in cases where individual components have less than a .0107 uniform probability of failure. As an example, von Neumann evaluated the probability of function failure where the NAND gates had a uniform failure probability of .005. With this individual gate failure probability, it required $N = 1000$ to achieve an overall failure probability of just .027 for a single multiplexed NAND function; with $N = 20,000$, this probability was reduced to $2.8 \times 10^{-19}$. Von Neumann stated that, using components with a failure probability of less than .0107, an arbitrary function failure probability could be achieved by making the wire bundles large enough.

The interest in this theory with regards to nanocomputing is twofold. First, systems designed with this approach can withstand a high probability of failure for their components. Second, nano-assembly techniques could provide the number of components needed to reach the redundancy levels required for usable systems. Of course, the high cost of redundancy greatly reduces the number of "usable" components provided by any nano-scale implementation technology. In this regard, von Neumann realized the impracticality of the approach for his time and acknowledged that it might be useful in the future, saying:

> This implies, that such techniques are impractical for present technologies of componentry (although this may perhaps not be true for certain conceivable technologies of the future), but they are not necessarily unreasonable (at least not on grounds of size alone) for the micro-componentry of the human nervous system. [49]

Subsequent analyses of NAND multiplexing have refined the bounds for the maximum fault probability for each device and the reliability of NAND multiplexing approaches. For instance, Evans and Pippenger in [14] illustrated that for "noisy" NAND gates that the maximum probability of failure for each component is $(3 - \sqrt{7})/4 \approx 0.08856$. Additionally, Han and Jonker in [20] performed some additional reliability analysis of NAND multiplexing and illustrated that using more restorative stages improved the performance of NAND multiplexing. They also suggested that, for circuits with many levels of logic, the restorative stages are not necessary. Considering this savings and using values of $N < 1000$, Han and Jonker also suggested that nano-chips with $10^{12}$ devices may result in $10^9 \sim 10^{10}$ effective devices by using NAND multiplexing.

In [38], Norman and others performed a subsequent analysis of using additional restorative stages to improve circuit reliability and have noted a flaw in [20]'s modeling of the "U" random permutation. This improved analysis has shown that Han and Jonker's bounds given in [20] are not always either upper or lower reliability bounds. Further, for smaller device fault rates, [38] shows that increasing the number of restorative stages does improve reliability, while for larger fault rates the increase in restorative stages can actually degrade reliability.

## Error-Control Coding

Another conventional technique to mask the presence of defects and faults is to use error-control coding[50, 31]. With this technique, the redundancy exists in how the data itself is encoded, not in replicating hardware having the same function. The extra bits required for error-control coding are used to help hardware distinguish between error-free data and data with errors. Ideally, if too many errors do not occur, the hardware can also use the encoding redundancy to locate and fix the data errors, if they exist. A variety of error-control codes have been developed, ranging from the well known single-error correcting (SEC) Hamming codes to Reed-Solomon and convolutional codes. This variety of codes are used to handle a variety of different error conditions: single-bit errors; multiple, independent errors; multiple, consecutive (or burst) errors; etc.

Error-control coding is commonly used in conventional computing, communications, and storage systems. For example, high-reliability server computers (and even some personal computers) typically use some form of error-correcting-code (ECC) RAM to fix single-bit errors in data words. Often in these schemes, 8 check bits are added to each 64-bit word in memory to provide single-error-correct/double-error-detect (SED/DED) protection. As another example, Reed-Solomon codes are an integral part of making compact discs a feasible storage medium where fingerprints, dust, and scratches can contribute to large burst errors in a data stream. In summary, error-correcting codes are frequently found protecting data internal to microprocessors, in caches, on busses, and other places where data is either communicated or stored in modern computing systems.

The costs of this approach to redundancy for defect and fault tolerance include: the additional hardware to both encode, decode, and correct data (more hardware); increased latencies or decreased circuit speed (more time); and increased power consumption (more power). Further, for this approach to work, the hardware used to encode, decode, and correct data is generally assumed to be reliable—a significant issue when applying this idea to nanocomputing.

## Reconfiguration

Often in conjunction with redundancy and self-assembly, reconfiguration has been explored as a defect and fault mitigation method for molecular-scale computers. The basic idea of reconfiguration is that the capability exists within a system to modify functionality after manufacture. Reconfiguration is a widely recognized defect and fault management technique in conventional electronics. Examples at the computer system level include the ability to de-activate chips or cores within a chip upon error diagnosis; ability to switch to spare bits for single cell failures in cache memories; to delete cache lines to map out bad bits; to bypass a cache or an entire memory card; and to mark I/O resources unavailable upon diagnosis of I/O failure. Diagnostic hardware must exist to detect the failure. Once a failure has been detected, the failed unit must be by-passed and, if a redundant resource exists, the redundant unit is activated and wired in. This process is illustrated in Figure 2.10, where the black node has failed and is replaced by the spare in its row. The approach is advocated in a nanocomputing architecture proposed by Han and Jonker [21] that combines

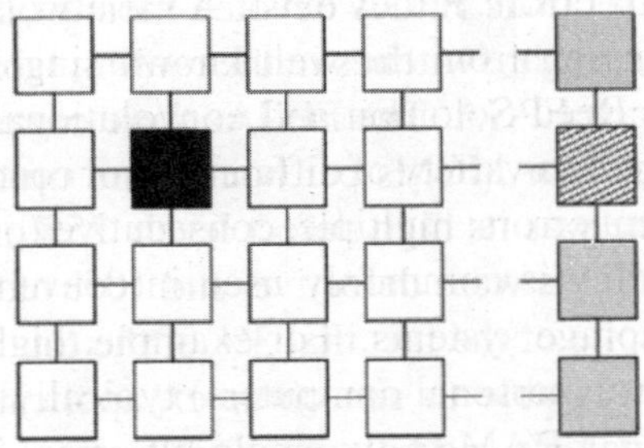

*Figure 2.10.* Fault Avoidance through Redundancy and Reconfiguration

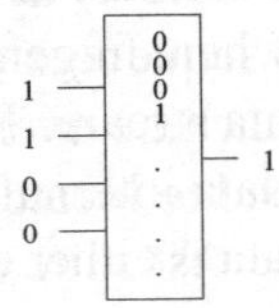

*Figure 2.11.* 4-input Look-Up Table

NAND multiplexing with additional redundancy for reconfiguration (see the Discussion portion of this section below).

While specific circuitry can be inserted into various components of a computer system to support dynamic reconfiguration of specific system resources, an alternative that has been proposed by several nanocomputing research groups is to build intrinsically reconfigurable architectures. Since it appears that a much higher degree of defects and faults must be tolerated in nanocomputers than traditional CMOS circuits, an emerging model of the nanocomputer as a massively replicated, inherently reconfigurable architecture is being studied [22, 21, 13, 11, 19].

**Defect Management on the Teramac.** An early experiment in defect-tolerant reconfigurable architectures was conducted by researchers at Hewlett-Packard in the early 1990's. The "Teramac" computer [1, 22] was composed of hundreds of SRAM-based field programmable gate arrays (FPGAs) designed and fabricated by HP. SRAM-based FPGAs generally use an $n$-bit look-up table (LUT) to determine a boolean function of $n$ bits. A 4-input LUT is shown in Figure 2.11. The 4 input lines form an address into a 1 bit $\times$ 16 deep memory. The illustration shows that LUT[3] = 1.

Since the LUT is stored in memory, any function of $n$ bits may be specified and, further, replaced by simply re-writing the LUT memory. FPGAs also contain a reconfigurable routing fabric so that LUTs may be connected to form larger circuits. The Teramac FPGA devoted most of its area to routing, using partially populated crossbar switches in the routing fabric. In fact, routing area comprised 90% of each FPGA. The FPGAs were organized 27 to a multi-chip module (MCM), with four MCMs per board, and up to 16 boards. A unique aspect of this machine was its high defect level: 75% of the FPGAs contained defects, half of the MCMs were defective, and 10% of the interconnect signals were unreliable. In spite of the fact that 3% of the total system resources were defective, Teramac was able to run complex applications such as magnetic resonance imagery analysis and volume visualization.

Reconfiguration was the enabling technology that allowed Teramac to compute despite the 3% defect rate. Test configurations ("signature generators") were run on the FPGAs to isolate defects in LUTs, wires, and switches. The signature generators created long sequences of pseudo-random bit strings and circulated them throughout the machine over a variety of paths. Observation of changes in the bit patterns helped to locate physical defects such as stuck-at conditions, opens, and shorts. The test suite established a database of defects. Circuits could then be physically mapped onto the remaining functional LUTs and routed through the remaining functional interconnect.

This same notion of using redundancy and reconfiguration to alleviate defects is suggested in a nano-scale Programmable Logic Array (PLA) architecture proposed by DeHon in [11]. In this architecture, crossed carbon nanotubes or nanowires are used as interconnect switching elements and as simple logic cells. Recognizing the high defect rate of these nano-scale wires, [11] suggests employing spares along with reconfiguration to create functional nanoarrays. Sparing can be used at the local wire level and also hierarchically at higher levels of organization. Likewise, Goldstein and Budiu also depend on reconfiguration in their nanoFabric programmable architecture [19] to achieve defect tolerance. For further discussion of the use of reconfiguration and sparing for nanocomputing, please see Chapter 7.

It should be noted that reconfiguring Teramac to run all the test vectors and create a defect database was a time-consuming operation, taking many hours to days to complete, depending on the size of the Teramac system. Once the database was created, the computer-aided design (CAD) tools could use it to place and route circuits.

This defect database was static, so that faults that might emerge during operation were not detected automatically. Only when the test procedure was re-applied would new faults be discovered. When the database was updated, all circuits would have to be re-compiled relative to the new database.

In [35], Mishra and Goldstein discuss a scalable testing methodology to find defects in reconfigurable devices. A metric of $k \cdot p$ defines the number of defective components in a test circuit, where $k$ is the number of components in the test circuit and $p$ is the probability that a component is defective. For a

Teramac-size machine in which $k \cdot p \ll 1$ (3% defect rate in the entire machine), a collection of test circuits is used so that a component in the reconfigurable system is a part of many different test circuits. By comparing the results from multiple test circuits, faulty components can be isolated. However, as $k \cdot p \gg 1$, the number of test circuits becomes unmanageably large. In this case, [35] proposes more powerful test circuits such as Reed-Solomon codes that can determine the number of defects in a component. Defect mapping then occurs in two phases. In the first phase, components are divided into "probably good" and "probably bad" groups. The former have $k \cdot p \leq 1$, so that Teramac-like test circuit coverage can be employed. The latter are discarded and are logically mapped out of the system. As with Teramac, this approach of defect mapping and reconfiguration depends on a rich interconnect and reconfiguration capability intrinsic to the system.

**Defect and Fault Management on Cell Matrix.** In contrast to a static defect discovery process employing test vectors, the Cell Matrix [13] architecture facilitates dynamic defect and fault discovery/recovery. The Cell Matrix (CM) is a fine-grained reconfigurable fabric composed of simple, homogeneous cells and nearest-neighbor interconnect. Like FPGAs, the CM cells are based on look-up tables (LUTs). There are no critical, irreplaceable elements whose failure could cause the entire system to fail. This homogeneity of cell structure and interconnect as well as the ability of the Cell Matrix to *self*-reconfigure makes the architecture inherently fault tolerant. Like Teramac, the Cell Matrix can function in the presence of a high degree of defects. Further, the Cell Matrix can continue to operate when faults occur by appropriately exploiting the Cell Matrix architecture and using application-level fault-tolerant design.

A cell can operate in one of two modes. In *Data* mode, a cell produces output as a function of its data inputs. In *Configuration* mode, a cell treats the input lines as a new configuration to store in the LUT memory. A cell's mode is a local property, allowing neighboring cells to operate in different modes. Any cell *A* can direct its neighbor *B* to go into *Configuration* mode simply by asserting the *Configure* signal from *A* to *B* (see Figure 2.12), which causes *B* to re-configure itself by loading its data inputs from *A* into the LUT memory. In this example, when the *Configure* input to *B* is asserted, the '1' bit input to cell *B* on its data channel overwrites a '0' in *B*'s LUT.

Thus each cell can receive configuration commands from adjacent cells, and can, in turn, send configuration commands to neighboring cells. This allows a cell or collection of cells to:

- monitor their neighbors' activities,
- detect erroneous behavior,
- disable defective neighboring cells, and

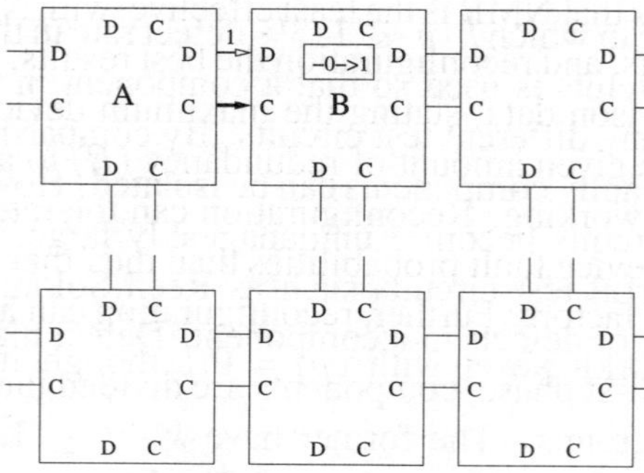


*Figure 2.12.* Cell Matrix Nearest Neighbor Reconfiguration

- relocate damaged segments of the circuit to other locations.

It should be noted that self-reconfigurability is a powerful capability that must be applied and managed very carefully. Erroneous or malicious use of self-reconfiguration can have widespread consequences on executing applications since it may be possible for one circuit to disrupt others through reconfiguration. Further, self-reconfigurability and the modification of circuits as they operate can add an additional level of complexity to application design and are, currently, not easy to analyze. As another issue, this architecture also imposes a substantial overhead to support self-reconfigurability—half of the interconnect lines are devoted to configuration control alone. However, despite these issues, the CM architecture and self-reconfiguration do allow autonomous, self-repairing circuits to be constructed from a simple, locally interconnected, homogeneous fabric [32, 12, 33]. For an additional discussion on Cell Matrix and related defect- and fault-tolerant strategies, please see Chapter 4.

## Discussion

Error detection, error masking, and reconfiguration are distinct methods for defect and fault detection and mitigation. In this section, we will discuss some of the research results of applying these techniques to nanocomputing. In the discussion, we will compare and contrast the results from different researchers. In addition, as these techniques are often used in tandem, we will describe research in using a combination of the techniques to improve the reliability of nanocomputers.

**Comparing NMR, NAND Multiplexing, and Reconfiguration.** In [37], Nikolić et al. compare three techniques: NMR, NAND multiplexing, and reconfiguration for different device fault rates. Assuming a chip with $10^{12}$ devices and with a requirement that the chip has a 90% probability of working, their analysis indicates that NMR is the least effective, with NAND multiplexing providing better results, and reconfiguration the best results. Table 2.1 provides some of their comparison data, stating the maximum device fault probability ($p_f$) allowable for the given amount of redundancy ($R$) to achieve a chip with a 90% probability of working. Reconfiguration can tolerate four to six orders of magnitude larger device fault probabilities than the other two techniques for the given redundancy factors. Further, reconfiguration can achieve a chip-level 90% probability of working even with a $p_f = 0.1$, though, it requires $R = 10^5$.

*Table 2.1.* Comparison of NMR, NAND Multiplexing, and Reconfiguration. The table provides the maximum device fault probability ($p_f$) given a redundancy factor, $R$, to achieve a 90% overall probability of working for a chip with $10^{12}$ devices. Data from [37].

| *Redundancy Technique* | *R* | *Maximum $p_f$ (approx.)* |
|---|---|---|
| NMR | 10 | $4 \times 10^{-9}$ |
| NAND Multiplexing | 10 | $3 \times 10^{-8}$ |
| Reconfiguration | 10 | $2 \times 10^{-3}$ |
| NMR | 100 | $4 \times 10^{-8}$ |
| NAND Multiplexing | 100 | $4 \times 10^{-7}$ |
| Reconfiguration | 100 | $2 \times 10^{-2}$ |
| NMR | 1000 | $1 \times 10^{-7}$ |
| NAND Multiplexing | 1000 | $5 \times 10^{-6}$ |
| Reconfiguration | 1000 | $3 \times 10^{-2}$ |

**Combining NAND Multiplexing and Reconfiguration.** In [21], Han and Jonker suggest a hybrid approach that combines NAND multiplexing with reconfiguration. Basically, they suggest a hierarchical approach which uses NAND multiplexing at the lowest level and then uses redundancy for reconfiguration at three additional implementation levels. To make the approach practical, a redundancy factor of $N = 3$ is suggested for implementing a bit slice of a 32-bit processor using NAND multiplexing with 11 logic levels. At the next level of hierarchy—the processor level, 16 spare bit slices (an additional 50%) are added for reconfiguration purposes to create the 32-bit processor. At the next level of the hierarchy, processors are combined into $32 \times 32$ arrays to form *clusters*. Within each cluster, four out of the 32 columns of processors are considered as spares. Finally, at the top level of the hierarchy—the chip level, the clusters are again organized into a $32 \times 32$ array, reserving four of the 32 columns for use as spares. Through this hierarchical approach, the researchers calculated that, for an individual gate fault probability of .01 and with $10^6$ devices per 32-bit processor, this approach would achieve a chip-level reliability of 99.8% with a total redundancy factor of less than 10, i.e., $3 \times \frac{3}{2} \times \frac{8}{7} \times \frac{8}{7} \times R_{other} < 10$ where $R_{other}$ is the redundancy of other necessary spare components. Their analysis assumes that faulty components can be effectively substituted with spare ones through reconfiguration. Further, they accounted for fault clustering, assuming that lower levels in the design hierarchy would have more correlated faults than at higher levels. So, with a system having $10^{12}$ devices and a high device fault rate (1%), they believe it would be possible to have a reliable system with $10^{11}$ effective devices—a large contrast with the large amount of redundancy ($N > 1,000$) required for von Neumann's original NAND multiplexing formulation or any of the other individual techniques listed in Table 2.1.

**Self-Correcting Logic Structures: NanoBoxes.** Another architectural approach using classical techniques to achieve usable nanocomputing systems is based on the NanoBox, as described in [25, 24]. A NanoBox is a look-up-table-based building block that uses redundancy through either TMR or error-control coding to correct any output errors produced by the LUT. Figure 2.13 illustrates what a 4-input, 1-output NanoBox might look like. Each NanoBox stores both the LUT function as well as the associated LUT check bits, which would be generated by the proposed CAD tool flow and then programmed into the NanoBox along with the LUT bits. The self-correcting LUT block does not actually correct the bits stored in the LUT memory or the check-bit memory, but simply corrects just the output, as needed. They assume that the errors seen are either transient or permanent and cannot be fixed, so they do not include circuitry to perform the fixes. Though this seems to contradict the current trend in SRAM FPGA reliability practice where upsets in the LUTs are assumed to be correctable and are corrected [9], it may be hard to determine when correcting the stored LUT and check-bit contents is beneficial, so not correcting the stored values may be reasonable.

In [25], KleinOsowski and Lilja perform several experiments to determine the best form of error correction for the NanoBox as well as the efficiency of the NanoBox for implementing control logic. Using a 90-nm, silicon-on-insulator CMOS process as way of implementing and evaluating the relative costs of the error correction schemes, the researchers built a 4-instruction arithmetic logic unit (ALU) and the control logic for an IBM Power4 floating-point-unit (FPU) controller using CMOS-based NanoBoxes. They compared the relative costs of using TMR, a Hamming code, a Hsiao code, and a Reed-Solomon code. Of the four, TMR was the least costly in terms of circuit area and power, being a factor of $2 - 3\times$ smaller than the best error control coding solution. Of the error control codes, Hamming was the least costly, while Reed-Solomon was the most costly in terms of area and power. They also demonstrated that the 4-input NanoBox was not very efficient in terms of memory usage for implementing the FPU controller—each of the controller's submodules used only 46–78% of

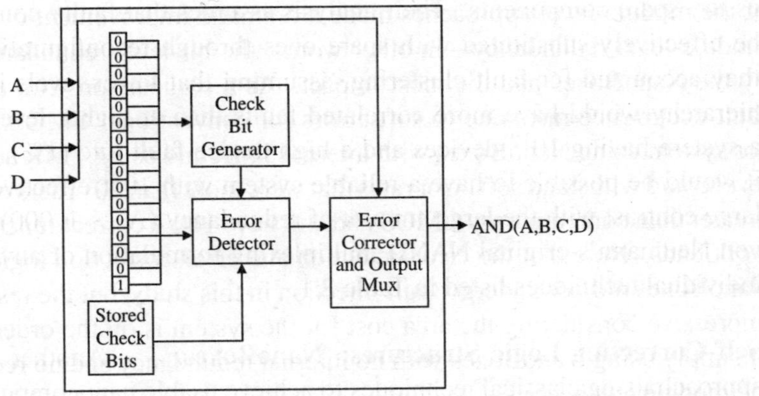


*Figure 2.13.* NanoBox, a Defect- and Fault-Tolerant Look-Up Table for Logic

the LUT memory bits for the LUTs that were utilized based on the results of using industry standard FPGA design mapping tools.

In [24], KleinOsowski et al. propose a defect and fault-tolerant architecture with multiple levels of redundancy and test the reliability of the architecture for high rates of injected faults. The architecture uses NanoBoxes at the bottom of the architectural hierarchy to provide bit-level defect and fault tolerance. Then, they implement a microprocessor using NanoBoxes. The microprocessor itself uses either time- or TMR-based redundancy for each of its internal modules (ALU, memory, controllers, etc.) to provide a second level of redundancy they call *module-level* redundancy. Finally, at the top of the hierarchy, several processors are arrayed in a two-dimensional grid where each processor can, if it fails, provide its workload to a neighboring processor to complete. Though nanotechnology is assumed to be the main method for implementing the processors themselves, conventional CMOS is used to implement the interconnection network for the 2-D array of processors.

Finally, in [24], the research group evaluates the defect and fault tolerance of the hierarchical architecture using a direct CMOS implementation of the processor as a base implementation and then implementing the architecture using NanoBoxes with no internal redundancy, with TMR, and with a Hamming code. At the processor level, they evaluated reliability using no redundancy, time-based redundancy (i.e., repeating operations), and TMR-based redundancy. By injecting faults into their circuits at the CMOS gate or LUT level (as appropriate), they determined that the most effective bit-level approach was TMR, which provides better than 60% correct computation when 9% of its internal nodes were injected with faults for their given processor workloads. The other techniques dropped below 60% at injected fault rates below 3%. As far as module-level redundancy, TMR performed the best, but none of the module-level approaches made a dramatic improvement over not using module-level redundancy—in other words, the bit-level redundancy provided by the NanoBoxes had the most impact. As a comparison with the expected fault occurrence rates due to soft errors for conventional CMOS, the processor system using NanoBoxes with internal TMR and using TMR at the module level could withstand soft error rates that were more than 20 orders of magnitude greater than contemporary CMOS and still have 100% correct function for their workloads. Note that the error-detection and error-correction logic within the NanoBoxes did not undergo fault injection in this study, but the results are still impressive considering the area cost for the system is on the order of 9× that of simply using NanoBoxes with no internal redundancy and no redundancy at the module level.

## 2.3 Non-Traditional Computing Models and Architectures

Beyond the application of more traditional forms for defect and fault tolerance (NMR, reconfiguration, error-control coding, etc.), several researchers are exploring other avenues for building reliable nanocomputing systems. This section will touch on several approaches for achieving defect and fault tolerance using techniques ranging from biologically inspired systems to new probabilistic system synthesis approaches.

### Neural Networks

Over the past twenty years, biologically-inspired artificial neural networks have been a popular and proven method for solving complex non-linear problems in a wide variety of application domains. Artificial neural networks abstract the properties of biological neurons and synapses, so that the synaptic interconnection of neurons along with the weights associated with a synapse form a distributed computation network. Nanoelectronics implementations of neural networks have been proposed, including some exploiting single-electron effects.

In [45], Rouw and Hoekstra identify two major challenges to nanoelectronic neural network fabrication. Neural networks are typically connection-rich, requiring each node to communicate across long distances. Nanoelectronic implementation favors local interconnection and short-distance communication. Neural networks using summation to determine connection weights are subject to errors due to the stochastic nature of single-electron transistors. To circumvent these limitations, Rouw et al. suggest

- local-interconnect-based linear topologies that exploit time delays and
- Hebbian learning and classical conditioning as the training methods.

The work of [17] uses single-electron latching switches as the basis of nano-scale binary weight, analog signal (BiWAS) synapses. Forwarding and branching latching switches are proposed, and these building blocks are used to design 2-D, square adaptive synaptic arrays. Experiments with free-growing arrays show that, for the connectivity typical of the cerebral cortex (each neural cell is connected to 10,000 others), interconnect density allows only a few neurons per $cm^2$. A more promising topology is based on nearest-neighbor communication in a 2-D mesh, in which neurons communicate on 4 axonic and 4 dendritic lines. The communication lines are interconnected by single-electron BiWAS switches. This scheme gives a neuronal density estimate of as high as $10^8$ neurons per $cm^2$.

The authors of [52] propose another novel neural network based on the stochastic nature of single-electron tunneling. A Boltzmann machine neural network contains bi-directionally connected nodes in which each node communicates with every other. A neuron has a binary output state that changes in response to inputs following a stochastic transition rule. All neurons operate in parallel, with each adjusting its state in response to state changes of the others. In [52], a digital oscillator is designed using a single-electron circuit that generates a randomly fluctuating 1/-1 binary stream required for Boltzmann machine operation. The authors do not address the problem of the massive interconnectivity required for such networks.

Neural networks seem attractive for nanoelectronic implementation due to intrinsic fault tolerance. Since computation is distributed through the array, the system may be insensitive to partial internal faults. On the other hand, since the computation is distributed, an error in one neuron or synapse potentially affects the whole network. It has been found that the degree of fault tolerance of a neural network is directly related to the degree of redundancy in the equilibrium solution to which it has been trained [41].

There has been a substantial body of research into methods of introducing fault tolerance into neural networks (see [41] for an excellent review, as well as [36] or [40]). One approach to increasing a neural network's fault tolerance modifies the learning procedure to force a neuron to tolerate larger variations in the input signals. A method that demonstrated enhanced fault tolerance for Gaussian radial basis functions forces a few nodes to zero (to simulate stuck-at-zero faults) and then trains the network. Alternatively, the neuron's output can be kept fixed to a given value once it has been found to be faulty. It is also possible to retrain a neural network, re-deriving synapse weights to compensate for a fault. All of these methods implicitly introduce redundancy into the network. Phatak and Koren in [40] prove that triple-modular redundancy is required for complete fault tolerance of feed-forward neural networks, thus, in the limit, imposing on the neural-net model the same requirements as traditional computing models.

## QCA Block Majority Gates

Quantum cellular automata (QCA) [47, 2] are a popular nanocomputing architecture being explored by the research community. In fact, Chapters 8.7 and 10 describe some of the challenges of and techniques for designing with QCA. Like many architectures at the nanometer scale, they have sensitivities to both manufacturing defects and environmental issues. In [16], Fijany and Toomarian analyze an implementation of the QCA majority gate and offer an alternative to the four-cell majority gate that is less sensitive to inter-cell alignment and cell failure. The QCA majority gate together with a QCA inverter chain (or NOT gate) provide a set of universal logic gates for building any logical function.

Figure 2.14 illustrates the conceptual, ideal QCA majority gate. Points $A$, $B$, and $C$ in the figure correspond to the inputs to the gate. In [16], the researchers describe the various types of alignment errors which are possible and point out that a failure of any of the cells will cause the gate to fail. During their studies, they used the University of Notre Dame's AQUINAS (A Quantum Interconnected Network Array Simulator) software to simulate the effects of misalignment and cell defects on the QCA logic gates. Figure 2.15 illustrates an example of a gate that would likely fail due to the $A$ input cell's misalignment.

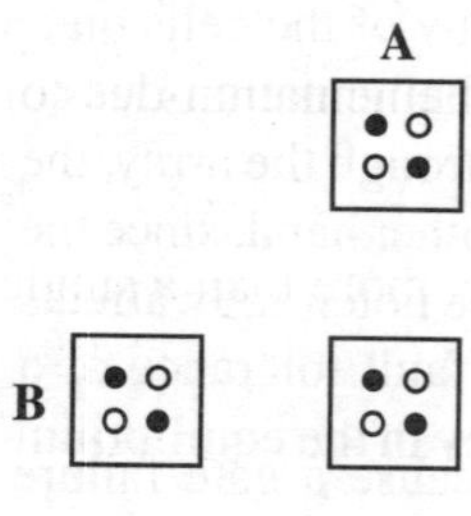


*Figure 2.14.* Ideal QCA Majority Gate

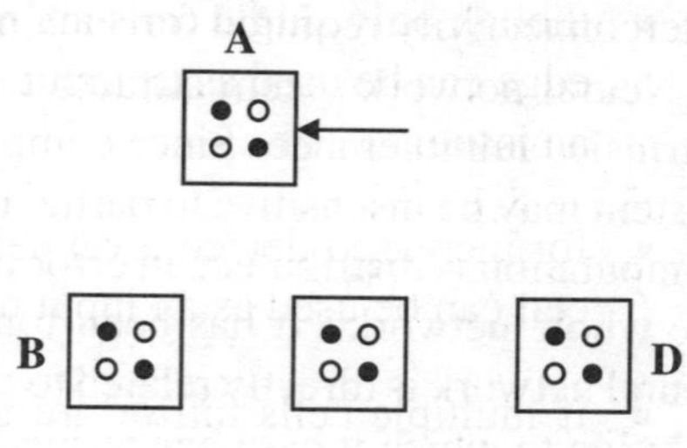


*Figure 2.15.* QCA Majority Gate with Misaligned Cell

Figure 2.16 illustrates the $11 \times 8$ block QCA majority gate proposed by Fijany and Toomarian, though, they do suggest that an array as small as $5 \times 4$ can be used as a block majority gate. The three gray QCA cells indicate possible inputs to the gate. Through their work, Fijany and Toomarian note that this block QCA gate is superior to the four-cell gate in several ways:

- the gate can function in the presence of many combinations of single-cell and/or multiple-cell failures or misalignments;

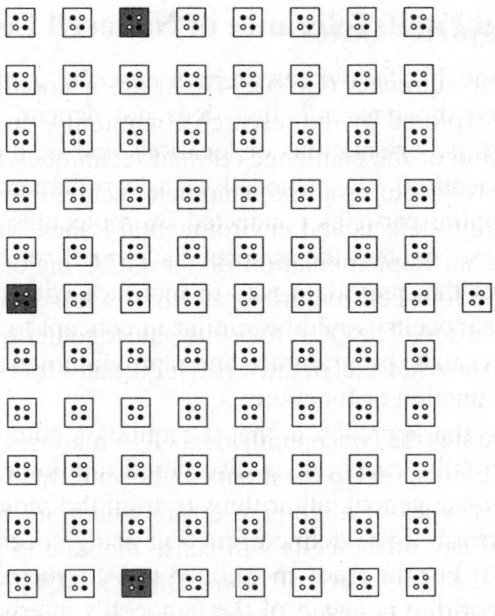

*Figure 2.16.* Block QCA Majority Gate

- though only single cells are highlighted as inputs, any of the cells on an edge can be used as an input, meaning that input cell alignment is less of an issue;
- for increased device-level defect and fault tolerance, more than a single cell can be used as an input on each side; and
- if multiple cells fail or are misaligned and, thus, cause a gate failure, using different input cells or more input cells on the edges can allow the gate to function again in some cases.

Clearly, the defect and fault tolerance provided by the redundancy in the array structure is significant. Unfortunately, the effects of misalignment and failures apparently are complex enough that it is not easy to give a simple reliability figure for the gate—simulations are generally required to determine if the gate will fail based on multiple variables (cell positions, cell failures, input cell locations, input widths, etc.).

The researchers also illustrated in [16] that the block majority gates can be used abutted so the output edge of one block majority gate can be placed next to the input edge of an adjoining gate to perform logic, assuming the gates are clocked with different clock phases. In this arrangement, the single output cell shown in Figure 2.16 is not a single point of failure, suggesting that the whole right edge of the block can be used effectively as the output.

## The Defect- and Fault-Tolerance of Nanocell Logic Gates

In [48], Tour and his fellow researchers describe a molecular computing structure, called the *nanocell*, that does not depend on the placement of molecules in precise orientations or locations and is, therefore, amenable to nano-scale fabrication. A nanocell is a two-dimensional network of self-assembled metallic particles connected by molecules that act as reprogrammable switches. Around the nanocell is a small set of lithographically defined access leads that provide a way to interface with the nanocell. In a sense, an array of nanocells would be similar in concept to an SRAM FPGA, where the nanocells can be programmed and reprogrammed after fabrication to perform a specific function or functions.

The downside to the approach is that it requires a complex programming scheme since, internally, each cell is disordered topologically. In their paper, the researchers use genetic algorithms to train the molecular switches in the nanocell to perform a pre-defined function using specific input and output leads to the cell. For the study mentioned in [48], the researchers assume that the genetic algorithm is aware of the nanocell's internal interconnection topology—something that would not be feasible for real systems but is a good first assumption for analysis. Despite the disorder of the internal switch network, they have been able to illustrate the training of the nanocells to perform as one-bit adders or as many as four independent two-input NAND gates.

Through simulating the training of nanocells, Tour and his colleagues noticed that, with a specific nanocell having only 14 molecular switches (a simple case used for thorough analysis) and a set I/O configuration, 13% of the possible switch states caused the cell to function as a NAND gate. For their complete set of 50 nanocells (each having between 5–16 switches per cell), 3%–19% of the switch states implemented the NAND function. This implies that the implementation does not depend on a single set of switch states, thus, providing a degree of device-level defect and fault tolerance.

Additionally, with larger nanocells having an average of 1826 switches, they did some additional tests to determine the defect tolerance of NAND gate logic in terms of ON-OFF output characteristics. With all switches in the ON position, the nanocell performed NAND logic with ON-to-OFF output current ratios of 20:1. When $> 60\%$ of the switches were turned off randomly, the minimum output ON-to-OFF current ratio dropped to 10:1, again illustrating a high tolerance to defects and faults at the switch and device level.

Lastly, they noticed that a large nanocell having 900 nanoparticles and 9000 switches could be trained as four independent two-input NAND gates at the corners of the cell. Between adjacent NAND gates, the switches were turned off, effectively isolating the gates from each other—another feature that can lead to better tolerance of device-level defects and faults when multiple gates are implemented by the same cell.

Similar to other architectures mentioned in this volume, the nanocell's device-level defect and fault tolerance come from redundant interconnections and the large number of "useful" logic states the cell provides.

## Markov Random Fields for Reliable Nanocomputation

Yet another novel approach to defect- and fault-tolerant nanocomputing design has been suggested by researchers at Brown University [4, 10]. The idea is to use probabilistic logic in the form of a Markov random network (MRN) to perform computations instead of depending on simple discrete-valued logic (see Chapter 5 for greater detail). By using probabilistic values for logic, the network can be optimized to maximize the probability that the outputs and intermediate results are correct, thus, accounting for defects and faults in the nano-scale implementation of the network.

In [10], the researchers describe conceptually how the MRN realization of Markov random fields (MRFs)[28] might be mapped to nanoelectronic structures based on carbon nanotubes (CNTs). The implementation of MRNs depends on three specific aspects: *weighted connections*, *clique energy summation*, and *probability maximization*. For their potential CNT implementation, the weighted connections are implemented by using multiple CNT paths for the same functional weighted input. The use of multiple nanotubes for connections clearly provides architectural-level defect and fault tolerance by allowing for some connection failures while still providing some fraction of the input value. As for clique energy summation, the Brown researchers propose the use of CMOS-style FETs based on CNTs where the gate of the FET is controlled by the summation of voltages provided by the nanotubes used for interconnection. Finally, they suggest that binary flip-flops can be used to fulfill the probability maximization function in the implementation, noting that more suitable persistent state mechanisms may be found as CNT device characteristics are explored further.

Based on this conceptual architecture, the researchers describe the design of the MRN interconnections in terms of topology and weighting. At the center of the design process is a technique that involves maximizing the probabilities of correct network-node operation by using the MRF formalism and the Belief Propagation algorithm[53]. In [4, 10], the researchers describe how the resulting design is tolerant to interconnection failures as well as discrete and continuous signal noise. The reader is referred to the discussion on this topic in Chapter 5 for more detail.

## 2.4 Tools

As materials scientists, chemists, and physicists grapple with device fabrication challenges, a variety of software tools have emerged to experiment with nano-scale computational structures through simulation, modeling, or analysis. Tools exist for analyzing nano-scale transistors [43], device physics modeling, molecular dynamics simulations, molecular bonding models, small molecule visualization, electron transport simulation, and many tools used in conventional computer-aided design (CAD) such as Spice [39] circuit simulation [42, 27].

Fewer tools exist to help study defect and fault behavior in fault-prone circuits, especially at the architectural and system levels. In this section, we describe recent advances in tools to understand and analyze error-prone computing architectures.

### Fault Simulation

An obvious technique to study faults in a system is to inject faults and then observe system behavior. This concept was exploited by the Space-Based Reconfigurable Computing project at the Los Alamos National Laboratory [7, 18]. In this work, SRAM-based FPGAs were used as compute engines in order to meet the size, weight, and power constraints of satellite-based processing. The work was motivated by a need to compute in the presence of on-orbit radiation effects without resorting to fully radiation-hardened electronics. In many ways, the problem mirrors the trade-offs between nanocomputing and conventional micro-scale computing: commercial components have many times the density of radiation-hardened electronics but suffer a high degree of faults in a radiation environment. In fact, it was determined that a collection of nine FPGAs would experience a transient error (single-event upset—SEU) 1.2 times/hour in low radiation zones and 9.6 times/hour in the presence of solar flares. These error rates are unacceptable in current computing environments.

In satellite-based processing, it is desirable to use commercial electronics for several reasons. Radiation hardened parts cost an order of magnitude more than conventional ones. The radiation hardened systems are too slow to do real-time data processing. In addition, the only fully radiation hardened FPGAs available cannot be reconfigured to hold different data processing algorithms. The available radiation-tolerant SRAM FPGAs (which are essentially commercial SRAM FPGAs fabricated on special wafers and are still susceptible to single-event upsets) use a configuration memory, so that the part may be repeatedly re-configured with new algorithms. In addition, the configuration data may be read out and repaired while the parts are active [9].

To explore the feasibility of using fault-prone, high-density devices for computing, a simulation—or more properly, emulation—environment has been developed. The emulator allows artificial injection of faults into an FPGA by

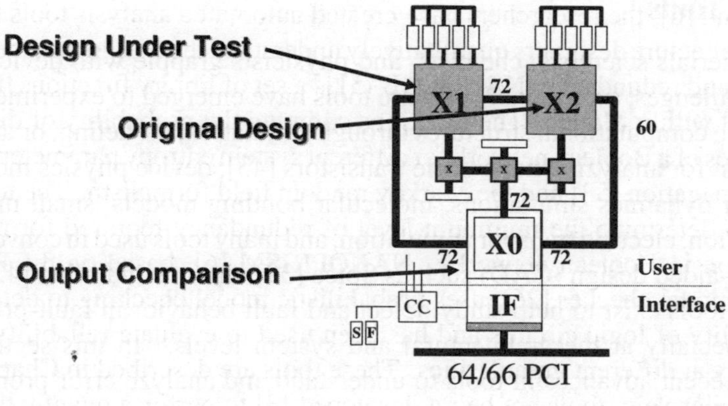


*Figure 2.17.* Transient Fault Emulation Testbed

dynamically reconfiguring the FPGA with corrupted configuration data. Figure 2.17 illustrates the mechanism. The emulator uses three FPGAs, all with a common clock. *X1* and *X2* initially hold identical designs. As the circuit is clocked, *X3* monitors outputs from *X1* and *X2* and signals the processor when outputs differ. During operation, *X1*'s configuration is selectively modified while the results from *X1* are compared to those from *X2* on a clock-by-clock basis. Through repeated and extensive testing, it is possible to correlate a single-bit upset in the configuration data with an output error, yielding for a specific circuit the probability of output failure attributable to each bit in the configuration. The emulator results have been compared to radiation testing in a cyclotron and show 97.6% correlation between output errors discovered through radiation testing and output errors predicted by the emulator [23].

This tool helps an application designer to understand the fault behavior of an application as well as where to insert redundancy or other error detection and correction circuitry to improve reliability. In a nanocomputing context, such a tool would be useful for characterizing an application's reliability for varying degrees of fault rates and types.

## Automated Trade-off Analysis of Nanocomputing Architectures

As discussed in detail in Section 2.2, redundancy is a standard technique for minimizing the effects of errors. However, since the redundant parts might themselves be error-prone, it is often not easy to determine the optimal level of redundancy beyond which reliability either does not increase or might actually decrease.

In [5] and [6], the researchers have created automated analysis tools to help micro-architecture designers quantitatively understand the design parameters of reliability and redundancy. NANOLAB [5] is a set of library functions that can be invoked within Matlab. The functions compute the probability of different output states of a Boolean network for different system entropy parameters using Belief Propagation [53] and the Markov random field formalism. The tool can be used to determine the minimum level of redundancy required for reliable computation in Boolean networks. NANOPRISM [6], based on the PRISM symbolic model checker [26], uses probabilistic model checking to determine the reliability of logic circuits and has been used to evaluate reliability using redundancy at different granularities. These tools are described in Chapter 6.

In a similar vein, tools are being developed [3] to perform quantitative statistical analysis of fault-tolerant QCA architectures [47]. While QCAs are resilient to certain perturbations in geometry and placement, they are sensitive to other errors such as non-symmetrical translation in position. The tools being developed integrate with the AQUINAS QCA simulator from the University of Notre Dame and perform quantitative statistical analysis of fault-tolerant QCA gate architectures.

## 2.5 Summary

Computing systems implemented with nanotechnology will need to employ defect- and fault-tolerant measures to improve their reliability due to the large number of factors that may lead to imperfect device fabrication as well as the increased susceptibility to environmentally induced faults when using nanometer-scale devices. For example, chemical assembly of molecular devices have only statistical yields and will not create all devices perfectly. As another example, single-electron transistors (SETs) and quantum-dot cellular automata (QCA) are susceptible to fluctuations in the background charge near the devices.

Researchers have approached this problem of reliability from many angles, using N-modular redundancy, NAND multiplexing, reconfiguration, error-control coding, artificial neural networks, and other novel architectures. All of the techniques use redundancy in some form to overcome defects and faults. The research results summarized here also suggest that many useful, yet strikingly different solutions may exist for tolerating defects and faults within nanocomputing systems.

Most of the work summarized has been performed at the architectural level for these nano systems and several claim to offer significant reliability (> 90%) for only an order of magnitude in redundancy. To solve the problem efficiently and effectively, we believe that defect and fault tolerance needs to be considered at all levels of the computing system: the device level, the architectural level, and the application level. For example, at the device level, a device's design parameters need to be considered to increase the probability that the device will operate properly and for a significant lifetime. Redundancies available at the device level should be exploited. At the architectural level, redundancy can be applied at various granularities and the trade-offs among system performance, cost, and reliability need to be weighed carefully. Further, many of the techniques being pursued can be applied throughout the architectural hierarchy: the gate level, the module level, and the system level. Finally, at the application level, there may be instances where the applications themselves can withstand some degree of noise in their operation. Requiring 100% reliability from the underlying computing hardware may not be absolutely necessary in those cases. Likewise, designing applications to be aware of system faults and to recover from them should be considered, despite the hopes that the underlying hardware will be able to mitigate the defect and fault issues.

Lastly, with the increased complexity of these computing systems as well as the increased complexity in designing the systems using nano-scale devices, designers will benefit from software tools that automate the mitigation of defects and faults as well as provide the ability to analyze systems for their reliability once designed. Without such tools, the costs of design and verification for these systems may be prohibitive and out of the reach of many designers.

## References

[1] R. Amerson, R. Carter, B. Culbertson, P. Kuekes, and G. Snider, *Teramac–configurable custom computing*, Proceedings of IEEE Workshop on FPGAs for Custom Computing Machines (Napa, CA) (D. A. Buell and K. L. Pocek, eds.), April 1995, pp. 32–38.

[2] Islamshah Amlani, Alexei O. Orlov, Geza Toth, Gary H. Bernstein, Craig S. Lent, and Gregory L. Snider, *Digital logic gate using quantum-dot cellular automata*, Science **284** (1999), 289–291.

[3] C. Duane Armstrong, William M. Humphreys, and Amir Fijany, *The design of fault tolerant quantum dot cellular automata based logic*, Proceedings of the 11th NASA VLSI Design Symposium (Coeur d'Alene, Idaho), NASA, May 2003, http://www.cambr.uidaho.edu/symposiums.

[4] R. Iris Bahar, Joseph Mundy, and Jie Chen, *A probabilistic-based design methodology for nanoscale computation*, Proceedings of the 2003 International Conference on Computer Aided Design (San Jose, CA), November 2003, pp. 480–486.

[5] Debayan Bhaduri and Sandeep Shukla, *Nanolab: A tool for evaluating reliability of defect-tolerant nano architectures*, IEEE Transactions on Nanotechnology (2004), To be published.

[6] ______, *Nanoprism: A tool for evaluating granularity vs. reliability trade-offs on nano architectures*, International Symposium in VLSI (ISVLSI 2004), Computer Society Press, 2004.

[7] Michael Caffrey, *A space-based reconfigurable radio*, Proceedings of the International Conference on Engineering of Reconfigurable Systems and Algorithms (ERSA) (Toomas P. Plaks and Peter M. Athanas, eds.), CSREA Press, June 2002, pp. 49–53.

[8] Carl Carmichael, *Triple module redundancy design techniques for Virtex FPGAs*, Tech. report, Xilinx Corporation, November 1, 2001, XAPP197 (v1.0).

[9] Carl Carmichael, Michael Caffrey, and Anthony Salazar, *Correcting single-event upsets through Virtex partial configuration*, Application Note XAPP216, Xilinx, San Jose, CA, June 2000.

[10] J. Chen, J. Mundya, Y. Bai, S.-M. C. Chan, P. Petrica, and R. I. Bahar, *A probabilistic approach to nano-computing*, Proceedings of the Second Workshop on Non-Silicon Computing (San Diego, CA), June 2003, http://www-2.cs.cmu.edu/ phoenix/nsc2, pp. Chen.1–8.

[11] Andre DeHon, *Array-based architecture for FET-based, nanoscale electronics*, IEEE Transactions on Nanotechnology **2** (2003), no. 1, 23–32.

[12] L. Durbeck and N. Macias, *Defect-tolerant, fine-grained parallel testing of a cell matrix*, Reconfigurable technology : FPGAs and reconfigurable processors for computing and communications IV (Boston, MA) (J. Schewel, P. James-Roxby, H. Schmit, and J. McHenry, eds.), Proceedings of the SPIE, vol. 4867, SPIE, July 2002, pp. 71–85.

[13] Lisa Durbeck and Nick Macias, *The cell matrix: An architecture for nanocomputing*, Nanotechnology (2001), 217–230.

[14] W. Evans and N. Pippenger, *On the maximum tolerable noise for reliable computation by formulas*, IEEE Transactions on Information Theory **44** (1998), 1299–1305.

[15] Joseph J. Fabula, Austin Lesea, Carl Carmichael, and Saar Drimer, *The NSEU response of static latch based FPGAs*, Proceedings of the 6th Annual International Conference on Military and Aerospace Programmable Logic Devices(MAPLD) (Washington, D.C.) (Richard Katz, ed.), NASA Office of Logic Design and AIAA, September 2003, pp. C5.1–23.

[16] Amir Fijany and Benny N. Toomarian, *New design for quantum dots cellular automata to obtain fault tolerant logic gates*, Journal of Nanoparticle Research **3** (2001), 27–37.

[17] Simon Fölling, Özgür Türel, and Konstantin Likharev, *Single-electron latching switches as nanoscale synapses*, Proceedings of the International Joint Conference on Neural Networks, July 2001, pp. 216–221.

[18] Maya Gokhale, Paul Graham, Eric Johnson, Nathan Rollins, and Michael Wirthlin, *Dynamic reconfiguration for management of radiation-induced faults in FPGAs*, 11th Reconfigurable Architectures Workshop (RAW 2004), IEEE Computer Society, April 2004.

[19] Seth Copen Goldstein and Mihai Budiu, *Nanofabrics: Spatial computing using molecular electronics*, Proceedings of the 28th International Symposium on Computer Architecture (Goteborg, Sweden), IEEE Computer Society, IEEE, June 2001, pp. 178–189.

[20] Jie Han and Pieter Jonker, *A system architecture solution for unreliable nanoelectronic devices*, IEEE Transactions on Nanotechnology **1** (2002), no. 4, 201–208.

[21] ______, *A defect- and fault-tolerant architecture for nanocomputers*, Nanotechnology (2003), 224–230.

[22] James Heath, Philip J. Kuekes, Gregory S. Snider, and R. Stanley Williams, *A defect-tolerant computer architecture: Opportunities for nanotechnology*, Science (1998), 1716–1721.

[23] Eric Johnson, Michael Caffrey, Paul Graham, Nathan Rollins, and Michael Wirthlin, *Accelerator validation of an FPGA SEU simulator*, IEEE Transactions on Nuclear Science **50** (2003), no. 6, 2147–2157.

[24] AJ KleinOsowski, Kevin KleinOsowski, Vijay Rangarajan, Priyadarshini Ranganath, and David J. Lilja, *The recursive NanoBox processor grid: A reliable system architecture for unreliable nanotechnology devices*, Proceedings of the IEEE International Conference on Dependable Systems and Networks (DSN) (Florence, Italy), IEEE Computer Society, IEEE Computer Society Press, June 2004, To be presented. Preprint available through http://www.arctic.umn.edu.

[25] AJ KleinOsowski and David J. Lilja, *The NanoBox Project: Exploring fabrics of self-correcting logic blocks for high defect rate molecular device technologies*, Proceedings of the IEEE Annual Symposium on VLSI (ISVLSI) (Lafayette, LA), IEEE Computer Society, IEEE Computer Society Press, February 2004, Preprint available through http://www.arctic.umn.edu.

[26] M. Kwiatkowska, G. Norman, and D. Parker, *Probabilistic symbolic model checker*, Proc. TOOLS '02 (T. Field, P. Harrison, J. Bradley, and U. Harder, eds.), LNCS, vol. 2324, Springer, 2002, pp. 200–204.

[27] Jiayong Le, Larry Pileggi, and Anirudh Devgan, *Circuit simulation of nanotechnology devices with non-monotonic I-V characteristics*, ICCAD 2003 (San Jose, CA), November 2003.

[28] S. Z. Li, *Markov random field modeling in computer vision*, Springer-Verlag, 1995.

[29] Konstantin K. Likharev, *Single-electron devices and their applications*, Proceedings of the IEEE **87** (1999), no. 4, 606–632.

[30] F. Lima, C. Carmichael, J. Fabula, R. Padovani, and R. Reis, *A fault injection analysis of Virtex FPGA TMR design methodology*, Proceedings of the 6th European Conference on Radiation and its Effects on Components and Systems (RADECS 2001), 2001.

[31] Shu Lin and Daniel J. Costello, *Error control coding: Fundamentals and applications*, Prentice Hall, Englewood Cliffs, NJ, 1983.

[32] N. Macias and L. Durbeck, *Self-assembling circuits with autonomous fault handling*, Proceedings of the 2002 NASA/DOD Conference on Evolvable Hardware (EH 02) (Alexandria, VA) (Adrian Stoica, Jason Lohn, Rich Katz, Didier Keymeulen, and Ricardo Salem Zebulum, eds.), NASA/-DARPA, IEEE, July 2002, pp. 46–55.

[33] N. J. Macias and L. J. K. Durbeck, *Adaptive methods for growing electronic circuits on an imperfect synthetic matrix*, Biosystems **73** (2004), no. 3, 173–204.

[34] R. Martel, V. Derycke, J. Appenzeller, S. Wind, and Ph. Avouris, *Carbon nanotube field-effect transistors and logic circuits*, Proceedings of the Design Automation Conference (New Orleans, LA), ACM/IEEE, ACM, June 2000, pp. 94–98.

[35] Mahim Mishra and Seth C. Goldstein, *Scalable defect tolerance for molecular electronics*, First Workshop on Non-Silicon Computing (Cambridge, MA), February 2002.

[36] Chalapathy Neti, Michael H. Schneider, and Eric D. Young, *Maximally fault tolerant neural networks*, IEEE Transactions on Neural Networks **3** (1992), no. 1, 14–23.

[37] K. Nikolić, A. Sadek, and M. Forshaw, *Fault-tolerant techniques for nanocomputers*, Nanotechnology **13** (2002), 357–362.

[38] Gethin Norman, David Parker, Marta Kwiatowska, and Sandeep K. Shukla, *Evaluating the reliability of defect-tolerant architectures for nanotechnology with probabilistic model checking*, Proceedings of the 17th International Conference on VLSI Design (VLSID'04) (Mumbai, India), IEEE Computer Society, IEEE, January 2004, pp. 907–912.

[39] Donald O. Peterson, *Simulation program with integrated circuit emphasis*, Sixteenth Midwest Symposium on Circuit Theory, 1973.

[40] Dhananjay Phatak and Israel Koren, *Complete and partial fault tolerance of feedforward neural nets*, IEEE Transactions on Neural Networks **6** (1995), no. 2, 446–456.

[41] Vincenzo Piuri, *Analysis of fault tolerance in artificial neural networks*, Journal of Parallel and Distributed Computing **61** (2001), 18–48.

[42] Arjit Raychowdhury, Saibal Mukhopadhyay, and Kaushik Roy, *Modeling of ballistic carbon nanotube field effect transistors for efficient circuit simulation*, ICCAD 2003 (San Jose, CA), November 2003.

[43] Zhibin Ren, *Nanomos*, http://nanohub.purdue.edu/NanoHub/tools/info/-nanomos.php (2004).

[44] Nathan Rollins, Michael Wirthlin, Michael Caffrey, and Paul Graham, *Evaluating TMR techniques in the presence of single event upsets*, Proceedings of the 6th Annual International Conference on Military and Aerospace Programmable Logic Devices(MAPLD) (Washington, D.C.) (Richard Katz, ed.), NASA Office of Logic Design and AIAA, September 2003, pp. P63.1–6.

[45] Eelco Rouw and Jaap Hoekstra, *Bio-inspired stochastic neural networks for nanoelectronics*, CP627, Computing Anticipatory Systems: CASYS 2001 - Fifth International Conference, American Institute of Physics, 2001.

[46] Daniel P. Siewiorek and Robert S. Swarz, *Reliable computer systems: Design and evaluation*, 2nd ed., Digital Press, Burlington, MA, 1992.

[47] G. L. Snider et al., *Quantum-dot cellular automata: Review and recent experiments*, Journal of Applied Physics **85** (1999), no. 8 15, 4283–4285.

[48] James M. Tour, William L. Van Zandt, Christopher P. Husband, Summer M. Husband, Lauren S. Wilson, Paul D. Franzon, and David P. Nackashi, *Nanocell logic gates for molecular computing*, IEEE Transactions on Nanotechnology **1** (2002), no. 2, 100–109.

[49] John von Neumann, *Probabilistic logics and the synthesis of reliable organisms from unreliable components*, Automata Studies (C. E. Shannon and J. McCarthy, eds.), Annals of Mathematics Studies, no. 34, Princeton University Press, 1956, pp. 43–98.

[50] Stephen B. Wicker, *Error control systems for digital communication and storage*, Prentice Hall, Englewood Cliffs, NJ, 1995.

[51] Michael J. Wirthlin, *Improving the reliability of FPGA circuits using triple-modular redundancy (TMR) and efficient voter placement*, Proceedings of the Twelfth ACM International Symposium on Field-Programmable Gate Arrays (Monterey, CA), ACM, ACM, February 2004, p. 252.

[52] Takashi Yamada, Masamichi Akazawa, Tetsuya Asai, and Yoshihito Amemiya, *Boltzmann machine neural network devices using single-electron tunneling*, Nanotechnology **12** (2001), 60–67.

[53] Jonathan S. Yedidia, William T. Freeman, and Yair Weiss, *Understanding belief propagation and its generalizations*, Exploring artificial intelligence

in the new millennium (Gerhard Lakemeyer, ed.), Morgan Kaufmann Publishers Inc., 2003, pp. 239–269.

# Chapter 3

# DEFECT TOLERANCE AT THE END OF THE ROADMAP

Mahim Mishra and Seth C. Goldstein
*Computer Science Department*
*Carnegie Mellon University*
mahim@cs.cmu.edu, seth@cs.cmu.edu

**Abstract** *As feature sizes shrink closer to single digit nanometer dimensions, defect tolerance will become increasingly important. This is true whether the chips are manufactured using top-down methods, such as photolithography, or bottom-up assembly processes such as Chemically Assembled Electronic Nanotechnology (CAEN). In this chapter, we examine the consequences of this increased rate of defects, and describe a defect tolerance methodology centered around reconfigurable devices, a scalable testing method, and dynamic place-and-route. We summarize some of our own results in this area as well as those of others, and enumerate some future research directions required to make nanometer-scale computing a reality.*

**Keywords:** nanotechnology, nanocomputing, end-of-roadmap CMOS, chemical assembly, defect tolerance, reconfiguration, testing

## Introduction

Future computing systems will inevitably be built using nanoelectronics, i.e., from devices and wires with feature sizes below thirty nanometers. The SIA roadmap [37] predicts that traditional silicon-based systems will have feature sizes of below 40nm within the decade. Additionally, there are advances being made in building computing systems using new technologies, such as molecular electronics [8]. Successfully harnessing nanoelectronics requires a new approach to building computing systems. While each technology has its own unique requirements, the small size of the individual devices and the large quantities of devices per chip are common to all nanoscale technologies.

The ever increasing improvement in computing performance is fueled by the ever increasing number and speed of available transistors, but it is driven by a hierarchy of abstractions. One plausible hierarchy is: Transistors → Logic Gates → Circuits → Blocks → ISA → Programs. Each abstraction layer hides the details of the layer below it, helping to control the complexity of designing and implementing systems with hundreds of millions of components. The layers also promote a separation of responsibilities, allowing independent progress to be made at different levels of the system. Currently, this hierarchy rests on certain assumptions about manufacturing, design, testing and verification. In particular it assumes that transistors and wires can be placed at will by the designer and will all work reliably. However, these assumptions break down at the nanoscale.

Perhaps the greatest impact of the nanoscale on electronics will be the reduced ability to arbitrarily determine the placement of the components of a system. The most extreme example of this is to be found in chemically assembled electronic nanotechnology (CAEN), a form of molecular electronics which uses bottom-up assembly to construct electronic circuits out of nanometer-scale devices. Large-scale molecular electronics requires some form of self-assembly [8]. When using self-assembly, individual devices and wires are first manufactured, and only later assembled into a circuit. While self-assembly promises to be a very economical process (compared with the cost of traditional semiconductor fabrication), it cannot be used to create the arbitrary patterns that can be formed using photolithography. Only simple, crystal-like structures can be created using self-assembly. Furthermore, defect densities of self-assembled circuits are projected to be orders of magnitude higher then in silicon-based devices. Thus, self-assembled circuits and architectures will have to be designed for defect tolerance.

To a lesser extent similar issues face traditional semiconductor technology as it continues to scale. The complexity of generating a reliable mask set which produces reliable chips is already limiting the ability to create arbitrary patterns of wires. This can be seen in the trend towards "structured" ASICs, which allow custom chips to share many of the same masks [44]. As devices scale down it is also harder to maintain constant characteristics for all the devices on a single chip [33]. Some argue that process variation will essentially eliminate the performance gains typically expected when feature sizes shrink [5]. These trends indicate that as feature sizes shrink even photolithographically manufactured chips will need to be crystal-like, i.e., built from very regular structures.

In order to implement useful reliable functionality on top of crystal-like structures, post-fabrication customization is required; this customization will be used for two purposes (1) to implement the desired functionality and (2) to eliminate the deleterious effects of the defects [11, 21]. Economics will also speed the movement towards customizable chips. As mask sets become more

| | | |
|---|---|---|
| Transistor | $\rightarrow$ | new molecular devices |
| Custom hardware | $\rightarrow$ | generic reconfigurable hardware |
| Defect free hardware | $\rightarrow$ | defect tolerance through reconfiguration |
| Synchronous circuits | $\rightarrow$ | asynchronous computation |
| Microprocessors | $\rightarrow$ | hybrid CPU + compiler-synthesized application-specific hardware |

*Figure 3.1.* Our proposal for how the abstractions used in devising computing systems should be changed.

expensive it becomes more economical to reuse the same chip for different tasks, i.e., to use programmable hardware (also called a reconfigurable fabric) such as field programmable gate arrays (FPGAs). A reconfigurable fabric is a network of processing elements connected by a programmable interconnect [22]. It can be programmed by determining how signals are routed on the interconnect. The desire to decrease time-to-market is also accelerating the trend towards using FPGAs for ever higher volume applications.

Defects in manufacturing have been, until now, primarily the concern of process engineers, not circuit designers or architects. In the era of nanoelectronics, delivering chips which can be viewed as defect free will likely be too expensive. In fact, this is happening already; for example, state-of-the-art FPGA chips with known defects can be purchased at a discount [47]. The "defective" chips can be used because the defects on the particular chip are determined not to affect the customer's design. In the future, defect tolerance will have to be designed in at the circuit and architectural level. In this chapter we discuss how defect tolerance can be achieved by combining reconfigurable fabrics with new tools. Reconfiguration provides defect tolerance by configuring the desired circuit around the defects, thus creating a reliable system from an unreliable substrate. Before the fabric is shipped its defects are mapped [31]. When the chip is used, the desired circuit is configured around the defects. The two main challenges are to develop architectures and tools which can find the defects quickly and then—in the field—quickly place-and-route (P&R) circuits around the defects. Final P&R needs to be done in the field so that a single configuration can be shipped for all devices, in spite of the fact that each device will have a different set of defects. For the remainder of this chapter, we will refer to such defect-tolerant reconfigurable fabrics made from future-generation technologies as *very large reconfigurable fabrics*, or *VLRFs*.

To summarize, our proposal for an alternative computer system architecture is based on dramatically different abstractions (outlined in Figure 3.1, and discussed in more detail in [20, 19]). In particular, the abstraction that a circuit is created at manufacturing time needs to be replaced by the ability to configure circuits at run-time; and the abstraction presented to upper layers of a defect-free, reliable computing fabric must give way to one where the defects are exposed to the upper layers – circuits, architectures, and software – and these upper layers must be actively involved in the objective of achieving defect tolerance.

## 3.1 Approaches for Achieving Defect Tolerance in the Nanometer Domain

The high defect densities in VLRFs require a completely new approach to manufacturing computing systems. No longer will it be possible to test a chip and throw it away if it has only a handful of defects, since we expect that every chip will have a significant number of defects. Instead, we must develop a method to use defective chips. The ability to tolerate defects in the final product in turn eases the requirements on the manufacturing process. In some sense, this introduces a new manufacturing paradigm: one which trades-off post-fabrication programming for cost and complexity at manufacturing time. In this section, we summarize some of the major approaches to defect tolerance used in computer systems of today, and which have been proposed for the technologies of tomorrow. This list, while by no means exhaustive, presents a good snapshot of the major research trends in this direction. Chapter 2 presents a more detailed survey of many of these approaches.

Modern memory chips and hard drives are able to achieve some degree of defect-tolerance by leveraging redundancy and post-manufacturing adaptiveness that allows them to substitute spare, working resources for defective ones. In large, high density memory chips, extra rows and columns are built into the chip. After manufacturing, a testing phase locates failing rows and columns, and these are replaced by the spare rows or columns by using a laser to burn a bypass path. Some modern operating systems go a step further: when they detect a memory error, a testing tool is run to detect the failing memory regions; the operating system then remembers not to use those regions when it stores data to memory. With VLRFs, techniques based on simple row or column replacement will not be sufficient: it is unlikely that a portion of the fabric of any appreciable size will be defect free. Moreover, these devices are being projected as a replacement not just for memories but also for logic, where simple techniques such as row-replacement will not work since logic is less regular.

Modern hard disks achieve defect tolerance by having a number of spare disk blocks. They ship with a map containing locations of all the bad disk blocks and the spare to be used in place of a bad block. During the lifetime of the hard drive, as more and more failing disk blocks are identified, the map is updated with this information. As we will see later, this approach is very similar in spirit to the one we are proposing for VLRFs.

One approach to achieve defect tolerance for logic would be to use techniques developed for fault-tolerant circuit design (e.g., [38, 34, 40]). Such circuit designs range from simple ones involving *triple-mode redundancy* or other relatively simple forms of majority logic to more complex circuits that perform computation in an alternative, sparse code space, so that a certain number of errors in the output can be corrected. However, the best such techniques available today require a significant amount of extra physical resources, result in a (non-negligible) slow-down of the computation, and they are also hard to automate well. Also, these circuits work reliably only if the number of defects are below a certain threshold. One major advantage of these techniques, however, is that they are able to tolerate defects that may occur during the operational lifetime of the fabric as well as transient faults, as long as the cumulative effect of these defects and faults remains smaller than the circuit's fault-tolerance threshold (for example, it can easily be shown that triple-mode-redundancy is able to enhance a system's reliability as long as the fault probability for each of the replicas remains below one-half).

A novel architectural approach for nano-scale computing is based on using probabilistic models of computation, such as the one proposed by Bahar et al. [2], and discussed in detail in Chapter 5. Their approach is based on Markov Random Fields and seeks to maximize the probability of correct boolean state configurations by minimizing the entropy of a suitable energy distribution that depends on neighboring nodes in the boolean network. Their architecture is based on an algorithm called *belief propagation*: starting from the network inputs, the known probabilities of the values of network nodes are used to compute probabilities on the values of their neighbors, and this process is carried out until the probabilities on the values of the outputs are known. This architecture is naturally defect and fault tolerant since computation happens in the probabilistic domain; however, it presents to us the challenge of adapting to a completely new architectural paradigm, and new programming models. The fault tolerance properties of this architecture are evaluated quantitatively by Shukla et al. [4] and described in more detail in Chapter 6.

Another paradigm that has been proposed for nano-scale computing are Quantum Cellular Automata [3] (see also Chapters 8.7 and 10 in this book). The fundamental component of a QCA architecture is the quantum dot, which is a nanoscopic site capable of holding a small, quantized coulombic charge. QCA devices perform logic and communication by arranging a number of quantum dots in linear and 2-dimensional patterns, and allowing the quantum dots to be affected by the charge state of their neighbors. On the one hand, their use of coulombic interactions give QCA architectures some robustness against small placement errors in the quantum dots; on the other, QCA-based devices proposed so far require more precise patterning and assembly than that required for mesh-based architectures, and hence are likely to be more fragile in the face of error-prone manufacturing processes. Some researchers have proposed using triple-mode-redundancy to increase the reliability of the wires and even the logic, or of architecting QCA logic gates in such a way to make them fault tolerant; there has also been some work in developing test strategies for QCA architectures (e.g., [41]).

A natural solution for achieving defect tolerance in VLRFs is suggested by looking at reconfigurable fabrics, e.g., Field-Programmable Gate Arrays (FPGAs). An FPGA is an interconnected set of programmable logic elements. Both the interconnect and logic elements may be programmed, or configured, to implement any circuit. The key idea behind defect tolerance in FPGAs is that reconfigurability allows one to find the defects and then to avoid them. On the theoretical side, Cole et al. [10] have shown that for fail-stop faults, an $N \times N$ reconfigurable array with $N^{1-\epsilon}$ worst-case faults can emulate a fault-free $N \times N$ array with constant slowdown. On the practical side, a number of approaches have been proposed and implemented for tolerating manufacturing defects and runtime faults in FPGAs and custom computing systems, as well as for run-time defect detection and reconfiguration to avoid defects [29, 18]. In the domain of custom computing systems, the Piperench reconfigurable processor [39] and more notably the Teramac custom computer [12, 23] had a notion of testing, defect-mapping and defect-avoidance built into them. Upto 75% of the FPGAs used in the Teramac had at least one defect; assembly was followed by a testing phase where the defects in the FPGAs were identified and mapped. Tools for generating FPGA configurations used this defect map to avoid defects.

The defect tolerance strategy we propose is similar to the one used for the Teramac: just like the Teramac's FPGAs, we expect that VLRFs will go through a post-fabrication testing phase at which point they will be configured for self-diagnosis. The result of the test phase will be a *defect map* which contains locations of all the defects. This map can be used by place-and-route tools to layout circuits on the fabric which avoid the defects. However, compared to the Teramac, the problem of implementing this defect-tolerance methodology on VLRFs is significantly harder because the Teramac used CMOS devices with defect rates much lower than those predicted for next-generation technologies. The enhancements we propose to make this process work for VLRFs are described in later sections.

## 3.2 Technology

This section reviews some recent results in Electronic Nanotechnology and Molecular Computing, as well as the basics of reconfigurable computing.

### Nanotechnology and Molecular Circuits

Significant progress has been made in developing molecular scale devices. Molecular scale FETs, negative differential resistors, diodes, and non-volatile switches are among the many devices that have been demonstrated. Advances have also been made in assembling devices and wires into larger circuits. Estimated densities are between 10x and 1000x greater than those predicted by the ITRS 2001 road-map for silicon-based CMOS circuits [8]. In addition to the increases in density, molecular electronics also promises to introduce devices with characteristics not found in silicon-based systems. One example is the non-volatile programmable switch, which holds its own state without using a memory cell, and can be configured using signal wires; such a switch has the same area as a wire-crossing. This contrasts with reconfigurable fabrics made today using standard CMOS, where a reconfigurable switch has the same area as a memory cell and is 2 to 3 orders of magnitude bigger than a wire-crossing.

The requirements imposed by the manufacturing process as well as the area advantages presented by the molecular reconfigurable crosspoints have prompted a number of researchers to propose regular, mesh-based reconfigurable architectures for VRLFs. One such architecture is the *nanoFabric* [21], which is fine-grained reconfigurable and is designed to overcome the limitations of self-assembly of molecular scale components. The basic unit of the nanoFabric is the programmable molecular switch, which can be configured either as a diode or as an open switch. As mentioned above, this molecular switch eliminates much of the overhead needed to support reconfiguration in traditional CMOS circuits, since the switch holds its own state and can be programmed without extra wires. These switches are organized into 2-D meshes called *nanoBlocks*, which can be configured to implement logic functions. The *nanoBlocks* in turn are organized into *clusters* which can be connected using *long lines* which run between the clusters. Within a cluster, each logic block is connected locally to 4 neighbors. In addition to the functionality of the logic blocks, the connections to the interconnect are also all reprogrammable. Another architecture based on mesh-like arrangements of active cross-points is the one proposed by DeHon [13], which uses arrays of configurable molecular FETs.

## Reconfigurable Hardware

In the last few years there has been a convergence in molecular-scale architectures towards reconfigurable platforms. Reconfigurable computing not only offers the promise of increased performance but it also amortizes the cost of chip manufacturing across many users by allowing circuits to be configured after they are fabricated.

Reconfigurable Hardware shares features of both custom hardware and microprocessors. Its computational performance is close to custom hardware, yet, because it is programmable, its flexibility approaches that of a processor. Because of their enormous potential as computational elements, there has been much research into using RH devices for computing (see [22] for an overview).

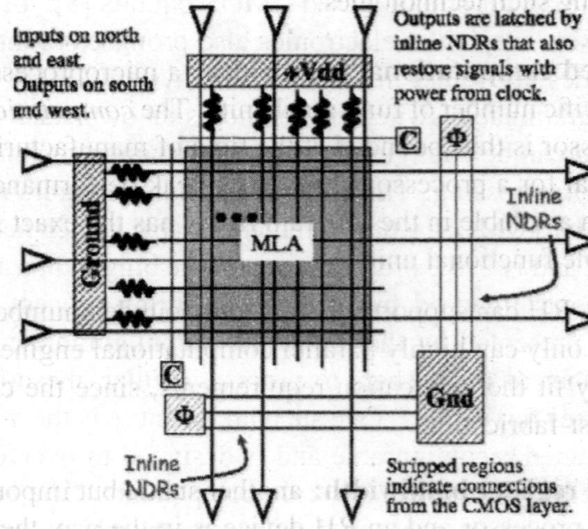


(a) A *nanoBlock* is the smallest configurable logic block of a *nanoFabric*.

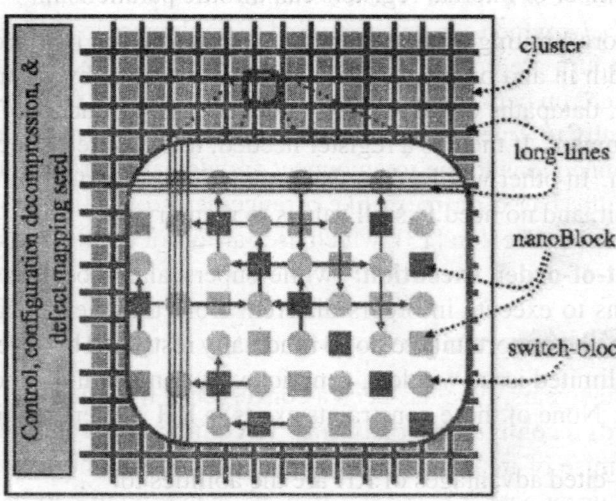


(b) A *nanoFabric* consists of many regularly tiled *nanoBlocks*, interspersed with routing resources.

*Figure 3.2.* The *nanoFabric*, an example of a reconfigurable CAEN-based architecture.

Several features of RH devices differentiate them fundamentally from processors, and are responsible for the extremely high performance of computing systems built using such technologies:

1 **Unbounded computational bandwidth:** a microprocessor is designed with a specific number of functional units. The *computational bandwidth* of a processor is thus bounded at the time of manufacturing. Moreover, it is unusual for a processor to reach its peak performance, because the parallelism available in the program rarely has the exact same profile as the available functional units.

   In contrast, RH can support a virtually unbounded number of functional units. Not only can highly parallel computational engines be built, they can exactly fit the application requirements, since the configuration is created post-fabrication.

2 **Unlimited register bandwidth:** another subtle but important difference between a processor and an RH device is in the way they handle intermediate computation results: processors have a predetermined number of registers. If the number of manipulated values exceeds the number of registers, then they have to be spilled into memory. Additionally, the fixed number of internal registers can throttle parallelism.

   Even more limiting is the number of register ports: in a processor the bandwidth in and out of the register file is bounded by a hard limit. In contrast, datapaths on RH directly connect the producers of a result to its consumers. If there is a register needed, it is inserted directly into the datapath. In other words, there is no monolithic register file, no register port limit, and no need to spill values to memory.

3 **Full out-of-order execution:** while superscalar processors allow instructions to execute in orders different from the one indicated by the program, the opportunity to do so is actually restricted by several factors, such as limited issue window, generic exception handling and structural hazards. None of these constraints exists in RH implementations.

Other often-cited advantages of RH are the abilities to:

4. Exploit all of an application's parallelism: task-based, data, instruction-level, pipeline, and bit-level,

5. Create customized function units and data-paths, matching the application's natural data size,

6. Eliminate a significant amount of control circuitry.

## Using VLRFs

There are two scenarios in which VLRFs can be used: (1) as factory-programmable devices configured by the manufacturer to emulate a processor or other computing device, and (2) as reconfigurable computing devices.

(1) In a manufacturer-configured device, user applications treat the device as a fixed processor (or potentially as a small number of different processors). Processor designers will use traditional CAD tools to create designs using standard cell libraries. These designs will then be mapped to a particular chip, taking into account the chip's defects. A finished product is therefore a VLRF chip and a ROM containing the configuration for that chip. In this scenario, the configurability of the VLRF is used only to accommodate a defect-prone manufacturing process. While this provides the significant benefits of reduced cost and increased densities, it ignores much of the potential in a VLRF. Since defect tolerance and limitations of the manufacturing process require that a VLRF be reconfigurable, why not exploit the reconfigurability to build application-specific processors?

(2) Reconfigurable fabrics offer high performance and efficiency because they can implement hardware matched to each application. However, this extra performance comes at the cost of significant work by the compiler. A conservative estimate for the number of configurable switches in a $1cm^2$ VLRF, including all the overhead for buffers, clock, power, etc., is on the order of $10^{11}$. Even assuming that a compiler manipulates only standard cells, the complexity of mapping a circuit design to a VLRF will be huge, creating a compilation scalability problem. Traditional approaches to place-and-route in particular will not scale to devices with billions of wires and devices.

In order to exploit the advantages listed above, we propose implementing the user application as a distributed dataflow machine, with only local communication between different parts of the machine and clean interfaces that are based on the request-response paradigm. This allows us to eliminate most global control and synchronization mechanisms found on the processors of today, while at the same time allowing us to synthesize hardware circuits directly from high-level descriptions.

## 3.3 Toolflow Required to Achieve Defect Tolerance

Our defect-tolerance approach is two-fold. First, we construct a map of the defects. Then, when configuring the device to implement a particular circuit, we avoid the defects by using only the good components of the device. Our approach requires three things from a reconfigurable device:

- it must be reprogrammable
- it must have a rich fine-grained interconnect

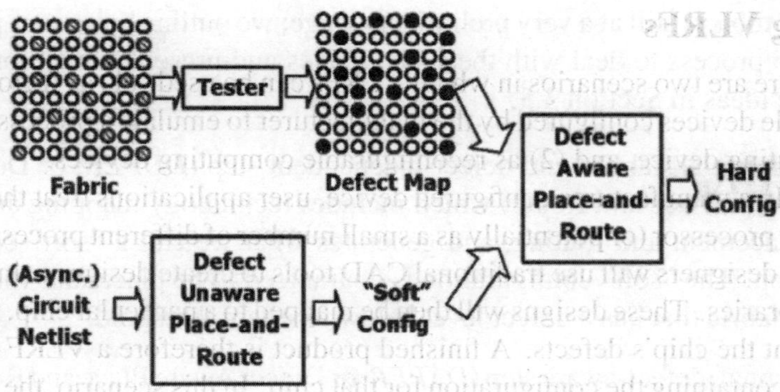


*Figure 3.3.* The tool-flow for using molecular reconfigurable fabrics for computation.

- it should allow us to implement a particular logic function in many different ways.

All three of these attributes are necessary for both defect detection and defect avoidance. During defect detection, we reprogram different test circuits on the device. Each different instance of a test structure gives us information about different sets of components on the device. The latter two attributes are most necessary during defect avoidance. They allow a particular circuit to be implemented without requiring us to use any of the defective components.

Dealing with high defect densities in VLRFs will require a new set of tools, such as fast testers to generate defect maps and place-and-route tools to convert circuit descriptions into fabric configurations taking into account the defect map for the fabric. Figure 3.3 depicts how these tools inter-operate. We briefly list the expected functionality from these tools below; the following sections contain more details on these ideas.

**Fabric testers** find and record defect locations. Defects can be tolerated by configuring circuits around them. A testing step first generates a defect map, much like the maps of defective blocks used for hard disk drives. This map may be a listing of all the defects in the fabric, or may be coarser-grained information such as a count of the number of defects in each portion of the fabric. We have done some initial algorithmic work that demonstrates that it is indeed possible to detect the defects in a large reconfigurable fabric with a defect density as high as 10%; Section 3.4 describes the details of our method and the results we have obtained.

**Place-and-route**: The place-and-route process generates fabric configurations that avoid using defective components. The place-and-route tools have to deal with the large size of the fabrics, as well as with the fact that each individual fabric has a unique set of defects and therefore requires some effort from the place-and-route tools to configure around its particular set of defects. Our work on these tools is still at a very preliminary stage; we outline below our proposed two-step process to deal with these challenges and present some more details on these ideas in Section 3.5.

1 An initial, fabric- and defect-independent step (*DUPER*, or Defect Unaware Place and Route) which includes technology mapping and global placement and routing, and generates what we call a "soft" configuration. The "soft" configuration is guaranteed to be place-and-routable on a fabric that has a defect level below a certain threshold.

2 A final, defect-aware step (*DAPER*, or Defect Aware Place and Route) that transforms the "soft" configuration into a "hard" configuration, taking into account the target fabric's defect map. At this step the final place-and-route is performed, using only non-defective resources. If the defect-map is imprecise or coarse-grained, this step requires the use of an on-the-fly defect mapper to pin-point the precise location of defects. The hard configuration is specific to each fabric.

We briefly summarize here the context in which we expect to use the above mentioned tools. Our expectation is that VLRFs will be used for general-purpose computation rather than only as a replacement for ASICs. In this scenario, the above mentioned tools will have to be part of a chain that allows the complete compilation of user programs from high-level programming languages down to hardware implementations that can carry out the desired functionality. The other important components of this hardware-compiler infrastructure will therefore have to include the following (see [20] for more details):

1 A hybrid processing fabric consisting of a classical CPU (manufactured using micro-scale technologies, if needed) and a large reconfigurable component. The CPU will serve to perform "housekeeping" tasks that are ideally not suited for execution on reconfigurable hardware, such as tasks that require a lot of interaction with micro-scale peripherals (for example, operating system tasks).

2 A compiler infrastructure that takes a high-level design description and compiles it down to a circuit description. Since creating the functionality of the circuit has been moved from manufacturing time to configuration time, the compilation encompasses tasks traditionally assigned to software compilers and to some tools in the CAD toolchain. The challenge of building a single tool spanning both domains presents us with the opportunity of reformulating the traditional interface between these two domains: the Instruction Set Architecture. Removing this rigid interface exposes a wealth of information to the compiler enabling novel classes

of optimizations. The disadvantage of this approach is the complexity between the source and target of the compilation process.

We propose the use of traditional, high-level software programming languages for programming VLRFs, as opposed to the traditional approach to programming reconfigurable hardware which uses hardware-description languages. The ability to use high-level programming languages will remove the burden from the users of these fabrics of having to learn hardware design methodologies. One such compiler is *CASH* [7, 6] which compiles ANSI C programs into collections of dataflow machines. The resulting dataflow machines are highly *composable*: they have simple interfaces using simple, timing-independent protocols. The natural substrate for implementing these dataflow machines are asynchronous (or globally-asynchronous, locally-synchronous) circuits. Asynchronous circuits also make it significantly easier to achieve rapid compilation and to meet our defect-tolerance requirements; we elaborate on this further in Section 3.5.

## 3.4 Testing

In this section we address the problem of finding the defects in high-density reconfigurable fabrics [31]. For our purposes here we limit ourselves to an abstract notion of defects and manifested faults: a defect is permanent and causes the defective resource to malfunction without affecting other surrounding resources. Also, a defect always manifests itself as a fault–it is not the case that faults are manifested in some situations and not in others depending on the operation context. Permanent stuck-open and stuck-at faults are examples of faults satisfying these conditions. We don't directly consider shorts, which may render a number of components connected to the shorted wires unusable; instead, a short manifests itself as failures in a number of fabric components located close to each other in a cluster, which our testing methods should be able to detect. We also do not consider defects which do not affect component functionality but only slightly affect parameters such as delay and power consumption. It should be noted that if the delay for any particular wire or logic component is too high, it will show up as a functional failure in our test circuits; for delays that are smaller, our proposal to use asynchronous or globally-asynchronous locally-synchronous circuits (see Section 3.5) should make it easier to tolerate these defects.

Although these assumptions may seem over-simplifying, they are not unrealistic for CAEN-based technologies (see [20] and references therein): the switches and logic elements will be made of molecules that should have fairly uniform operational characteristics, and most defects will occur because molecules fail to make contact or to align properly during the assembly process.

Also, shorts can be made very rare because for CAEN-based fabrics, we can engineer the molecules and assembly processes to be highly biased towards opens. This is likely to cause a higher overall defect rate, but a single short is more harmful than a handful of opens. Finally, we do not consider faults that may occur during the operational lifetime of the fabric, although some of our ideas for rapid defect location and reconfiguration to avoid defects have applicability for tolerating defects that occur in the field as well.

## Background

VLSI testing is a much-studied area of research. A large number of testing strategies and design methodologies have been proposed over the years to improve the speed and accuracy of VLSI testing, and hence to enhance manufacturing yield [1]. Most such techniques have been designed around the assumption that a single, or at most very few faults exist in the portion of the circuit under test. The problem we wish to tackle is significantly harder, since a large fraction of the resources under test may be defective. A key advantage we enjoy over traditional VLSI testing is that since the fabric is reconfigurable, we have the freedom to implement a circuit of choice to carry out the testing, rather than being limited to passing input vectors to the fabricated circuit.

Testing and defect tolerance are widely studied problems for FPGAs and custom computing systems. A number of testing methods have been proposed for particular FPGA architectures (e.g., [26, 15, 25, 48]), as well as many FPGA architectures designed with Design-for-Testability considerations [9, 30].

Our testing and analysis techniques have resonances with a large body of work in Statistics and Information Theory on *Group Testing* [16], which is a collection of techniques for finding members of a population which satisfy a particular property (i.e., which are "defective" in our setting). Our work is based on certain aspects of *non-adaptive, probabilistic group testing*. Different flavors of this technique have been applied to a variety of problems [14, 45, 28]. However, none of the problems discussed in the group testing literature have constraints as hard as ours: they have lower defect rates and assume that members of the population can be tested individually; while testing VLRFs, however, accessing individual fabric components will probably not be possible, because the on-fabric routing and switching resources used to carry signals to and from the particular component may themselves be defective, and also because with the abundance of resources, testing each component individually will be prohibitively expensive.

## Fabric Architecture

The particular architecture of the reconfigurable device is not essential to our defect detection or defect avoidance algorithms. In modeling the specifics of the algorithm we assume an architecture similar to an island-style FPGA, i.e., a mesh of interconnect resources surrounding islands of reconfigurable logic. Examples of such an architecture include the *nanoFabric* (see Section 3.2) and the architecture proposed by DeHon [13]. Both these architectures satisfy all three characteristics necessary for defect tolerance as enumerated in Section 3.3.

## Proposed Testing Method

We propose configuring sets of fabric components into test circuits whose output is used to infer the defect status of individual components. Unlike the dedicated Built-in Self Test structures often incorporated in current digital designs, the test circuits placed on the fabric during this self-diagnosis phase will utilize resources that will be available later for normal fabric operation; our expectation is that testing in this way should not require any dedicated fabric resources, and so supporting such a testing methodology should not incur either an area or a delay penalty.

For the description in this section, we use an abstract notion of a "defect" and a "fabric component". A defect is assumed to be permanent, and to cause some incorrectness in the output of a circuit using that defective component (see above). When performing the tests on a real fabric, the test-circuits used will have to be specialized according to the type of defects being diagnosed - shorts, opens, wire-breakages etc. What we mean by a fabric component is also left unspecified. It will depend on the design of the fabric, on the granularity of reconfigurability, and on how much of the fabric resources we are willing to sacrifice to achieve quick testing. Depending on these factors, a "component" may be one or more simple logic gates, a small configurable mesh of active cross-points, or a look-up table; the on-fabric interconnect resources are also considered "components" in the sense that they are configurable and may be defective.

As an example, consider the situation in Figure 3.4(a). Five components are configured into one test-circuit, so that defects in one or more circuit components would cause the circuit output to be incorrect. By comparing the circuit's output with the correct result, it can be determined if any of the circuit's components were defective. In the first run, the circuits are configured vertically, and test circuit 2 detects a defect. In the next run, the circuits are configured horizontally, and test circuit 3 fails. Since no other errors are detected, we can say that the component at the intersection of these two circuits is defective, and all others are good. This testing method, used for the Teramac [12], relies on the fact that most of the constructed test circuits will be defect-free. Hence, all their components can be assumed to be good. However, if the defect rate is higher, as it is likely to be in VLRFs made with future-generation technologies, this method no longer produces very good results. For example, in Figure 3.4(b),

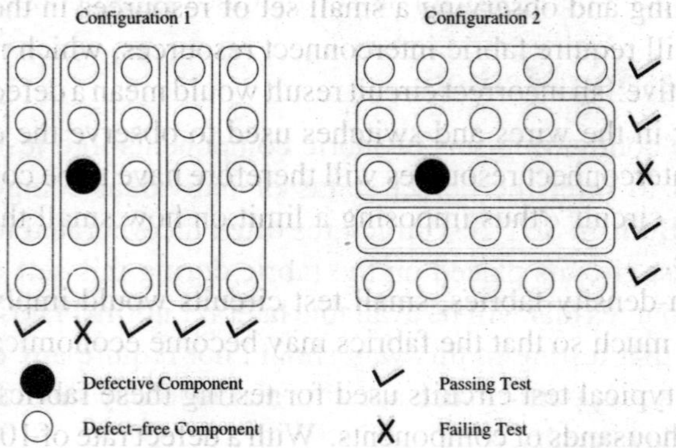


(a) For small defect rates, precise defect location is possible, and the false-positive rate (defect-free components identified as being defective) is low.

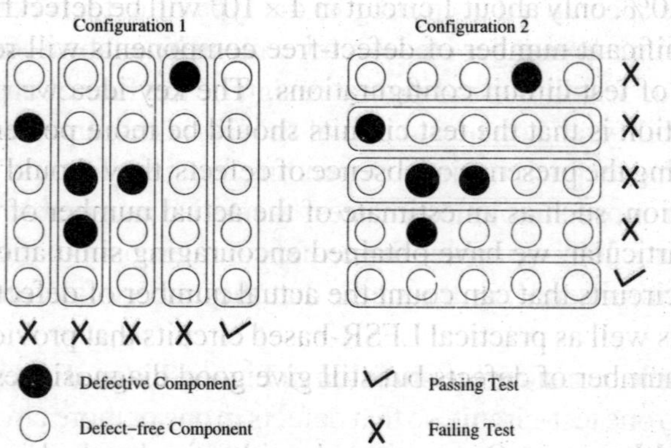


(b) For a high defect rate, most test circuits fail and the false-positive rate is high.

*Figure 3.4.* Examples showing how defective components can be identified using two different test-circuit configurations, for fabrics with small and high rates of defects. Any component that is part of a passing test is assumed to be free of defects.

only one vertical and one horizontal circuit return the correct result, and we gain no information about defect locations in the rest of the fabric. Also, the test circuits used will have a much larger number of components than in the simple examples shown here. This will be true for two reasons:

1. Controlling and observing a small set of resources in the interior of the fabric will require fabric interconnect resources, which may themselves be defective: an incorrect circuit result would mean a defect in the circuit's parts, or in the wires and switches used to observe the circuit's output. These interconnect resources will therefore have to be considered part of the "test circuit", thus imposing a limit on how small these circuits can be made.
2. For high-density fabrics, small test circuits would imply a long testing time, so much so that the fabrics may become economically unviable.

We believe typical test circuits used for testing these fabrics will have hundreds or even thousands of components. With a defect rate of 10%, a test-circuit with as few as a hundred components is expected to have 10 defects. This implies that the simple technique used in the examples in Figure 3.4 will not work at all, since an overwhelming majority of circuits will contain some defects. For example, it is easy to see analytically that for a test-circuit size of 100 and a defect rate of 10%, only about 1 circuit in $4 \times 10^4$ will be defect free. Therefore, locating a significant number of defect-free components will require an enormous number of test circuit configurations. The key idea we propose to deal with this situation is that the test circuits should be more powerful: instead of simply indicating the presence or absence of defects, they should provide us with more information, such as an estimate of the actual number of defective components. In particular, we have obtained encouraging simulation results using idealized test circuits that can count the actual number of defective constituent components, as well as practical LFSR-based circuits that provide approximate counts of the number of defects but still give good diagnosis results.

**Testing Algorithm.** Our testing algorithm is sketched in Figure 3.5. It consists of two phases: the *probability assignment* phase (lines 1–10) and the *defect location* phase (lines 11–22). The *probability assignment* phase assigns each component a probability of being defective, and discards the components which have a high probability. This should result in a large fraction of defective components being identified and eliminated from further testing. The remaining components are now likely to have a small enough defect rate that they can be tested in the *defect location* phase using the simple method employed in the example of Figure 3.4 to identify all the defect free components. In each phase, the fabric components are configured into test circuits in a particular orientation, or *tiling*; since each circuit uses only a small number of components, many such circuits can be configured in parallel, or *tiled*, across the fabric. For example,

```
   // Probability Assignment Phase
1  mark all fabric components not suspect
2  for iteration from 1 to N1 do
3    while probabilities not stable do
4      for all fabric components marked not suspect do
5        configure components into type-1 test-circuits using
         a particular tiling
6        compute defect probability for each component using
         circuit results from current iteration
7      done
8    done
9    mark components with high defect probability as suspect
10 done

   // Defect Location phase
11 for iteration from 1 to N2 do
12   while results improve do
13     for all fabric components marked not suspect or
       not defective do
14       configure components into type-2 test-circuits using
         a particular tiling
15       for all circuits with correct output do
16         mark all circuit components not defective
17       done
18     done
19   done
20   mark some suspect components not suspect
21 done

22 Mark all remaining components as defective
```

*Figure 3.5.* Proposed testing algorithm.

the circuits in Figure 3.4 are arranged in two tilings, vertical and horizontal. For now, we assume that arbitrary tilings are possible; this will generally not be the case in a real fabric with limited connectivity. Later on, we discuss some implications of limited fabric connectivity and ways to deal with it.

In the *probability assignment* phase, test circuits are repeatedly configured on the fabric, and circuit results are analyzed to compute the probability of being defective for each component (*while* loop of lines 3–8). When these probabilities have stabilized, components with a high probability are discarded, or marked `suspect`. These `suspect` components are not made a part of further test-circuits. This whole process is repeated a fixed number of times ($N_1, N_1 \geq 1$) or until a point of diminishing return is reached (*for* loop of lines 2–10). Note that this phase is probabilistic, and many good components will be misidentified as `suspect`. The purpose of running the *for* loop multiple times is to get a finer resolution of defect-probabilities for the components and thus to minimize the misidentification rate. The remaining points of interest here are the *type-1* test-circuits used in line 5, and the method for computing defect probabilities in line 6. Both of these are described in greater detail in subsequent subsections.

After the *probability assignment* phase, a number of components have been marked `suspect` and are not included in the further test-circuits. For the remaining components, the defect density is expected to be low enough that a substantial number of test-circuits will be defect-free. These defect-free components are identified in the *defect-location* phase (lines 11–22). In this phase, the circuits used (*type-2 test circuits*, line 14) return a wrong answer in the presence of a defect, but are not required to provide any more information. Components of error-free circuits are marked `not defective`. This whole process is repeated till no more good components are identified (*while* loop, lines 11–19). At this point, some of the components previously marked `suspect` are added back (line 20), and the whole process is repeated; this is done a total of $N_2$ times. The purpose of this is to try to reduce the number of components misidentified as `suspect` by the *probability-assignment* phase.

We define a quality metric, *recovery*, to evaluate our algorithm. Recovery is the percentage of defect-free components which our algorithm identifies as such. For example, if the fabric has a 10% defect rate, and the algorithm identifies 45% of the components as defect free, the recovery is 50% (45/(100-10)×100%). The recovery value is usually less than 100% because the algorithm has many false positives, i.e., good components that are identified as bad. In general, recovery will depend on the type of test-circuits used, number of tests run and the rigor of the post-testing analysis. There are a number of trade-offs that can be made between testing time and recovery, for example by adjusting the value of $N_1$ and $N_2$, or by changing the termination condition for the *while* loops on lines 3 and 12. In particular, the value of $N_2$ has been set to more than 1 in our simulations and the step on line 20 has been added in an effort to maximize recovery. It should be noted that if the test circuits used in the *defect location* phase can detect all modeled defects, this method of defect-mapping will never produce false negatives (i.e., bad components which are identified as good): all the components that the algorithm says are good will actually be good.

Another important trade-off is the obliviousness or non-adaptiveness of the test-circuit generation. By this, we mean that the results of previous tests are not used to generate new circuits. In the algorithm as it stands, a small amount of re-routing of test circuits is required after each iteration of the *for* loops in the two phases, when some components are discarded or added back. This re-routing is unlikely to require much effort since only routes to bypass certain components need to be included in pre-computed tile configurations; even this small amount of routing effort can be traded-off against recovery by adjusting $N_1$ and $N_2$.

**Some Candidate Test-Circuits.** We need test-circuits that can give us some notion of the number of defects in the circuit's constituent components. We have considered two kinds of circuits for this purpose: idealized *counter* circuits that can actually count the number of defects, and *none-some-many* circuits, which can tell us, with reasonable certainty, if the circuit had none, some or many defective components.

**Counter circuits**: These are idealized circuits that can count the number of defects, upto a certain threshold. For example, if the circuit's threshold is $t$, it can tell us if there are 0, 1, 2, ..., $t$ or more than $t$ defects in the circuit's components. Naturally, circuits with a higher threshold are more powerful and give better recovery results than those with a lower threshold. Although it is extremely difficult to practically realize such counter circuits with high thresholds, our simulation results (discussed later) show that even circuits with a threshold of 1 (i.e., which can only tell us if there were 0, 1 or more defects) can give significant recovery for moderate defect rates. For certain defect types, designing such low-threshold counter circuits may be possible.

**None-some-many circuits**: These circuits are a weaker version of the counter circuits described above. They can tell us, with some degree of accuracy, if the circuit contained *none*, *some* or *many* defects. We have designed simple circuits based on Linear-Feedback Shift Registers (LFSR) that can give us such information. These test circuits operate as follows:

- A set of components are organized into an LFSR, which is provided with an initial input and run autonomously for a while. Its signature is then matched against the correct output. If the signature matches, the circuit is assumed to contain no defects. Note that this does not mean the circuit is defect-free, because a correct output may have been obtained because of aliasing.
- If the large LFSR produced a signature mismatch, it is split into some number of smaller LFSRs (say 4), which are again provided initial inputs and run autonomously. If less than half of the smaller LFSRs have a signature mismatch, the entire circuit is assumed to contain *some* defects; if there are more mismatches, the circuit is assumed to contain *many* defects.

Breaking up a larger LFSR into some smaller ones should require minimal reconfiguration, since only a small number of forward and feedback connections need to be redirected. We have found that the two-step schema above works better than using smaller LFSRs from the start because smaller LFSRs have a higher aliasing probability. Also, in the initial stages of testing, most circuits will be found to contain some or many defects. However, as components are marked `suspect` and removed from subsequent testing steps, many of the large LFSRs will turn out to be defect-free, obviating the need to run the smaller LFSR circuits. This should speed up the testing. An alternative to LFSRs would be circuits based on linear cellular automata [36], which have the advantage of not having long feedback connections. However, low aliasing probabilities are harder to achieve with cellular automata and they are also harder to design since they need more fabric resources.

**Analysing Circuit Outputs.** Once the results of the test circuits have been obtained, they are used to determine the probability of each individual component being defective. We have considered two methods for doing this analysis:

**"Sorting" analysis**: Let a component $c$ be part of $n$ different circuits. Based on the results of these $n$ circuits, we calculate a *fault-value* for the component, as follows: if the test circuits are *counter* circuits, the fault-value is simply the sum of the number of defects in each of the $n$ circuits. If the circuit is an LFSR-based *none-some-many* circuit, we assign numerical weights to each result (e.g., 2 to *many* defects, 1 to *some* and 0 to *none*) and sum up all $n$ weights for the component. Once this calculation has been performed for all components under test, they are sorted according to their fault-values and components with higher fault-values are assigned a higher probability of being defective. This method involves simple calculations and places no specific restrictions on the shape or nature of the tilings.

**Bayesian analysis**: Again, let a component $c$ be a part of $n$ different test circuits. Let $p$ be the *a priori* known defect rate in the fabric, obtained through some initial testing or from knowledge of the manufacturing process. Let $a_1, a_2$, ..., $a_n$ represent numerical results for each of these circuits (these can be actual defect counts for *counter* circuits, or numerical weights for the *none-some-many* circuits as described above). We need to find the posterior probability of component $c$ being defective given our knowledge of the circuit results. Let $A$ be the event that $c$ is good, and let $B$ be the event of obtaining the circuit results that we have obtained for the $n$ circuits. Therefore, we need to find $P(A|B)$.

Now, from Bayes' rule,

$$P(A|B) = \frac{P(A \cap B)}{P(B)}$$
$$= \frac{P(A \cap B)}{P(A \cap B) + P(\overline{A} \cap B)}$$

If $c$ is the only component that the $n$ circuits share, this equation simplifies to the following:

*Figure 3.6.* A schematic representation of how testing will proceed in a wave-like manner through the fabric. The black area is tested and configured as a tester by the external tester; each darker-shaded area then tests and configures a lighter-shaded neighbor.

$$P(A|B) = \frac{1}{1 + \frac{(1-p)^{n-1} n^n}{p^{n-1}(n-a_1)(n-a_2)...(n-a_n)}}$$

This equation is solved for each component to obtain its probability of being good (the probability of being bad, which is required by the algorithm, is simply this value subtracted from 1). The simple closed-form expression obtained above holds true only if all the circuits that a component is part of have only that component in common. It can easily be shown that in general, the amount of computation grows exponentially with the number of components that two different circuits can share. This is a significant limitation of the method: on a fabric with limited routing resources, it severely restricts the number and type of tilings that are practically realizable. Although the sorting analysis gives results inferior to the Bayesian analysis, it does not suffer from this limitation. We are currently exploring ways to combine these two techniques to get around this while maintaining high recovery.

**Scaling The Testing Process with Fabric Size.** A short testing time is crucial for the usability and low cost of these fabrics, so it is important to ensure that the testing procedure scales with fabric size. We shall begin by analyzing the testing strategy above to see how long it takes to run.

The size of a test-circuit will depend on the granularity of access the fabric provides us–in general, smaller circuits provide more accurate information but are harder to realize. Let the circuit size be $k$. Then, since the circuits in a tiling are independent of each other, a $k \times k$ piece of fabric can be configured to run $k$ test circuits in parallel. Let the average number of iterations of the two *while* loops in lines 3–8 and 12–19 of the algorithm (Figure 3.4) be $x$ and

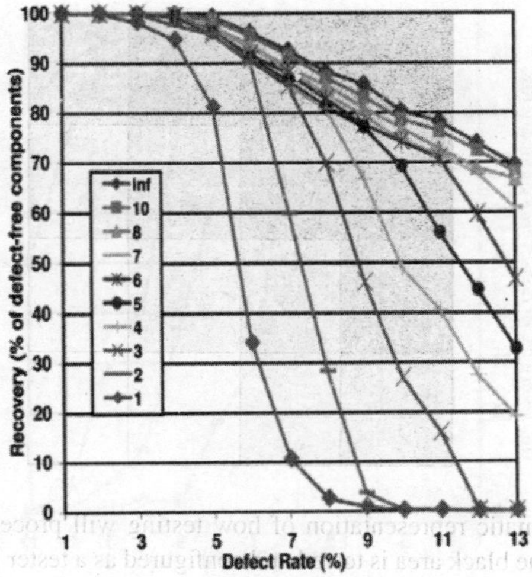


*Figure 3.7.* Recovery vs. defect rate for *counter* circuits using *sorting* analysis of circuit results. Each line represents a circuit with a different upper bound for the number of defects that can be counted. *inf* stands for infinity.

$y$ respectively ($x$ and $y$ will depend on the termination condition used for the two loops). Then, the total number of tilings used equals $N_1x + N_2y$. We have observed empirically that if $d$ is the defect rate, the number of tilings required before recovery stops improving scales as $O(k \times d)$. Therefore, for testing $k \times k$ components, we require $O(k \times d)$ fabric reconfigurations. If the fabric is larger then $k \times k$ it can be split into many sections of size $k \times k$, each of which can be tested separately.

We envisage that the reconfigurability of the fabric can be leveraged to reduce the time spent on an external tester significantly. Once a part of the fabric is tested and defect-mapped, it can be configured to act as a tester for the other parts. Also, there is nothing to prevent us from having multiple testers active simultaneously. In such a scenario, the first area to be tested tests its adjacent ones, which test their adjacent ones and so on, and the testing can move in a wave through the fabric (see Figure 3.6). For large fabrics, multiple such waves may grow out from different externally-tested areas. Now, as the fabric size increases, testing time grows linearly with the distance this wave has to traverse through the fabric, which is proportional to the length of the fabric's edge, and to the square root of the number of components in the fabric.

## Evaluation

We performed simulations of our algorithm to determine its efficacy and to evaluate the two types of test-circuits and analysis methods described above. These simulations were carried out using a very abstract notion of the fabric

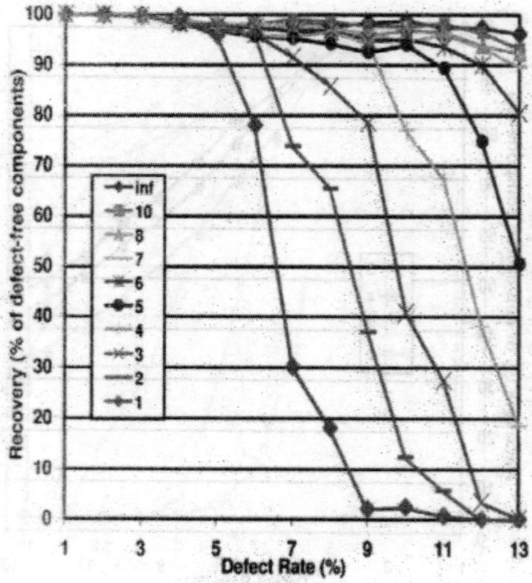


*Figure 3.8.* Recovery vs. defect rate for *counter* circuits using *Bayesian* analysis of circuit results. Each line represents a circuit with a different upper bound for the number of defects that can be counted. *inf* stands for infinity.

architecture and defect model: the fabric was assumed to consist of a large number of "components" arranged in a rectangular array with practically unlimited routing resources to connect them together. This allowed us to configure arbitrary test-circuit tilings onto the fabric, which will in general not be feasible on a real fabric. It was also assumed that suitable test-circuits were available to identify each of the different types of defects that can occur on the fabric. Although this abstracts away all the details of the fabric architecture and device failure model, our results are still a fairly good indication of the level of recovery achievable in such a high-defect-rate regime.

The simulations were carried out using test-circuits that had about 100 components each, and with fabrics that had defect rates ranging from 1 to 13 percent. What this actually means is that our analysis methods were tested for an average of 1 to 13 defects per test circuit, and our results are valid if test circuits used on actual fabrics have a defect count approximately in this range. Therefore, the size of the test-circuit will have to be adjusted according to the defect rate of the fabric: fabrics with a small defect rate can be tested using larger test circuits, but fabrics with higher defect rates will require smaller circuits and therefore the architecture will need to provide more fine-grained access to fabric internals.

Figures 3.7 and 3.8 present simulation results for *counter* circuits using, respectively, the *sorting* and *Bayesian* analysis. Figures 3.9 and 3.10 present results for the LFSR-based *none-some-many* circuits. For all of the above simulations, we assume that the defects have a random, uniform distribution throughout the fabric. Experience with VLSI fabrication has shown that defects

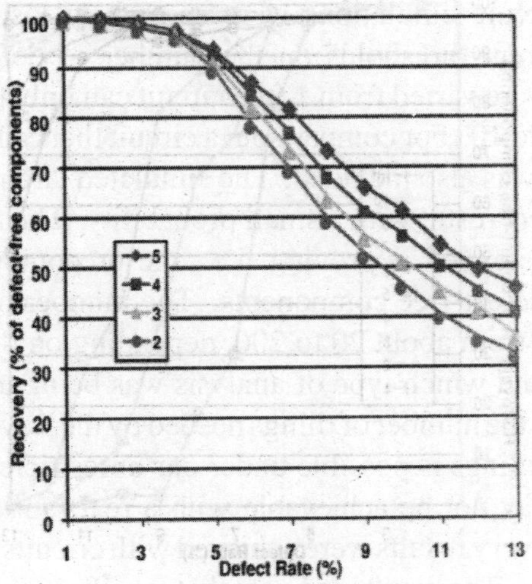


*Figure 3.9.* Recovery vs. defect rate for *none-some-many* circuits using *sorting* analysis of circuit results. The label for each curve represents the number of smaller LFSRs that the larger LFSR is split into.

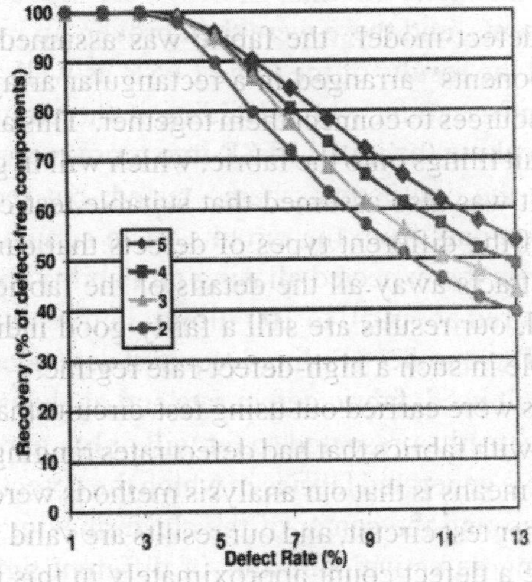


*Figure 3.10.* Recovery vs. defect rate for *none-some-many* circuits using *Bayesian* analysis of circuit results. The label for each curve represents the number of smaller LFSRs that the larger LFSR is split into.

are often clustered rather than uniformly scattered. However, the assumption of randomly scattered defects is pessimistic compared to clustered defects: if the defects are clustered, a larger number of circuits are expected to be defect-free, and hence defect diagnosis can be expected to be easier, as we show later in this section.

The *counter* circuit simulations were carried out using a number of test circuits with different thresholds on the number of defects they can count. These thresholds were varied from 1 (the circuit can only tell us if there were 0, 1 or more defects) to 10. For comparison, a circuit that can count an unbounded number of defects was also included. The simulated circuits were also allowed to return an incorrect result with a small probability. The test cycle was carried on till the results stopped improving, i.e., the precomputed tilings failed to identify any new defect-free components. The number of tilings required for each circuit varied from about 20 to 200, depending on the type of circuit, the fabric defect rate and which type of analysis was being used (sorting analysis needed about twice the number of tilings needed by the Bayesian analysis). Such a large number of tilings is possible under our unrestricted fabric connectivity assumption, but may not be achievable with a real fabric. For both types of analysis, good recovery results were achieved with circuits having relatively low counting thresholds. The results indicate that significant recovery is achievable if the test-circuit can count about a third of the expected number of defects per circuit (e.g., a circuit that can count 3 or 4 defects gives good results if the average number of defects per circuit is less than 10).

For the *none-some-many* circuits, LFSRs were first configured using about a hundred components, and these were then broken up into 2, 3, 4 or 5 smaller LFSRs (recall that the results of the large LFSR as well as each of the smaller pieces are taken into account while deciding if the circuit has none, some or many defects). Breaking the initial LFSR into more pieces gives more accurate information about the components being tested, but requires more effort to get inputs to and outputs from the circuits. The simulated circuits produced incorrect results with the expected aliasing probability for circuits of that size. These circuits required about 50 to 100 tilings, before results stopped showing an improvement. The number of tilings required depended in the expected way on the type of circuit used, fabric defect rate and analysis technique.

To show that the results presented so far will only improve in the presence of clustered defects, we generated fabrics in which the defects occurred in clusters, with components around the center of the cluster having a normal probability distribution of being defective. Clusters of different tightness were obtained by varying the standard deviation of the normal distribution. These results are presented in Figure 3.11: all the fabrics had a defect rate of 9%; the bars to the left represent larger standard deviations and hence larger clusters, while those to the right represent tighter clusters. The leftmost bars correspond to a standard deviation of infinity, which is essentially the uniform distribution. We simulated counter circuits with thresholds of 2 and 3 and Bayesian analysis. As the clusters get tighter, the recovery results for the circuits with threshold 2 improve dramatically. For the threshold 3 counter, the results were good to start with, but still show a small improvement.

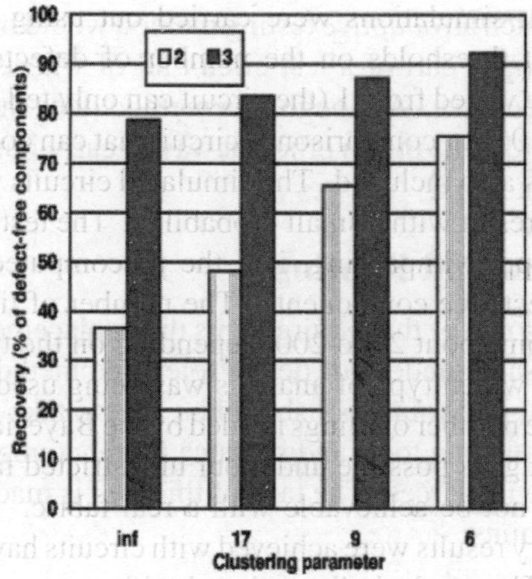


*Figure 3.11.* Recovery results for clustered defects, using counter circuits of thresholds 2 and 3 and Bayesian analysis. The horizontal axis shows different values of a clustering parameter, which corresponds to the tightness of the clusters. Larger values represent looser clusters: infinity (*inf*) represents unclustered, uniformly distributed defects, or the results in Figure 3.8.

To summarize, the more sophisticated Bayesian analysis has significantly better recovery than the simpler sorting analysis. Also, simulations using the sorting analysis technique needed about twice the number of reconfigurations needed by the Bayesian analysis. However, the sorting analysis is much easier to implement, particularly because it places far fewer restrictions on the types of tilings that can be used (recall that the Bayesian analysis required that any two test circuits have at most one component in common). We are presently investigating other methods of analysing circuit results, including a combination of sorting and Bayesian analysis, to reduce the number of reconfigurations required as much as possible while not sacrificing recovery.

## Testing: Discussion

We have presented a testing algorithm that can obtain a reasonably high level of recovery of fabric components, using test-circuits that provide more information about the defect status of their components than simply the presence or absence of defects. Our algorithm attempts to minimize the amount of rerouting required at test time, since rerouting is a slow and computationally expensive procedure. Therefore, we restrict ourselves to using pre-computed tiling configurations which undergo a minimal amount of adaptation at test time. Of course, once the algorithm has identified some fabric resources as being defect free, test circuits can be generated which incrementally test fabric resources whose defect status is unknown, potentially achieving close to 100% recovery.

A naive example of such an *adaptive* testing method would be one that tests each component by making it part of a test-circuit all of whose other components are known to be defect-free. Although this method does the required job, it is very inefficient, and identifying an efficient way to carry out this final testing is a focus of our current work.

## 3.5 Placement and Routing

This section lists some of the requirements that a placement and routing tool must satisfy to be usable under our defect tolerance methodology. To briefly summarize, the place-and-route tool should be able to take in a map of defects (as generated by our testing tool) and produce fabric configurations that avoid using these defective resources. The task of this tool is made more difficult by the following constraints:

1 As opposed to current hardware design methodologies where the place and route tools are run once per design, our approach would require running them once per fabric. This is because each fabric would have a defect map that is unique to it, requiring some place and route effort to avoid the defects. To make this process practical, we will have to find ways to minimize the per-fabric effort, while still achieving the desired functionality.

2 Adjusting the design to avoid defects may have an unpredictable effect on the timing of various signals. It has to be ensured that in spite of these adjustments, the implemented circuits are able to meet all the timing and functionality guarantees.

3 Since we envision using VLRFs for general-purpose computation rather than only as ASIC replacements, the execution time of the place-and-route tool should be comparable to software compilation and installation runtimes–measured in seconds and minutes, rather than in hours and days, as is the case for current hardware design or reconfigurable logic toolflows. This should be true in spite of significantly more available resources compared to current hardware design tools.

4 For significantly high defect rates, the complete defect map may not be available. This is because storing a detailed defect map for the full fabric may require resources comparable to those available on the fabric itself.

In the next few subsections, we propose techniques that may be used to deal with each of these problems.

## Two-phase Place-and-route Process

As shown in Figure 3.3, we propose splitting the place-and-route process into two steps, a *Defect Unaware Place and Route (DUPER)* step which is performed once per-design by the application developer, followed by a *Defect Aware Place and Route (DAPER)* step that is performed once per-fabric by the end user. The DUPER phase will produce a "soft" configuration: this will contain enough place and route information so that it can be quickly finalized by the DAPER phase to produce a "hard" configuration, that can be loaded onto the fabric. To give the user the impression that his application starts up in time comparable to current software installation times, the DAPER phase should be quick and lightweight, and as much place-and-route functionality as possible should be handled by the DUPER phase.

The DUPER phase will be akin to global placement and routing performed by place and route tools of today. For example, if a quadratic-placement based method is used (e.g., [42, 17]), the DUPER phase will correspond to the initial quadratic optimization phase, while the final legalization will occur in the second, DAPER phase. If a recursive partitioning-based placement method is used (e.g., [27, 43]), the partitioning of the design into small groups of logic blocks can occur in the DUPER phase, while final placement for the groups will happen in the DAPER phase. Iterative improvement methods, such as those based on simulated annealing (e.g., [35]), are not likely to be useful for our purposes, since they commit the placement to be legal fairly early in the place-and-route process. Similarly, global routing to assign nets to channels (or perhaps small groups of channels) will be done in the DUPER phase, while the nets will be assigned to individual wires in those channels in the DAPER phase.

We expect that the placement and routing tools for the DAPER phase will be implemented to run on the VLRF itself, so that a companion compute system is not required to execute these layout tasks. This may require modifications to the layout algorithms, to make them amenable to a parallel, distributed implementation on the VLRF; for example, Wrighton et al. [46] describe a distributed and parallel placement algorithm based on simulated annealing that can be implemented on a reconfigurable fabric and runs up to 3 orders of magnitude faster than standard tools, with a small degradation in placement quality. Also, the fabric can be architected suitably to assist in placement and routing, reducing runtime even further. One such approach described by Huang et al. [24] designs the interconnect network in such a way that it is able to help significantly with the task of identifying possible paths between two nodes, and congestion levels along those paths. Their methods are able to obtain 2-3 orders of magnitude speedup for the routing phase while seeing only a few percent degradation in routing quality compared to standard routing tools.

The configuration produced by the above process may not be well optimized with respect to area or wirelength, which translates into slower execution. However, the reconfigurability of the fabric can be leveraged to improve on this situation: while the computing system is lying idle, it can perform background optimizations on the designs that have been "installed" on it, tweaking the configuration to extract better performance. The next time the user loads up the configuration, it will be faster and more compact. This background, post-install optimization may very well be carried out by iterative methods.

## Hierarchical Fabric Architecture

To make the task of defect-mapping and place-and-route simpler, we propose architecting the reconfigurable fabric in a hierarchical manner. The fabric will be divided into a number of smaller regions, each of which will consist of a small number of basic configurable blocks. An example of such a hierarchical design is the *nanoFabric* (Figure 3.2): the individual *nanoBlocks* are arranged into *clusters*, which are tiled regularly across the plane to obtain the complete fabric.

While testing the fabric as well as performing place and route, we can apply a threshold on the number of defects in each region to determine whether that region is usable. If a region has only a few defects, the defect map will contain detailed information for it and the DAPER phase will be able to utilize the region; if the region has too many defects, the whole region may be marked unusable.

## The Need for Asynchronous Circuits

Synchronous circuits present two complications to our methodology: they increase place-and-route time since timing closure is one of the most difficult and time-consuming objectives for CAD tools to achieve, and it is difficult to guarantee timing closure will be achieved at all after the DAPER phase, since the results of this phase are dependent on the unique defect map for each fabric.

To alleviate both these problems, we propose using asynchronous or globally asynchronous locally synchronous (GALS) circuits [32]. If the fabric architecture is hierarchical (see above), one possible implementation is to make each region (or *cluster* in a *nanoFabric*) synchronous, and make inter-region communication asynchronous in nature. This will make it easy for design tools to deal with regions that have been deemed unusable because of a very high number of defects, by allowing asynchronous inter-regions signals to be routed around them. For each individual synchronous region, the clock can be given enough slack so that the small adjustments required to avoid defects does not cause a timing violation. Using a GALS rather than a fully asynchronous design also minimizes the overhead caused by asynchronous signaling and handshaking protocols.

## Hierarchical, Multi-step Testing

Making the defect map hierarchical as suggested above – making it have precise defect information for regions with a relatively small number of defects while marking the regions with more defects completely unusable – helps to compact the defect map a little. However, this map may still be too large to ship with the fabric – the complete map may require storage that is larger than the fabric itself.

One way around this problem is to include the complete defect map for only a small portion of the fabric. This portion should be large enough to accommodate most small and medium-sized applications, so that the rest of the fabric needs to be used only rarely for the very large applications. A small testing tool is also shipped with the fabric, and enough defect information about the remainder of the fabric is included in the map (such as the count of the defects in each region, or statistical information on the nature of the defects) so that this tester can very quickly test and map the defects in that portion. When a large application needs to be configured on the fabric, the DAPER tool runs the tester to obtain this defect information before using that part of the fabric. This testing and configuration can happen in an incremental, on-demand manner, so that part of the application is up and running, while another part is still being configured.

## 3.6 Summary

Next generation manufacturing technologies are expected to achieve extremely high device densities, yielding computational fabrics with many billions of components. However, this boon comes at the cost of large defect densities. In order to make the entire process economical it is important that the chips be defect tolerant. One possible approach to defect tolerance is to use a reconfigurable architecture for the computing devices, coupled with a testing methodology that can locate the defects in the reconfigurable fabric and a place-and-route process that can implement user circuits on the fabric while avoiding the defects.

In this chapter, we have proposed a toolflow that can meet the requirements for a reconfiguration-based defect tolerance strategy. We have described our own work which demonstrates that it is indeed possible to quickly and reliably locate the defects in a large reconfigurable fabric with a defect rate as high as 10%. The testing strategy we propose is based on powerful test-circuits and analysis methods which collate results from many different test-circuits to determine whether a particular fabric component is defective or not. We have also enumerated some of the challenges that layout algorithms will need to meet to enable them to configure circuits on the fabric while going around the identified defects. We have proposed solutions to some of these problems; in particular, we propose using a two-stage place-and-route process, hierarchical fabric architecture, multi-stage testing, and implementing user functionality as asynchronous circuits.

## 3.7 Acknowledgments

This research is funded in part by the National Science Foundation under Grant No. CCR-9876248 and by Darpa under contract #MDA972-01-03-0005.

## References

[1] Vishwani D. Agrawal and Sharad C. Seth, *Tutorial: Test generation for vlsi chips*, Computer Society Press, 1730 Massachusetts Avenue, N.W., Washington, DC 20036-1903, 1988.

[2] R. Iris Bahar, Joseph Mundy, and Jie Chen, *A probabilistic-based design methodology for nanoscale computation*, Proceedings of the International Conference on Computer Aided Design (ICCAD-2004) (San Jose, CA), November 7–11 2004, pp. 480–486.

[3] Gary H. Bernstein, *Quantum-dot cellular automata: Computing by field polarization*, Proceedigs of the 2003 Design Automation Conference (DAC) (Anaheim, CA), June 2–6, 2003, pp. 268–273.

[4] Debayan Bhaduri and Sandeep Shukla, *NANOLAB: A tool for evaluating reliability of defect-tolerant nano architectures*, Tech. Report 2003-09, Formal Engineering Research using Methods, Abstractions and Transformations (FERMAT) Lab, Virginia Tech, 2003.

[5] Keith A. Bowman and James D. Meindl, *Impact of die-to-die and within-die parameter fluctuations on the maximum clock frequency distribution for gigascale integration*, IEEE Journal of Solid State Circuits **37** (2002), no. 2, 183–190.

[6] Mihai Budiu, *Spatial computation*, Ph.D. thesis, Carnegie Mellon University, Computer Science Department, December 2003, Technical report CMU-CS-03-217.

[7] Mihai Budiu and Seth Copen Goldstein, *Compiling application-specific hardware*, Proceedings of the International Conference on Field Programmable Logic and Applications (FPL) (Montpellier (La Grande-Motte), France), September 2002, pp. 853–863.

[8] M. Butts, A. DeHon, and S. Goldstein, *Molecular electronics: Devices, systems and tools for gigagate , gigabit chips*, Proceedings of the 2002 IEEE/ACM International Conference on Computer Aided Design (ICCAD) (San Jose, CA), November 2002, pp. 433–440.

[9] X.T. Chen, W.K. Huang, F. Lombardi, and X. Sun, *A row-based FPGA for single and multiple stuck-at fault detection*, Proceedings of the IEEE International Workshop on Defect and Fault Tolerance in VLSI Systems (Lafayette, LA), November 13-15 1995, pp. 225–233.

[10] Richard J. Cole, Bruce M. Maggs, and Ramesh K. Sitaraman, *Reconfiguring arrays with faults part I: worst-case faults*, SIAM Journal on Computing **26** (1997), no. 6, 1581–1611.

[11] C. P. Collier, E. W. Wong, M. Belohradský, F. M. Raymo, J. F. Stoddart, P. J. Kuekes, R. S. Williams, and J. R. Heath, *Electronically configurable molecular-based logic gates*, Science **285** (1999), 391–394.

[12] B. Culbertson, R. Amerson, R. Carter, P. Kuekes, and G. Snider, *Defect tolerance on the Teramac custom computer*, Proceedings of the 1997 IEEE Symposium on FPGA's for Custom Computing Machines (FCCM '97) (Napa Valley, CA), April 16-18 1997.

[13] André DeHon, *Array-based architecture for FET-based, nanoscale electronics*, IEEE Transactions on Nanotechnology **2** (2003), no. 1, 23–32.

[14] R. Dorfman, *The detection of defective members of large populations*, Annals of Mathematical Statistics **14** (1943), 436–440.

[15] Abderrahim Doumar, Satoshi Kaneko, and Hideo Ito, *Defect and fault tolerance FPGAs by in shifting the configuration data*, Proceedings of the International Symposium on Defect and Fault Tolerance in VLSI Systems (Albuquerque, NM, USA), November 1-3 1999, pp. 377–385.

[16] Ding-Zhu Du and Frank K. Hwang, *Combinatorial group testing and its applications*, second ed., Series on Applied Mathematics, vol. 12, World Scientific, New York, 2000.

[17] Hans Eisenmann and Frank M. Johannes, *Generic global placement and floorplanning*, Proceedings of the 35th Design Automation Conference (DAC) (San Francisco, CA), June 1998, pp. 269–274.

[18] John Emmert, Charles Stroud, Brandon Skaggs, and Miron Abramovici, *Dynamic fault tolerance in FPGAs via partial reconfiguration*, Proceedings of the IEEE Symposium on Field-Programmable Custom Computing Machines (FCCM 2000) (Napa Valley, CA), April 2000, pp. 165–174.

[19] Seth Goldstein, Mihai Budiu, Mahim Mishra, and Girish Venkataramani, *Reconfigurable computing and electronic nanotechnology*, Proceedings of the IEEE 14th International Conference on Application-specific Systems, Architectures and Processors (ASAP 2003) (The Hague, Netherlands), June 24–26 2003.

[20] Seth C. Goldstein and Mihai Budiu, *Molecules, gates, circuits, computers*, Molecular Nanoelectronics (Mark A. Reed and Takhee Lee, eds.), Amer-

ican Scientific Publishers, 25650 North Lewis Way, Stevenson Ranch, California 91381- 1439, USA, May 2003.

[21] Seth Copen Goldstein and Mihai Budiu, *Nanofabrics: Spatial computing using molecular electronics*, Proceedings of the 28th Annual International Symposium on Computer Architecture (ISCA 2001) (Göteborg, Sweden), July 2001, pp. 178–191.

[22] Reiner Hartenstein, *A decade of research on reconfigurable architectures - a visionary retrospective*, Design, Automation and test in Europe (DATE) (Munich, Germany), March 2001.

[23] James R. Heath, Philip J. Kuekes, Gregory S. Snider, and R. Stanley Williams, *A defect-tolerant computer architecture: Opportunities for nanotechnology*, Science **280** (1998), 1716–1721.

[24] Randy Huang, John Wawrzynek, and André DeHon, *Stochastic, spatial routing for hypergraphs, trees, and meshes*, Proceedings of the International Symposium on Field-Programmable Gate Arrays (FPGA 2003) (Monterey, CA), Februaury 23–25 2003, pp. 78–87.

[25] W. K. Huang, X. T. Chen, and F. Lombardi, *On the diagnosis of programmable interconnect systems: Theory and application*, Proceedings of the 14th VLSI Test Symposium (Princeton, NJ), April 28-May 1 1996, pp. 204–209.

[26] Tomoo Inoue, Satoshi Miyazaki, and Hideo Fujiwara, *Universal fault diagnosis for lookup table FPGAs*, IEEE Design and Test of Computers **15** (1998), no. 1, 39–44.

[27] Jurgen M. Kleinhans, Georg Sigl, and Frank M. Johannes, *Gordian: A new global optimization/ rectangle dissection method for cell placement*, Proceedings of the 1988 IEEE/ACM International Conference on Computer-aided Design (Santa Clara, CA), November 7–10 1988, pp. 506–509.

[28] E. Knill, W. J. Bruno, and D. C. Torney, *Non-adaptive group testing in the presence of errors*, Tech. Report LAUR-95-2040, Los Alamos National Laboratory, Los Alamos, NM, September 1996.

[29] John Lach, William H. Mangione-Smith, and Miodrag Potkonjak, *Low overhead fault-tolerant FPGA systems*, IEEE Transactions on Very Large Scale Integrated (VLSI) Systems **6** (1998), no. 2, 212–221.

[30] P.K. Lala, A. Singh, and A. Walker, *A CMOS-based logic cell for the implementation of self-checking FPGAs*, Proceedings of the International Symposium on Defect and Fault Tolerance in VLSI Systems (Albuquerque, NM, USA), November 1-3 1999, pp. 238–246.

[31] Mahim Mishra and Seth C. Goldstein, *Defect tolerance at the end of the roadmap*, Proceedings of the International Test Conference (ITC), 2003 (Charlotte, NC), Sep 30 – Oct 2 2003, pp. 1201–1211.

[32] J. Muttersbach, T. Villiger, and W. Fichtner, *Practical design of globally-asynchronous locally-synchronous systems*, Proceedings of the International Symposium on Advanced Research in Asynchronous Circuits and Systems (ASYNC) (Eilat, Israel), April 2000, pp. 52–59.

[33] Sanil Nassif, *Delay variability: Sources, impacts and trends*, Proceedings of the ISSCC, 2000, pp. 368–9.

[34] Nicholas Pippenger, *Developments in "The synthesis of reliable organisms from unreliable components"*, Proceedings of Symposia in Pure Mathematics **50** (1990), 311–324.

[35] Carl Sechen and Alberto Sangiovanni-Vincentelli, *The TimberWolf placement and routing package*, IEEE Journal of Solid-State Circuits **SC-20** (1985), no. 2, 510–522.

[36] Micaela Serra, Terry Slater, Jon C. Munzio, and D. Michael Miller, *The analysis of one-dimensional linear cellular automata and their aliasing properties*, IEEE Transactions on Computer Aided Design **9** (1990), no. 7, 767–778.

[37] *International technology roadmap for semiconductors*, Sematech, available online from http://public.itrs.net, 2001.

[38] Daniel P. Siewiorek and Robert S. Swarz, *Reliable computer systems*, second ed., Digital Press, Burlington, MA, 1992.

[39] Steven K. Sinha, Peter M. Karmachik, and Seth C. Goldstein, *Tunable fault tolerance for runtime reconfigurable architectures*, Proceedings of the IEEE Symposium on Field-Programmable Custom Computing Machines (FCCM 2000) (Napa Valley, CA), April 2000, pp. 185–192.

[40] Daniel A. Spielman, *Highly fault-tolerant parallel computation*, Proceedings of the 37th Annual IEEE Conference on Foundations of Computer Science (FOCS'96) (Burlington, VA, USA), Oct. 14–16 1996, pp. 154–163.

[41] Mehdi Baradaran Tahoori, Mariam Momenzadeh, Jing Huang, and Fabrizio Lombardi, *Defects and faults in quantum cellular automata at nano scale*, Procceings of the 22nd VLSI Test Symposium (VTS 2004) (Napa Valley, CA), April 25–29, 2004.

[42] Ren-Song Tsay, Ernest S. Kuh, and Chi-Ping Hsu, *Proud: A sea-of-gates placement algorithm*, IEEE Design & Test of Computers **5** (1988), no. 6, 44–56.

[43] Maogang Wang, Xiaojian Yang, and Majid Sarrafzadeh, *Dragon2000: standard-cell placement tool for large industry circuits*, Proceedings of the 2000 IEEE/ACM International Conference on Computer-aided Design (ICCAD) (San Jose, CA), November 5–9 2000, pp. 260–263.

[44] Ron Wilson, *Structured ASICs arrive*, EETimes **May 5, 2003** (2003).

[45] Jack K. Wolf, *Born again group testing: Multiaccess communications*, IEEE Transactions on Information Theory **IT-31** (1985), no. 2, 185–191.

[46] Michael Wrighton and André M. DeHon, *Hardware-assited simulated annealing with application to fast FPGA placement*, Proceedings of the International Symposium on Field-Programmable Gate Arrays (FPGA 2003) (Monterey, CA), Februaury 23–25 2003, pp. 33–42.

[47] Xilinx, Inc., *Virtex-II Series EasyPath: Frequently Asked Questions*, http://www.xilinx.com/publications/products/v2/faq/faq101_easypath.pdf, January 2003.

[48] Yinlei Yu, Jian Xu, Wei Kang Huang, and Fabrizio Lombardi, *Minimizing the number of programming steps for diagnosis of interconnect faults in FPGAs*, Proceedings of the 8th Asian Test Symposium (Shanghai, China), November 1999, pp. 357–362.

# Chapter 4

# OBTAINING QUADRILLION-TRANSISTOR LOGIC SYSTEMS DESPITE IMPERFECT MANUFACTURE, HARDWARE FAILURE, AND INCOMPLETE SYSTEM SPECIFICATION

Lisa J. K. Durbeck
*Cell Matrix Corporation*
*Blacksburg, VA, USA*
ld@cellmatrix.com

Nicholas J. Macias
*Cell Matrix Corporation*
*Blacksburg, VA, USA*
nmacias@cellmatrix.com

**Abstract** New approaches to manufacturing low-level logic—switches, wires, gates—are under development that are stark departures from current techniques, and may drastically advance logic system manufacture. At some point in the future, possibly within 20 years, logic designers may have access to a billion times more switches than they do now. It is sometimes useful to allow larger milestones such as this to determine some of the directions of contemporary research. What questions must be answered so that we sooner and more gracefully reach this milestone at which logic systems contain a billion times more components? Some problems include how to design, implement, maintain, and control such large systems so that the increase in complexity yields a similar increase in performance. When logic systems contain $10^{17}$ switches or components, it will be prohibitively difficult or expensive to manufacture them perfectly. Also, the handling and correction of operating errors will consume a lot of system resources. We believe these tendencies can be minimized by the introduction of a low-cost redundancy so that, in essence, if one switch or transistor fails, the one next to it can take over for it. This reduces effective hardware size by a factor in exchange for a way both to use imperfect manufacturing techniques, and, through similar means, maintain the system during its life cycle. It may also be possible to use similar basic principles for a more complex problem, designing a system that can



*S.K. Shukla and R.I. Bahar (eds.), Nano, Quantum and Molecular Computing*, 109-132.

catch and compensate for operating errors, but with low enough cost in time and resources to allow incorporation into all large systems. We suggest that such a system will be a distributed, parallel system or mode of operation in which systems failure detection is a hierarchical set of increasingly simple, local tasks run while the system is running. Work toward answering these questions appears to also yield some useful ways to approach a more general question, of constructing systems when their structure and function cannot be completely predetermined.

**Keywords:** integrated systems verification, integrated systems design, fault tolerance, fault isolation, fault handling, Cell Matrix, cellular automata, nanotechnology, electronics, transistor, Von Neumann, Drexler, CPU/memory architecture

## 4.1 Four Areas for New Research

Hardware and logic designs have come a long way. The transistors used in a modern single-chip CPU are several hundred million times smaller than the original transistor built in 1947. If a contemporary CPU were built with the original transistor technology, it would take up a space of roughly one square kilometer. Current ways to produce logic designs pack many more transistors into hardware than their predecessors ten years ago, and ten to twenty years from now there may be ways to produce hardware devices with a billion times more transistors or switches. Note that such an increase in fabrication density does not lie on most current technology road maps, such as ITRS. This prediction is instead based on the expectation that researchers will uncover fundamentally different technologies that cause a sudden jump in device density. At the end of the curve following Moore's Law, we may find that process technology begins a completely new curve.

There has been and continues to be strong economic incentive for miniaturization of logic designs and electronics. Although for some products this has been used to simply reduce the footprint, designers have also been freed to create larger and more complex designs as transistor density has increased.

How complex will designs be with a billion times more capacity available? And what of the fact that some of the work to uncover replacements for the field effect transistor is being done in scientific disciplines in which three dimensional structures are not at all unusual– ten to twenty years from now there could be ways to produce three dimensional hardware for logic designs.

Technical breakthroughs over the next ten to twenty years could come gradually, but may instead exhibit sudden leaps in progress as problems are solved, discoveries are made. The path will depend on many variables: the nature and timing of future breakthroughs, whether they combine to form a complete production method, and how rapidly these new ideas are put into practice. In addition to having much larger switch counts and much smaller package sizes for logic designers to work with, there may be production means, and product parameters, very different from the ones in use today. There are many hard and interesting research questions at this junction that are appropriately addressed both in academia and industry. They include:

1 getting involved in the analysis of nascent switch-production methods to model how well they fit the engineering requirements of integrated circuits and what the outcomes could be from using a particular new method. Considerations include the number of switches per unit area or volume inside the device (density), total number of switches inside the device (volume), operating condition limitations, operating speeds, power requirements, production costs, and the reliability of production method and of the product during its lifecycle. This analysis will help guide the directions of research in switch production methods, determine how a particular method is best applied, and also help in comparing and contrasting differing production methods;

2 coming up with designs, and design tool capacity, for effective use of $10^{17}$ transistors or switches, such as designs that will scale up gracefully or even seamlessly as density increases, and ways to produce designs that readily lead to production of larger switch counts;

3 coming up with more powerful design and verification tools to handle logic designs with many orders of magnitude greater scope and complexity, a topic which Hsiao et. al address in Chapter 11;

4 coming up with more flexible processes for the product path, from definition of a new product's requirements, through logic design, test and verification, and implementation in hardware. Processes should be flexible enough to permit things like:

   (a) co-development of design, test and build, with none of these steps assumed to be fixed;

   (b) development of design and verification that exploit a new fabrication method's strengths and minimize its weaknesses; and

   (c) blurring of test and build into a more iterative process that takes the imperfect nature of a build process into account, rather than assuming that a perfect build is possible or normative while an imperfect build is unusable.

We have done some work related to these four areas of inquiry. The question we are interested in is what logic designs are suitable for billions to trillions of times larger and more complex designs, particularly if you remove the assumptions of perfect hardware and completely predetermined usage. We propose a two layer hierarchy, in order to decouple logic and manufacturing issues. Figure 4.1 gives some idea of this approach and its effect on the production of logic systems. For a given logic design X, we build a two-layer logic design, with the lower layer Y representing what is actually built in hardware, and the upper layer either representing the logic design X, or representing a means by which X can be constructed, or often a representation of a bit of both, such as a hardware library containing the components of X and a means to copy them and lay them out onto the hardware layer Y. The lower layer Y contains the logic design needed to either represent X in an efficient manner, or to construct X, after which it represents X in an efficient manner. This means we have direct control over the hardware layout of Y, but not X, which means we can control the placement of X's gates and wires on Y, but this is one step removed from the native hardware and its placement of transistors or switches and wires. In exchange, we gain the capacity to have parts of X defined, or redefined, at runtime, when information about the hardware's imperfections is known. Similarly, we can use this capacity to cause the layout of X to be affected by other information at any time during the system's useful life, and can use it to do things like optimize X or its implementation for the situation in which the product is used, such as the specific inputs it gets, use patterns, changes in use or inputs, environmental conditions, damage incurred throughout its life, etc.

The lower layer Y of this two-level design is a homogeneous structure with local-only interconnect, one that appears to be a good fit with the expected strengths and weaknesses of the revolutionary, post-field-effect transistor production methods under development. An illustration of Y's structure is provided in Figure 4.2. Layer Y's specification leads to hardware with every square or cubic millimeter packed with simple building blocks that can implement transistors/switches and wires, or small-scale- integration-sized gates, logic blocks, and wires. We call these building blocks logic cells, illustrated in Figures 4.3 and 4.4. Logic cells combine low level signals and produce outputs, and individual cells' functions can be combined to make more complex logic blocks and functions such as memory, state machines, multipliers, floating point units, and so on: any digital logic design. However, they also have this additional property of supporting non-static, partially-predetermined functions. We achieve this partly from the design of the logic cells, and partly from the functional directives we provide them. That is, we assign them not a static but a dynamic function, analogous to switch statements in programming languages (if X then do A, else B). Then we base their function on the inputs they receive, and organize groups of cells together to do useful work as dynamic circuits. The Y layer of hardware can be thought of as smart transistors and wires that can move around and change system function to suit new directives, and the new capability of dynamically changing function can be thought of as a system that can process and modify both data and its own logic, its own circuits, its own layout. The hardware layer Y is the physical and logical substrate upon which these traditional static and new dynamic logic designs exist. The logic design layer X is the set of current logic designs plus designs that take advantage of the ability to change the logic and its layout during its operation, for example, circuits that take in, process, and modify circuits.

We have explored the use of this new capability to create systems that determine on their own how to best utilize their hardware resources, and to create systems that lay out perfect circuits on damaged hardware [2, 3, 5]; to create self-replicating and self-modifying systems [2, 3, 5, 10, 11]; to create an expanding counter that extends itself when it detects impending overflow conditions [11]; to create orderly copying of libraries of logic blocks onto Y hardware [10, 11]; to create mechanisms to test enhanced wires and transistors for faults and report them [2]; to create mechanisms to isolate faulty transistors and wires and demonstrate that they cannot affect the rest of the hardware [3]; to organize X,Y systems that efficiently perform computations via highly distributed, highly parallel functions [1, 12]. We have also ensured that traditional capabilities are well-served: it is easy to lay out a simple circuit onto Y hardware and can be done via straightforward engineering practice, for example, compilation of circuit schematics [14] or HDL [15]. We have fully defined the layer Y with a complete and simple specification [6, 8] and have initial, unsophisticated and fairly low-level ways to get X onto Y [14] and are working our way up to more sophisticated tools for the use of higher level representations of logic designs.

Our work to date can provide researchers with a convenient framework in which to develop research programs for these four areas of inquiry. In this chapter we describe and address one particular problem associated with so many orders of magnitude more switches per unit area/volume, which will give the reader a better idea of the problems that cause the need for new research in these four areas, and will provide a better idea of how to use our framework to approach them as well.

Figure 4.1 shows a high level view of how the production of a particular logic design might be performed using the Cell Matrix computing architecture we have designed and the X,Y approach described here. The approach is similar to that used for reconfigurable devices today, with several important unique aspects that are marked in the figure by asterisks. Time proceeds roughly to the right in two parallel, independent tracks, and then down the right hand side of the diagram. The construction of the X and Y layers are performed independently, shown by the first two rows of events. The product is produced by the configuration of the Cell Matrix hardware in Layer Y according to the Logic System Design X, shown as steps proceeding down the right column of events. Because the implementation of X onto hardware is a post-manufacturing process, there is opportunity to perform useful functions in a post-manufacturing process labeled with a star, such as mapping and avoiding hardware defects, and

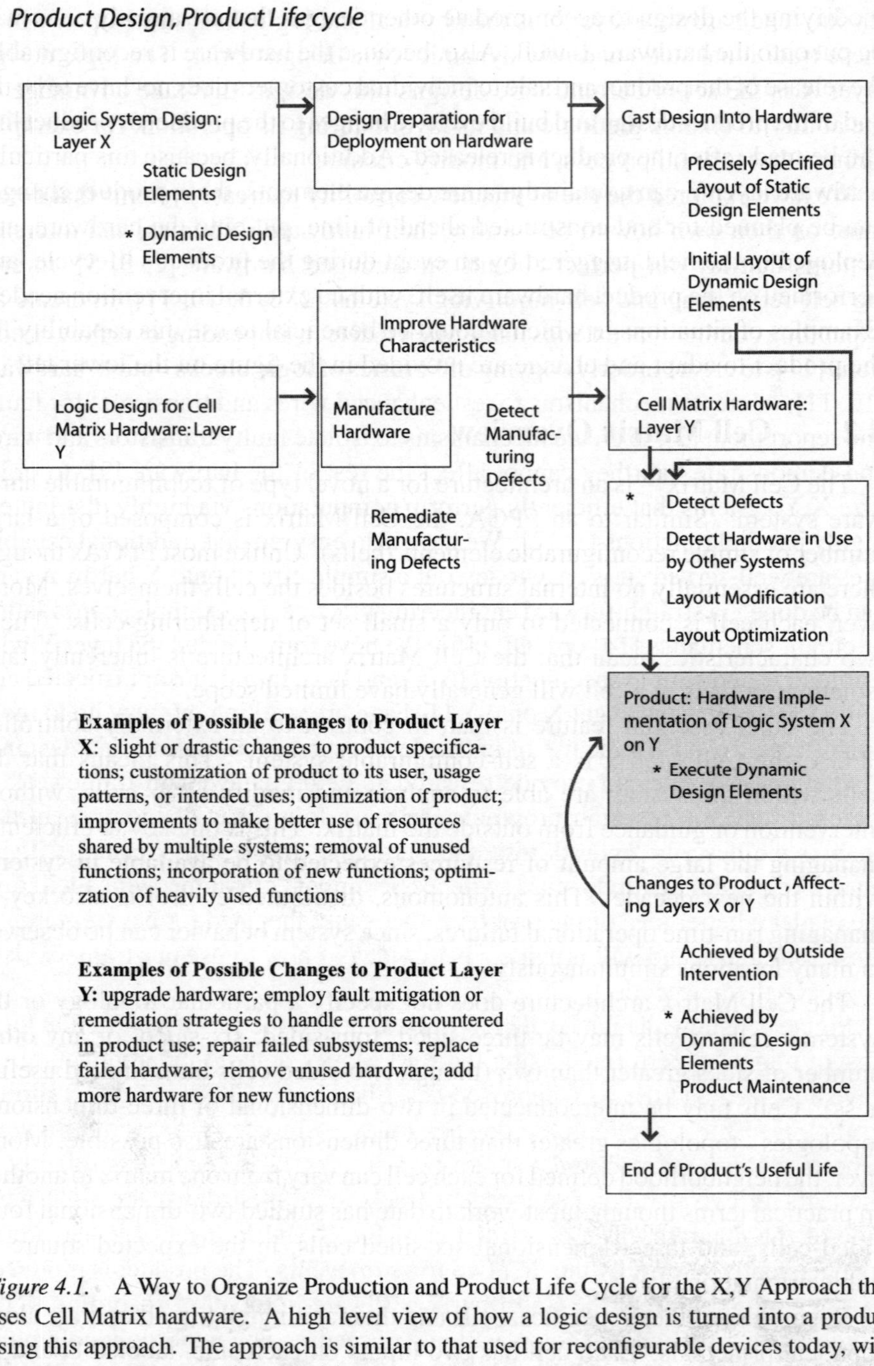


*Figure 4.1.* A Way to Organize Production and Product Life Cycle for the X,Y Approach that uses Cell Matrix hardware. A high level view of how a logic design is turned into a product using this approach. The approach is similar to that used for reconfigurable devices today, with the unique aspects marked by asterisks.

modifying the design to accommodate other designs that are already on or will be put onto the hardware as well. Also, because the hardware is reconfigurable, the release of the product and sale to individual customers does not have to be the end of the product design and build cycle. Changes to the product or product line can be made after the product is released. Additionally, because this particular hardware layer Y can contain dynamic design elements, these product changes can be planned for and constructed ahead of time, put onto the hardware, and deployed in the field, triggered by an event during the product's lifecycle, and performed on the product hardware itself, with no external intervention needed. Examples of situations in which it could be beneficial to use this capability for the product to adapt and change are provided in the figure on the lower left.

## 4.2 Cell Matrix Overview

The Cell Matrix™is an architecture for a novel type of reconfigurable hardware system. Similar to an FPGA, the Cell Matrix is composed of a large number of simple reconfigurable elements (cells). Unlike most FPGAs though, there are essentially no internal structures besides the cells themselves. Moreover, each cell is connected to only a small set of neighboring cells. These two characteristics mean that the Cell Matrix architecture is inherently fault isolating: defects in a cell will generally have limited scope.

The other essential feature is that, in contrast to an externally-controlled FPGA, the Cell Matrix is a self-configurable system. This means that the cells within the system are able to analyze and modify other cells, without intervention or guidance from outside the matrix. This is one key to efficiently managing the large amount of resources expected to be available in systems within the next decade. This autonomous, distributed control is also key to managing run-time operational failures, since system behavior can be observed in many locations simultaneously.

The Cell Matrix architecture does not specify a particular topology or the system's cells. Cells may be three-sided, four-sided, six-sided, or any other number of sides greater than two (though two-sided cells have limited usefulness). Cells may be interconnected in two-dimensional or three-dimensional topologies - topologies greater than three dimensions are also possible. Moreover, the neighborhood defined for each cell can vary from one matrix to another. In practical terms though, most work to date has studied two-dimensional four-sided cells, and three-dimensional six-sided cells, in the expected square or cubic orientation.

Regardless of these specifics, cells and their arising matrix all operate along identical principals. Each cell has two inputs on each side, labeled D and C. Each cell has a corresponding set of outputs (D and C). Cells are interconnected according to the matrix-wide definition of a neighborhood, with inputs and

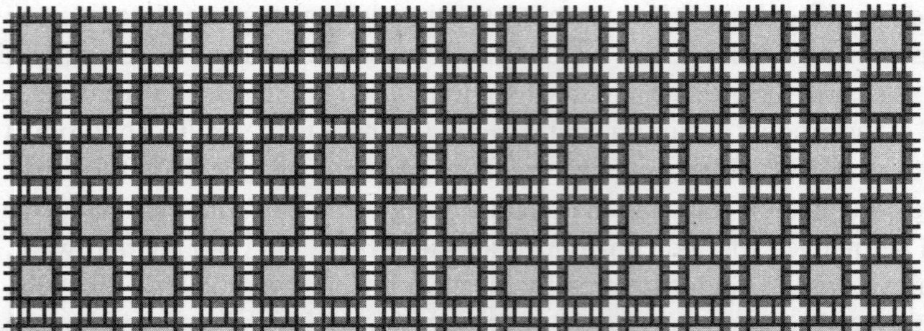

*Figure 4.2.* Hardware Layer Y definition. The hardware is a densely packed substrate containing a regular mesh of identical, very fine-grained SSI-scale processors that are connected in a local, nearest-neighbor scheme. The definition of the processors includes cases for connections with n local neighbors, 3, 4, 6, or other convenient local connection schemes; here, the processors are shown as 4-sided cells each connected to their four neighbors. The content of each of these processor cells is detailed in Figure 4.3.

outputs connected in the obvious fashion. Additionally, each cell contains an internal lookup table (LUT). The LUT maps every possible combination of D inputs to a set of C and D outputs.

Each cell within the matrix operates in one of two ways, depending on the *mode* in which the cell is operating. If a cell is in *D mode*, it continually samples its D inputs, looks up a set of outputs in its internal LUT, and sends those LUT values to its C and D outputs. Note that this happens continuously, without any clocking or synchronization.

If, instead, a cell is in *C Mode*, it samples its D inputs as specified by a system-wide clock, and loads the sampled bits into its internal LUT. Moreover,

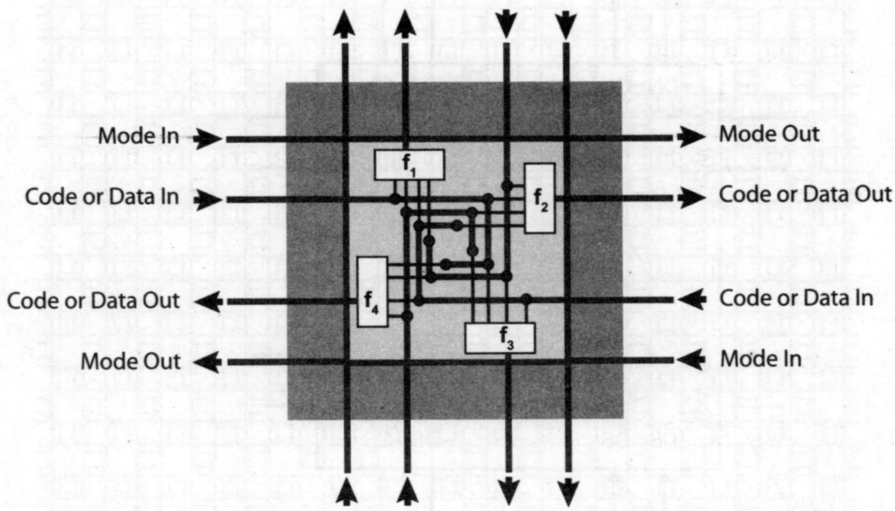


*Figure 4.3.* Structure of each logic processor cell in the hardware Layer Y from 4.2. A four-sided cell is shown. As shown by the input signal labels on the left of the figure, each side provides a data input to the cell, and receives a data output (which can also be used to output the cell's content, or code), as well as a Mode output that can be used to put the neighboring cell into a programmable control mode called C mode, during which the neighbor's content or code is changed. The input signals are combined to produce any logic function of four inputs and eight outputs, labeled $f_4$, that is delivered to the left side of the cell. The other three sides' outputs are other logic functions $f_1$ through $f_3$ that are based on any of the inputs to the cell. The functions $f_1$ through $f_4$ are settable: they are reset when the "Mode In" is set high, at which point the "Code or Data In" line is used to input a new code for the functions. The cell is completely symmetric in its function. A different view of the cell structure and function is provided in Figure 4.4.

before a LUT bit is overwritten, it is sent to the cell's D outputs. C Mode is thus the mode in which a cell's LUT is read or written, while D Mode is the mode in which a cell is able to perform data processing functions via it's LUT.

Finally, a cell's mode is specified by its C inputs. If any C inputs are asserted, then the cell is in C Mode. Otherwise the cell is in D Mode. Three key consequences of this are:

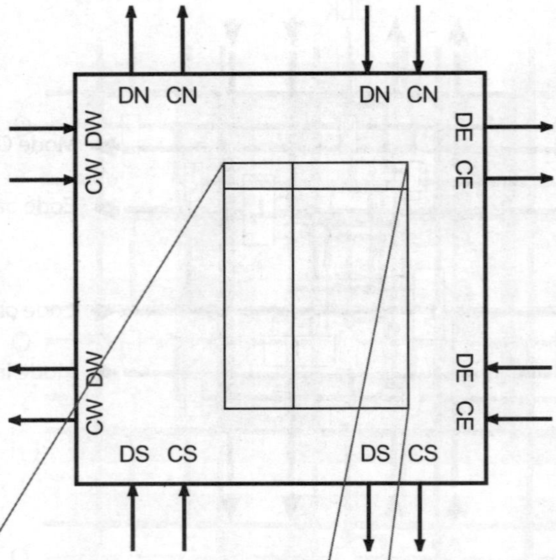


| DN | DS | DW | DE | CN | CS | CW | CE | DN | DS | DW | DE |
|---|---|---|---|---|---|---|---|---|---|---|---|
| 0 | 0 | 0 | 0 | 0 | 0 | 0 | 0 | 0 | 0 | 0 | 0 |
| 0 | 0 | 0 | 1 | 0 | 0 | 0 | 0 | 0 | 0 | 0 | 0 |
| 0 | 0 | 1 | 0 | 0 | 0 | 0 | 0 | 0 | 0 | 0 | 1 |
| 0 | 0 | 1 | 1 | 0 | 0 | 0 | 0 | 0 | 0 | 0 | 1 |
| 0 | 1 | 0 | 0 | 0 | 0 | 0 | 0 | 0 | 0 | 0 | 0 |
| 0 | 1 | 0 | 1 | 0 | 0 | 0 | 0 | 0 | 0 | 0 | 0 |
| 0 | 1 | 1 | 0 | 0 | 0 | 0 | 0 | 0 | 0 | 0 | 1 |
| 0 | 1 | 1 | 1 | 0 | 0 | 0 | 0 | 0 | 0 | 0 | 1 |
| 1 | 0 | 0 | 0 | 0 | 0 | 0 | 0 | 0 | 0 | 0 | 1 |
| 1 | 0 | 0 | 1 | 0 | 0 | 0 | 0 | 0 | 0 | 0 | 1 |
| 1 | 0 | 1 | 0 | 0 | 0 | 0 | 0 | 0 | 0 | 0 | 1 |
| 1 | 0 | 1 | 1 | 0 | 0 | 0 | 0 | 0 | 0 | 0 | 1 |
| 1 | 1 | 0 | 0 | 0 | 0 | 0 | 0 | 0 | 0 | 0 | 1 |
| 1 | 1 | 0 | 1 | 0 | 0 | 0 | 0 | 0 | 0 | 0 | 1 |
| 1 | 1 | 1 | 0 | 0 | 0 | 0 | 0 | 0 | 0 | 0 | 1 |
| 1 | 1 | 1 | 1 | 0 | 0 | 0 | 0 | 0 | 0 | 0 | 1 |

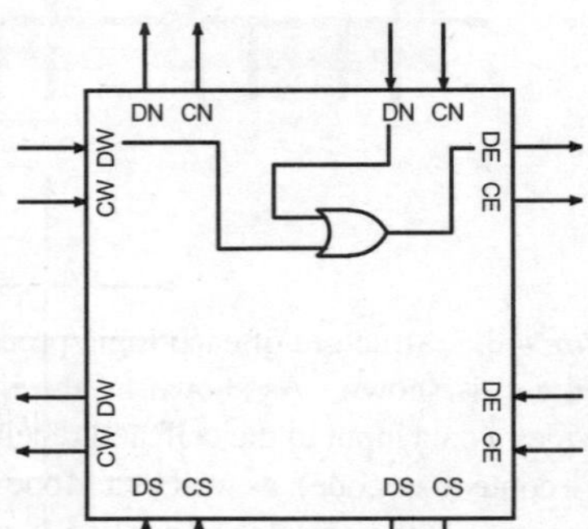


*Figure 4.4.* Logic Cell Structure and Function. The Signals in and out of the cell are shown, with the sides of the cells labeled as North, East, South, and West. The label "DN" at the top indicates the Data or Code signals sent in or out the North side of the cell, and "CN" indicates the Control Mode signals sent in or out of the North side of the cell. The lookup table that dictates cell function is shown and enlarged on the lower left. This particular lookup table configuration corresponds with the logical OR function shown on the lower right. A multi-cell lookup table configuration is shown in Figure 4.5.

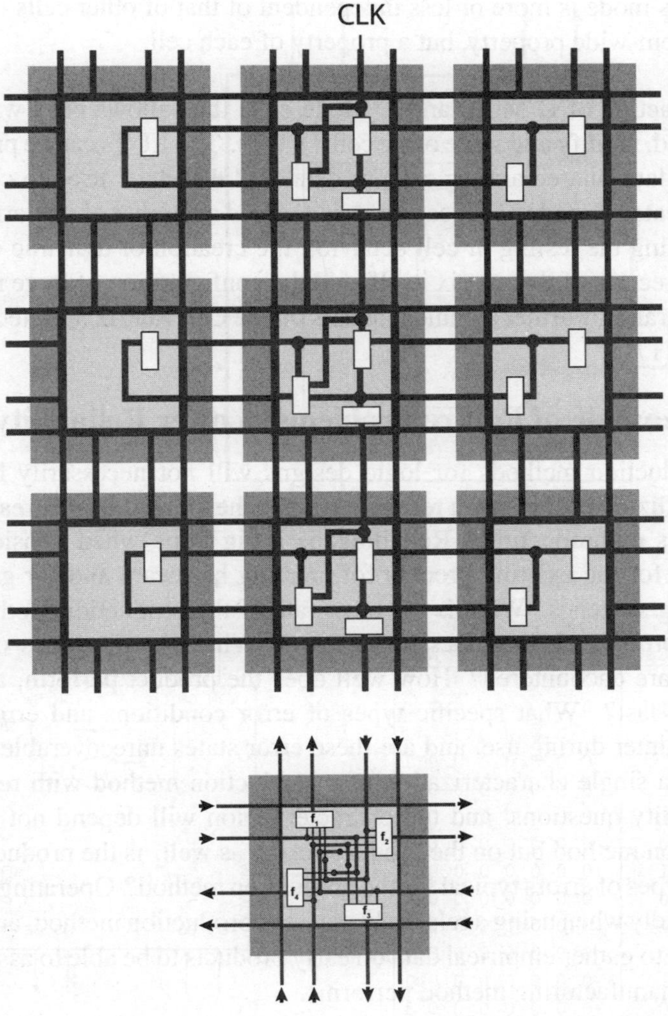


*Figure 4.5.* Six logic cells set up to function as a 3-bit counter. The same functional view from Figure 4.3 is shown in the inset at the bottom of the figure. Only the functions $f_1$ through $f_4$ that are used for this circuit are displayed in the upper $3 \times 3$ matrix of logic cells. A Clock signal CLK is provided at the top, and the bits Q are produced on the right. Many of the functions $f_1$ through $f_4$ are in this case used for simple wires to propagate signals; others in the center column of cells are used to perform boolean logic needed for the counter.

1 the mode of any cell C1 can be specified by any of its neighbors C2 (since C2 has a C output connected to one of C1's C inputs);

2 the mode of a cell can change over time, since the value of its C inputs can change over time; and

3 a cell's mode is more or less independent of that of other cells - it is not a system-wide property, but a property of each cell.

The interaction of D Mode and C Mode cells thus allows cells within the matrix to read, modify and write other cells' LUTs. The LUTs can be processed as ordinary data, shared among cells as data, and then later used to configure cells, i.e., treated as code. This can be used to yield a number of powerful functions, including the testing of cell behavior, the creation of dynamic circuitry under the direction of the matrix itself, and the configuration of large numbers of cells in parallel. Further technical details on the Cell Matrix architecture can be found in [17].

## 4.3 Example of Future Problems: Lower Reliability

New production methods for logic designs will not necessarily have the benefit of utilizing past gains in reliability, since they may deviate greatly from past methods of production. Reliability is a big issue when considering a replacement for the existing process for creating hardware and for gradually miniaturizing switches. What is the error rate of the production method, and how many product defects does it produce? What specific types of errors and defects are encountered? How well does the product perform, and how long does it last? What specific types of error conditions and error states does it encounter during use, and are these error states unrecoverable? There may not be a single characterization of a production method with respect to these reliability questions, and the characterization will depend not only on the production method but on the product design as well: is the product robust against the types of errors typical for the production method? Operating failures are highly likely when using a relatively untested production method, and it will be important to gather empirical data on early products to be able to assess how well a new manufacturing method performs.

What the reader should be interested in is not just being on the receiving end of new hardware manufacturing methods, but also participating in their development, and their improvement. A new manufacturing process' expected density and volume are important, as are its expected cost and operating parameters such as speed, temperature, and power. However, reliability is also a big issue when considering a replacement for existing methods of manufacture. It seems unlikely that any radically different production method will have reliability even close to the current methods of producing field-effect transistors, because they will likely be too different to be able to incorporate the last five decades of improvements in transistor production. And yet they may have highly desirable traits in terms of volume or density, or operating parameters, and so we will want to be able to find a way to overcome their reliability problems enough to begin using them, and gradually improve them until they are reasonably reliable. Even if they eventually reach the reliability of current techniques, their reliability will forever be challenged by the tendency to utilize more switches in designs as density improves, what we are calling here increased volume. Therefore, even if the defect or operating error rates remained fixed, the number of defects or operating errors will continue to rise with volume. With volumes approaching a billion times today's volumes, operating errors will likely be commonplace in products as they are used.

Even with a manufacturing method that exhibits a small defect rate, such as a one in a million defect rate, logic designs with $10^{10}$ to $10^{17}$ transistors will be hard to manufacture perfectly, and, during their operation, will exhibit a mean time between failures that is markedly decreased, and reciprocally related to hardware volume, or system size. How are logic designers to handle this? Can designers continue to assume that the manufacturing process will produce perfect implementations of their designs that will also operate perfectly? Must logic designs be intimately aware of and tied to the physical constraints of the manufacturing method and its typical operating errors? Must each contemporary design be redone to increase its robustness against manufacturing defects and operating errors? A good goal for research in these areas would be to solve these reliability problems and yet allow logic designers to have to change as little as possible about what they do now.

With our two-layer approach there is opportunity to greatly minimize what logic designers have to know about the manufacturing method and its imperfections. This is because the view of the hardware is abstracted up one level above the physical layer, and is handled later in time, post-manufacture. There may be errors in the manufacture of the Y layer, but designs are laid onto the Y layer in a later step, which has the advantage of knowing what defects are present in the Y layer. Current research with our collaborators is intended to provide ways to handle operating errors that can exist within, underneath, alongside, or above the desired logic design X, using the dynamic capabilities of our architecture. The intention is that these capabilities can simply be evoked by the use of base-level logic blocks with extra capabilities for fault handling, dynamic system modification, and other needed functions. Although in some cases this will be an overly reductive approach, it is possible to present a view of the hardware as perfect to the logic designer, and delegate the handling of imperfections to a post-processing step performed on the design. This approach also supports the basic premise of managing complexity through encapsulation.

The focus of the discussion below will be on how two aspects of this approach–redundant hardware and dynamic functions–lead to ways to tolerate the reliability problems of larger, more complex systems built using immature production technology.

## Manufacturing Defects

Manufacturing methods must be refined and improved, but perfection is a difficult goal to hold, particularly when the desire to continue to miniaturize switches persists in driving manufacturing onward to new challenges. There are at least five conceivable ways to handle defects in the construction of logic designs in hardware:

- discard any hardware that is not perfect;
- repair, or remove and replace, the individual defects;
- build redundancy into the hardware and a means to use only perfect resources within the hardware;
- use logic designs for Layer X that can function despite defects or runtime faults; or
- perfect the manufacturing technique so that it creates no defects.

The tact typically taken today is to place the burden largely on the manufacturing technique to provide perfect hardware, and to discard any hardware it constructs that is not perfect. We argue, however, that certainly during the development of revolutionary new fabrication techniques, and quite possibly long after their refinement, this will not be the most cost-effective option to choose. The other approaches above should also be considered, such as building redundancy into the hardware. Incorporating redundant copies of hardware components is used widely today to effect fault tolerant systems. This option is readily available for use in our approach, although it takes different forms for the different layers. Again, there are two layers to a logic design in our approach, the normal logic design X, and the lower level implementation layer Y. It is possible to incorporate redundancy into the logic systems design and implementation, as it is done today for systems onboard satellites and spaceships, via modifications to the logic design layer X. The lower layer Y can also be used quite effectively to safeguard products against manufacturing defects. The basic mechanisms offered above for safeguarding a logic system's perfect function against manufacturing defects are to either enlarge the system hardware to include redundant copies of resources, or to go in and repair, remove, or replace defects. Layer Y can achieve both of these models. How is does this is the next topic of discussion.

To define our use of terminology, we are using the general term errors to refer to any sort of nonoptimal function, and looking only at one cause of errors, hardware failures of some kind. We use the term manufacturing defect for those failures that are turned up during initial testing of the hardware, and operating errors for all other hardware failures that turn up later in the product life cycle. We note here that hardware failures' effects can be persistent or transitory, periodic or aperiodic, and can appear predictable or unpredictable in their response to testing.

## Redundant Copies of Hardware

One benefit of our approach is that some measure of both fault tolerance and redundancy is automatically provided for all logic designs X by the nature of layer Y. Y contains low-cost redundancy already, because the logic cells themselves are resources that can be used to implement transistors, wires, flip flops, truth tables, gates, logic blocks, state machines, or any other digital circuit component. If one cell is bad, a design layout tool can use the one next to it. This is a great way to provide redundancy, because the system designer does not have to decide how much redundancy to put in ahead of time, or where exactly to focus the extra resources (three copies of this subsystem, four of that); instead, resources are pulled from a general pool, and used as cleverly as the design layout tool or diagnostic system is designed to utilize them. The problem is thus reduced to the question of how much larger to make the hardware than would be strictly needed for perfect manufacture, which could potentially be answered statistically, or via over engineering. This approach would then require an additional processing step in the design flow that modifies the placement of logic and wires using knowledge of hardware defects. A first version of such a tool for this X,Y approach was created by Dimitri Yatsenko [15]. We have also demonstrated the Y layer's capacity to make the necessary determinations on its own as to what hardware is good and where to put the gates and wires in a circuit definition it receives [2, 3, 5], which serves as an example of how to use dynamically interpreted, non-predetermined directives.

Nearly all circuits in use today cease to function properly when they incur any sort of damage. Any error in the creation of its transistors or wires is usually fatal to a circuit hardware implementation. Our cells are a much more fault tolerant circuit design because an error in the production of one cell is generally limited in effect to the immediate neighborhood of the cell. Because the X,Y Layered approach described here lays a logic design X out on Y hardware in a post-manufacturing step, we can attack this problem of manufacturing defects from a different angle, invoking several different ways to ensure that a defective piece of hardware is not intimately and irrevocably tied to the implementation of a critical piece of logic. Figure 4.1 indicates a number of functions that can be performed during the layout of the logic design onto the hardware in the righthand side box above the gray Product box.

No particular piece of hardware is necessary in layer Y, aside from the receipt of power to each logic cell. There are no differentiated cells, no buses, no external memories, no specialized structures. In fact there is no heterogeneity at all: the hardware is simply a densely packed surface or volume of logic cells connected only to their nearest neighbors. layer X can be put onto layer Y in a post-manufacturing step, at a point in time after Y has been tested for defects, and after an inventory of defect type and location is made. At that point in time, the layout of layer X can safely be determined so that it uses only good hardware. Or, if errors are encountered at runtime, there is support for dynamically re-allocating Y's hardware at runtime instead. This change to design flow insulates logic designs from hardware defects. It prevents logic system designers from having to change their current view that they will receive perfect implementations of their logic designs. In a later section on operating errors, we describe how error avoidance can be done at runtime, to deal with errors that arise in systems later, post production, from parts that start to fail, or from environmental effects.

## SCANDISK for Logic Designs

One way to achieve this design flow process that can lay out perfect systems on top of an imperfect Y layer is to implement something analogous to what is used for memory, or disks, today, such as a SCANDISK type of process that checks each region of hardware and constructs a map of all bad regions, which is then used during writes to strictly prevent the copying of data into bad regions of hardware. The system that performs the scanning must itself be nondefective, and it should have the goal of marking as little extra hardware off-limits as is possible. We have constructed such a process for scanning the Y layer and reporting failed regions of hardware [2, 5]. The process has the capability of marking off $n \times m$ regions of cells as defective if any one of the cells fails any one of the tests it performs; most tests set the logic cell up to perform a specific function, then tests it with a set of input values to make sure it provides the expected outputs for the given inputs. The size of the region tested and marked is set to whatever is convenient, if testing is done by an external system proceeding from an edge of the hardware, and to a square region of $44 \times 44$ cells if testing is done by autonomous agents set up on the hardware, with most of that space required for replicating the testing apparatus and the ability to mark sectors to all locations where the testing is being done. Testing begins at an edge of a defined region, and scans the whole region using only already-tested hardware. Because of the distributed, local nature of signal processing in the Y layer's structure and function, we were also able to set up the test process so that it ran efficiently, testing many independent regions of hardware at the same time [3].

## Isolation of Good Hardware from the Effects of Bad Hardware

If the intention is to use hardware despite the presence of manufacturing defects, then no defective hardware can be used in the layout of logic designs, and the defects present in the hardware must not be allowed to affect the logic design's function. How we approach this in the X,Y approach is that the layout of layer X on layer Y must be one that avoids placing any logic or wires within defective regions. We also explicitly prevent defective hardware from interfering with logic system function.

This preventive step could be done by either physically removing the defects from layer Y's hardware, or by logically removing them from layer Y's functioning. This effectively stops the spread of defects and guarantees that they will not alter the operation of logic design X in any way.

Layer Y has the desirable property that defects tend to remain localized near the source of the defect. This is largely due to the physical organization of layer Y, which is fine-grained cells connected only locally, to their nearest neighbor cells. We were able to validate this expectation by looking at failed cells' effects on silicon chips, and we have also studied failure modes by configuring reconfigurable chips with explicit errors in the definition of layer Y. Because there are no specialized structures within the hardware, and no distant connections, a defect is limited in its ability to spread. A cell C1 can fail, and for some types of failure modes, this cell's failure can send signals that affect the outputs of good neighboring cells C2, rendering them unusable. However, analysis of these scenarios suggests that there is only a very slight possibility that cells C2 will in turn affect their neighbors C3. The extent of the region affected by a specific defect will generally be one cell, and in some cases also its neighbors, and in a few of those cases also their neighbors, for a total of two connected levels outward from the defective cell.

This situation is a promising one, because it means that defects are naturally limited in their locality and effects. We go further, however, to ensure that defects do not affect good hardware. We do this by finding and then explicitly isolating faulty cells, as illustrated in Figure 4.6. They are explicitly isolated by setting the logic up in all neighboring cells to completely ignore any signals received from the offending side of the cell. This is described more fully in several sources [6, 5]; however, the effect is to construct a functional/logical wall around the defective region, by specifying the functions of the good cells at the boundary to ignore anything they receive from the defective region: it doesn't matter what they receive, they do not use it. This approach uses one set of cells surrounding a defective region to guarantee that no signals escape out of the wall around the defective region.

Figure 4.6 shows how this guard wall is erected. In the figure, it is assumed that analysis of each cell for defects has already been performed, and the center cell has been determined to be defective. Layout tools need to have knowledge of both the defects and the guard walls and are responsible for ensuring that they do not attempt to lay down a part in this region: if they try, they will fail, because the guard wall is already in place and irrevocable. They may use only the unutilized resources and sides within the outermost level of the guard wall. The defective cell is logically isolated from the functioning of the rest of the matrix by ensuring that its outputs will be explicitly ignored. This is done by putting its immediate neighbors, the plus shape of dark gray cells around the center, permanently into C mode, which causes them to send out low signals on the rest of their output lines no matter what signals they receive. They are put into C mode by the light gray cells, as indicated by the 1 on the $C_{i}n$ lines of the dark gray cells. This strategy completely contains the signals from the defective center cell. It uses up the four neighboring cells completely, and the edge of one of each neighbor one level further out in the adjacency that is used to send the C mode signal. The smaller figure below it shows that the guard wall grows compactly around the defect with a two-cell defective region.

## Detection of Bad Hardware

Before defects can be examined and explicitly walled off, they must be found and pinpointed as precisely as possible. Because cells have the ability to exchange data with other cells, as well as the ability to change a neighboring cell's function, it is possible for one cell to perform a series of tests on a neighboring cell. For example, a cell can be configured to always output 0, and by then verifying that the output is zero, one can confirm the output is not stuck-at-1. Similarly, a cell can be configured to always output 1, to detect a stuck-at-0 fault in the output. By configuring a cell as a wire that outputs its input, one can partially verify the cell's configuration mechanism. Configuring a cell as an inverter allows one to check for a short between the input and the output of the cell, as well as to further test the configuration mechanism of the cell. More complex test patterns can be used to further exercise the cell-under-test. Using these basic concepts, we have developed ways to test for defects at a granularity of one cell, and the passage or failure of one test by that cell [2].

Most defects can be detected by testing cell function, and extensive testing, while time-consuming, can be organized as a distributed, parallel set of local processes, so that many cells can be tested at each iteration of a testing regime (ideally an increasing number at each iteration to reduce the order of magnitude). We have constructed a distributed, parallel, order $\sqrt{n}$ means to access logic cells using only already-verified hardware to do so, and to supply tests of their functioning, and reporting of any failures [2]. These test patterns can then be

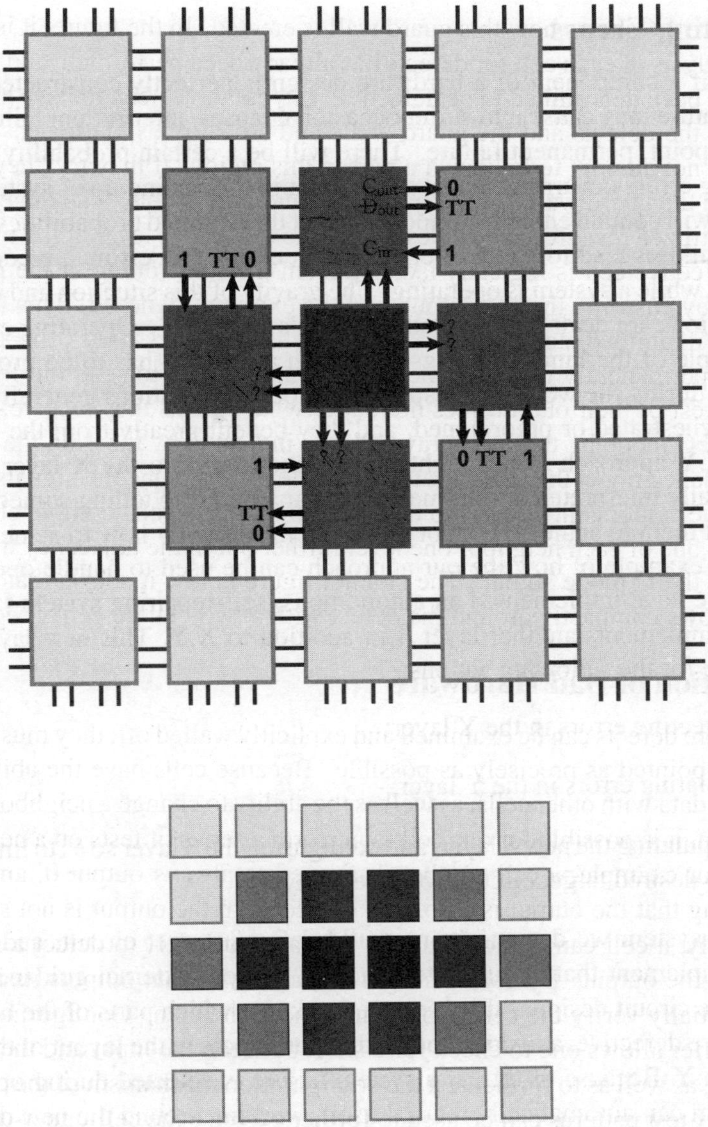


*Figure 4.6.* The Construction of a Guard Wall, and inset example of a guard wall for a 2-cell-wide defect. Black cells are defective, dark gray cells are used to isolate the defective cells' signals, and light gray cells are used to orchestrate this isolation.

customized to the expected defect types for a specific hardware manufacturing process.

## Operating Errors

Even if a component of a hardware design is perfectly constructed, its inherent nature may cause it to suffer occasional lapses, intermittent failures, and at some point, permanent failure. There will be a certain probability of these operating errors occurring, and at some point in the scaling up of system sizes, systems will contain enough components that the summed probabilities of component failures lead to a rate of errors high enough that errors are completely common while a system is operating. The gravity of this situation and potential remediations are dealt with in Andre Dehon's chapter 7 . Operating errors are an example of the kinds of things for which a solution has to be provided at runtime, during the working lifespan of a product. Solutions generally cannot be preorchestrated or preordained, and they benefit greatly from the capacity of our X,Y approach to put higher level functions onto the X layer that are dynamically interpreted at runtime. This is analogous to telling someone what you want them to achieve, but not telling them explicitly how to achieve it.

As an example of how the our approach can be used to handle operational errors, we have implemented an autonomous, self-repairing system [5]. This system implements another layer S in addition to X,Y. This new layer is responsible for the following actions:

- detecting errors in the Y layer;
- isolating errors in the Y layer;
- populating the new S layer; and configuring the S layer so as to implement the desired target circuit from the X layer.

In the system we designed, one could define a target circuit, and layer S would implement that circuit automatically, without external guidance. This means the circuit designer does not need to know which parts of the hardware layer Y are defective, and does not need to participate in the layout of the target circuit on Y. Because of this, if a system failure is detected during operation, the system can automatically rebuild itself, working around the new defective areas.

This is, in a way, nothing more than an automatic place-and-route system, but with a few key differences:

- the place and route is being run in the same hardware, in a parallel, distributed fashion;
- the algorithm can be implemented on hardware containing defects;
- the hardware first self-organizes to create the circuitry for running the place and route; and

- the same hardware that implements the place and route algorithm will eventually implement the final target circuit.

Thus, by having layer S, the system no longer has to find, isolate and work around defects for **any** application. Instead, it solves these problems for a particular problem, that of the place and route controller. Once that controller has been implemented, **it** is then responsible for implementing the final target circuit. Since the place and route system is now aware of the location of faults in the hardware, it is a relatively simple task to avoid them in implementing the target circuit.

## Repairing, Removing, or Replacing Defective Hardware

Another approach to manufacturing defects is not to accept them and handle them, but to go back in after manufacture and testing, and physically remove or replace them, or to effect repairs to them. This second defect-handling process could be done by the same process that built the hardware, by a sort of "try again" model, or by another process that specializes in removal and repairs, or by a combination thereof.

Because of the physical organization of the Y layer, it is amenable to either the replacement or removal of hardware. Missing hardware is no different from unused hardware, except in that it can never be called upon for use. Our layouts of layer X designs often have empty space within them now, to make them easier to view, or to add to or amend later. Also, the size of the hardware does not matter to the Y layer. It can be arbitrarily big in its dimensions: on the Y layer there are no presumptions or structures dependent on size, such as data buses and address spaces. And the hardware can be grown or shrunk at a later date. We have demonstrated the ability to make and use a larger piece of Y hardware by simply joining two pieces of Y hardware, by a simple physical process of simply joining cells' neighbor wires along the edges of two pieces of hardware. Removed hardware would simply register as defective hardware via the detection mechanisms described above, which can be combined with layout procedures in a manner similar to that for defects, so that the layout of layer X does not include the use of any missing hardware or holes in the hardware.

The checking of repairs can also be handled by the testing and defect handling mechanisms described above, such as the analogue to SCANDISK. If the repairs are good enough, the cell passes all tests and is treated like any other good hardware; if they are insufficient, the cell gets treated appropriately as a defective cell. Iterating between repair and this process would be a good way to determine when the repair job has succeeded: the cell or region would no longer appear on the defect list.

Efforts to perfect the manufacturing technique, even if not met with complete success, are also desirable because they result in the highest density and volume for a fixed manufacturing process. This effort may also help to reduce operating errors. The construction of a simpler target is often easier to perfect. Accordingly, the construction of the Y layer is easier to perfect because it is simply a task of repeatedly constructing the same, small design for the logic cell, and repeating the same interconnection pattern throughout.

## 4.4 Summary, Conclusions

Revolutionary, not evolutionary, new manufacturing processes are a hot research topic. With the capacity to vastly increase the number of switches designers can use in logic designs, and with the likelihood that these new manufacturing processes will be too different from the current techniques to borrow many advancements from them, their introduction may well usher in an age where reliability cannot be guaranteed by the manufacturer and must be handled upstream in the design flow as well, and, to a much larger degree than is true today, downstream in the product also. Reliability must be directly addressed, on a per-design basis with logical/physical co-design where necessary and appropriate, and with a broader brushstroke wherever possible.

The broader brushstroke we advocate upstream, for design systems, is to address reliability using a two-layer, multi-step layout process we dub X,Y, where a convenient, tightly packed hardware design Y is built with low-cost, low-level redundancy. Then the hardware is tested, and then used to implement specific logic designs built up from SSI level logic components, wires, or state machines. This approach is much less dense than would be a direct hardware representation of the logic design X. However, it makes it possible to routinely use imperfectly manufactured hardware, which means that it is straightforward to scale hardware up to much larger designs without paying a large penalty in manufacturing yields, significantly increasing applications while simultaneously lowering costs. And it also means that a new manufacturing process can be put into production sooner, before it is perfected, particularly for small logic designs that can fit on the potentially small amount of good hardware in a device that early versions of a process will produce. Layout of a design onto faulty hardware would most likely need to be done by an automated process, since each piece of hardware will have different defects. For one-offs or layouts onto a specific piece of hardware, however, design tools could build fault locations into the design checks and simply disallow the placement of logic on particular regions of the hardware space. This capability could serve in a facility for diagnosing and then repairing customers' systems.

The manufacturing process must of course be improved as much as possible, and its unreliability duly minimized, but the design system will likely be left with something still imperfect, and must take it from there. Accommodations, such as these suggested here, should be developed so that hardware that is imperfect can still be used if it is found to have acceptable flaws or acceptable reliability problems. This increases the effective yield of a manufacturing process by finding a way to use imperfect hardware. For instance, a manufacturer could use each piece of Y layer hardware it produces for whichever of the logic designs it is producing - whichever one automatic tools can find a way to fit.

Downstream, reliability hits due to operating errors will plague products containing massive numbers of switches. There are ways that have been developed to handle this within logic designs themselves such as subsystem redundancy and voting. Again we advocate using a broader brushstroke wherever possible, rather than a per-design fault handling strategy. The broader brushstroke we advocate downstream for operating errors is finding ways to, as cheaply and painlessly as possible, provide logic systems designers with efficiently self-analyzing, efficiently self-modifying systems that automatically handle low-level operating errors locally. This capability is supported in the system architecture we have developed by using the dynamic, local processing capabilities of Layer Y. This allows the management of some of the complex system problems that arise with scaling up in a more encapsulated way, and can be used for various approaches. What approaches will work best is a wide area of research, and there may be different ideal ways to do this for different kinds of applications. However, we have included several ideas here, such as allowing the placement of logic blocks and routing of signals to continue in some sense during system operation, with logic designs that can shift and move their logic to avoid using failing hardware, or investing designs with the ability to test their higher and lower level systems for operating errors, and giving them the ability to make repairs to themselves, possibly by cutting the wires to a subsystem and wiring up a new copy of it elsewhere on the hardware. The main benefit to this approach is that rather than trying to predict all failure modes or failure situations and invest all systems with a game plan for them, each circumstance is handled if and when it arises, by a more general approach, and the hardware is eventually fine-tuned to its own quirkiness, usage patterns, and environment. The vagaries of intermittent failures do not lend themselves to a static solution, but rather a solution as dynamic as the problem itself.

## References

[1] Durbeck L and Macias N 2001 The Cell Matrix: an architecture for nanocomputing. Nanotechnology vol 12 pp 217-30 (Bristol, Philadelphia: Institute of Physics Publishing)

[2] Durbeck L and Macias N 2002 Defect-tolerant, fine-grained parallel testing of a Cell Matrix Proc. SPIE ITCom 2002 Series 4867 ed J Schewel, P James-Roxby, H Schmit and J McHenry pp 71-85

[3] Macias N and Durbeck L 2002 Self-Assembling Circuits with Autonomous Fault Handling Proc. The 2002 NASA/DOD Conference on Evolvable Hardware ed A Stoica, J Lohn, R Katz, D Keymeulen and R Salem Zebulum pp 46-55

[4] Redl F X, Cho K-S, Murray C B and O'Brien S 2003 Three-dimensional Binary Superlattices of Magnetic Nanocrystals and Semiconductor Quantum Dots. Nature, 423, pp. 968 - 971 (26 June 2003).

[5] Macias N and Durbeck L 2004 Adaptive methods for growing electronic circuits on an imperfect synthetic matrix Biosystems Volume 73 Issue 3 (Elsevier ireland Ltd.), Pages 173-204, March 2004.

[6] Macias N, Raju M and Henry L 1998 Self-Reconfigurable Parallel Processor Made From Regularly-Connected Self-Dual Code/Data Processing Cells. US Patent 5,886,537

[7] Culbertson B, Amerson R, Carter R, Kuekes P and Snider G, Defect Tolerance on the Teramac Custom Computer, Proceedings of the 1997 IEEE Symposium on FPGA's for Custom Computing Machines, pp. 116-123.

[8] Durbeck L and Macias N 2001 Self-configurable parallel processing system made from self-dual code/data processing cells utilizing a non-shifting memory. US Patent 6,222,381

[9] Drexler K E 1992 Nanosystems: molecular machinery, manufacturing, and computation. Wiley Interscience.

[10] Macias N 1999 Ring Around the PIG: A Parallel GA with Only Local Interactions Coupled with a Self-Reconfigurable Hardware Platform to Implement an O(1) Evolutionary Cycle for EHW. Proc. 1999 Congress on Evolutionary Computation pp 1067-75

[11] Macias N 1999 The PIG Paradigm: The Design and Use of a Massively Parallel Fine Grained Self-Reconfigurable Infinitely Scalable Architecture. Proc. The First NASA/DOD Workshop on Evolvable Hardware ed A Stoica, D Keymeulen and J Lohn pp 175-80

[12] http://cellmatrix.com/entryway/products/pub/ForesightPoster.pdf

[13] von Neumann J 1966 Theory of Self-Reproducing Automata. University of Illinois Press, Urbana, Illinois.

[14] http://cellmatrix.com/entryway/products/software/layoutEditor.html

[15] http://cellmatrix.com/entryway/products/pub/yatsenko2003.pdf

[16] http://www.xilinx.com

[17] http://www.cellmatrix.com

# Chapter 5

# A PROBABILISTIC-BASED DESIGN FOR NANOSCALE COMPUTATION

R. Iris Bahar, Jie Chen, and Joseph Mundy
*Division of Engineering*
*Brown University*
*Providence, RI 02912*

**Abstract** As current silicon-based techniques fast approach their practical limits, the investigation of nanoscale electronics, devices and system architectures becomes a central research priority. It is expected that nanoarchitectures will confront devices and interconnections with high inherent defect rates, which motivates the search for new architectural paradigms. In this chapter, we exam probabilistic-based design methodologies for designing nanoscale computer architectures based on Markov Random Fields (MRF) The MRF can express arbitrary logic circuits and logic operation is achieved by maximizing the probability of state configurations in the logic network. Maximizing state probability is equivalent to minimizing a form of energy that depends on neighboring nodes in the network. Once we develop a library of elementary logic components, we can link them together to build desired architectures. Overall, the probabilistic-based design can dynamically adapt to structural and signal-based faults.

## 5.1 Introduction

During the past decades, we have witnessed the boom of microelectronics. The whole semiconductor industry thrives on the miniaturization of microscale devices based on "Moore's Law". That is, every couple of years, there is a double of the number of transistors on a single chip. However, silicon-based devices are fast approaching their practical limits, and Moore's Law will no longer be sustainable. For instance, short-channel effect can cause direct tunneling between source and drain, or between gate and source/drain when we keep scaling the device down below $40nm$ [14][1]. Other limitations come from lithography, high-field effects, etc. As a result, many alternatives to silicon-based devices are being explored for the basis of developing new nanoelectronic systems. In the process, it is expected that the past approach



*S.K. Shukla and R.I. Bahar (eds.), Nano, Quantum and Molecular Computing*, 133-156.

of using global interconnections and assuming error-free computation may no longer be possible, thereby presenting new challenges to computer engineers. It is likely that nanoscale computing will be dominated by communication, where processing is based on redundant and adaptive pathways of error-prone connections.

Limitations of previously mentioned architectures can be classified into three types:

- Redundant architectures, such as Nanofabrics, are programmed to work *around* manufacturing defects. Our view is that the defect density will be too high to permit this strategy. Too much space and power will be wasted on testing and redundant devices.
- Architectures such as quantum cellular arrays currently must operate at low temperature to overcome thermal noise and may not soon achieve room-temperature operation. This will very likely prevent them having a serious impact on the mainstream computing domain. Also, cellular arrays exhibit high communication costs for computations involving global constraints.
- Neural network style architectures require training, and it is difficult to analyze or optimize their performance according to engineering principles. It is not clear how their behavior generalizes to new computational examples.

## Anticipated Architecture Characteristics

Up to now, the fabrication of nanocircuits has been limited to a few devices intended to demonstrate simple logic or memory operations. There are no actual data to measure the characteristics of large networks of devices. However it is possible to pose two likely characteristics that will have to be confronted in the development of computational architectures that use these devices.

1 A high and dynamic failure process: It can be expected that a significant fraction of the devices and their interconnections will fail. These failures will occur both during fabrication and at a steady rate after fabrication, thus precluding a single test and repair strategy.

2 Operation near the thermal limit: As device sizes shrink, the energy difference between logic states will approach the thermal limit. Thus, the very nature of computation will have to be probabilistic in nature, reflecting the uncertainty inherent in thermodynamics.

The first characteristic is a simple extrapolation of current device fabrication experience. The smaller the device dimensions become, the more phenomena can interfere with correct operation. It seems likely that architectures will have to cope with device and connection failure rates of 10% or more. At the same time, the nature of connections and devices will be based on mechanisms that can easily mutate over time, such as chemical reactions or fusing and bridging of connections.

The second conclusion can be arrived at in several ways. Perhaps the most direct is to consider the evolution of current CMOS technology towards smaller device sizes, perhaps using silicon nanowire devices. Current processor chips are nearing 100 million transistors and dissipate over 100 Watts. It can be assumed that this power level is already near the practical limit in terms of battery life and dangerous external temperatures. The natural evolutionary forces that drive the number of gates per chip and clock rate upward will decrease logic transition energy limits to within a few orders of magnitude of $k_bT$, where $k_b$ is Boltzmann's constant, $1.38 \times 10^{-23} J/^{\circ}K$, and $T$ is absolute temperature in $^{\circ}K$. At room temperature, $k_bT = 0.026$ electron-volts. Current devices still consume energy several orders of magnitude above this thermal limit.

Another approach to the same conclusion is the study of the ultimate limit on the energy required for computation, starting with the paradox of Maxwell's demon [3] and ultimately clarified by the development of information theory. It can be shown that the energy cost of computation cannot be reduced below $(\ln 2)k_bT$ per bit. This basic result is derived from the necessary increase in randomness as information is lost during computation. For example, the input to an AND gate has four possible states while the output has only two. The evolution of device technology will relentlessly drive towards this limit, requiring approaches that can confront randomness in logic state.

## Nano-architecture Approaches

To date, architecture research has taken two approaches. The first approach simply increases existing machine resources while the second approach uses modular and hierarchical architectures to improve the performance of traditional single-thread architectures. In particular, the highly regular, locally connected, peripherally interfaced, data-parallel architectures of the second approach offer a good match to the characteristics of nanoelectronic devices [2].

It will be necessary to evolve new architectural concepts in order to cope with the high level of device failure and randomness of logic states anticipated by the reasoning just proposed. Current architectural studies aimed at nanoscale systems are focused primarily on the first issue, device failure.

There are two basic approaches being proposed to deal with significant device failure rates: testing and routing around failures; and designing with redundant logic in the form of error correction. Illustrative of the first approach are the architectures of DeHon [5] and Goldstein and Budiu [8]. They propose designing in extra circuit elements that can be used to supplement failed devices and connections. A major issue is the testability of the network and the ability to confront continuous failures over the life of the device.

The second approach was suggested in the pioneering work by Von Neumann [15] where he used majority logic gates as a primitive building block and randomizing networks to prevent clusters of failures from overwhelming the fault tolerance of the majority logic. The architecture proposed in [10] is based on this approach. See Chapter 2 for more discuss about these approaches.

One architectural approach that can provide continuous adaptation to errors is based on neural network structures [6]. The synaptic weights of the neural network are implemented using multiple connection paths and the summation is provided by conventional CMOS differential amplifier nodes. The connections are adaptively configured using single electron switching devices. It is proposed that useful computation could arise by training the network with a series of required input-output pairs. Another work we have recently become aware of attempts to build Bayesian networks with subthreshold CMOS circuits [12]. The paper focuses specifically on the sum-product algorithm useful for solving problems in artificial intelligence, signal processing, and communications.

The approach taken in this chapter has some similarity to the architecture proposed in [6], but is based on Markov Random Fields (MRF). The MRF provides a formal probabilistic framework so that computation can be directly embedded in a network with immunity to both device and connection failures. Since logic states are computed probabilistically the computation is also robust to the logic signal fluctuations that will arise as operation approaches the thermal limit of computation. Furthermore, the MRF is general and directly programmed without learning.

In a computing system errors may occur either in devices or connections; however, we do not distinguish between them. Instead, we have *structure-based* and *signal-based* faults. Nanoscale devices contain a large number of defects or structural errors, which fluctuate on time scales comparable to the computation cycle. The signal-based type of error is directly accounted for in the probability maximization process inherent in the MRF processing. The following sections describe how the MRF and BP framework can be used to handle both types of faults.

## 5.2 MRF Design for Structural-based Faults

**Overview of the Markov Random Field.** The Markov Random Field (MRF) [11][4][13] has been widely used in computer vision, physics, computational biology and communications and is proposed here as a model for uncertain and noisy computation. The MRF represents the relationship between a set of random variables, $\mathbf{X} = \{X_1, X_2, \ldots, X_n\}$. Each $X_i$ can take on values from a range set $\mathcal{L}$. In some MRF treatments, the random variables are called *sites* and the set $\mathcal{L}$ is called the *label set*. The notation presented here is based on the treatment of Geman and Kochanek [7].

The joint probability of variable assignments is denoted as,

$$p(x_1, x_2, \ldots, x_n) = p(X_1 = x_1, X_2 = x_2, \ldots, X_n = x_n), \tag{5.1}$$

where $x_i \in \mathcal{L}$. The conditional probability of a particular variable, say $x_2$, is in general,

$$p(x_2|x_1, x_3, \ldots, x_n) = \frac{p(x_1, x_2, \ldots, x_n)}{p(x_1, x_3, \ldots, x_n)} \tag{5.2}$$

In this case, the random variable set $\mathbf{X}$ is completely statistically interdependent. If all the random variables are independent then,

$$p(x_1, x_2, \ldots, x_n) = p(x_1)p(x_2)\ldots p(x_n), \text{and} \tag{5.3}$$

$$p(x_2|x_1, x_3, \ldots, x_n) = p(x_2). \tag{5.4}$$

The MRF defines the concept of a *neighborhood*, $\mathcal{N}_i$ to represent the conditional dependence of a variable, $X_i$, on a subset of $\mathbf{X}$. The neighborhood can vary from complete dependence (the entire set $\mathbf{X}$) to complete independence (the null set). As an example, suppose the neighborhood of $X_2$ is $\mathcal{N}_2 = \{X_3, X_5, X_7\}$. Then,

$$p(x_2|x_1, x_3, \ldots, x_n) = p(x_2|x_3, x_5, x_7) \tag{5.5}$$

The formal definition of a Markov random field can now be stated. Let $\mathbf{X}$ be a family of random variables defined over a set of values from $\mathcal{L}$. $\mathbf{X}$ is said to be a Markov random field on $\mathcal{L}$ with respect to a neighborhood system $\mathcal{N}$ if and only if the following two conditions hold:

$$p(x_i) > 0, \quad \forall X_i \in \mathbf{X} \qquad (Positivity) \tag{5.6}$$

$$p(x_i|\{\mathbf{X} - x_i\}) = p(x_i|\mathcal{N}_i) \quad (Markovianity) \tag{5.7}$$

A remarkable key property of Markov random fields is that $p(x_i|\mathcal{N}_i)$ can always be expressed in terms of a function of the *cliques* formed from a site and its neighborhood. In this context the sites are considered to be nodes in a graph and the conditional dependencies between nodes are the graph edges. Recall that a clique in a graph is a set of nodes where each node in the set has edges to the all other nodes. This graph interpretation of a MRF neighborhood is illustrated in Figure 5.1. In this interpretation the edges of the graph indicate elements of a neighborhood. The influence of each clique on the probability of the entire set of random variables can be expressed in terms of a set of terms, $\mathcal{U}_c$, called *clique energy* functions. The variable, $c$, indexes the cliques over the

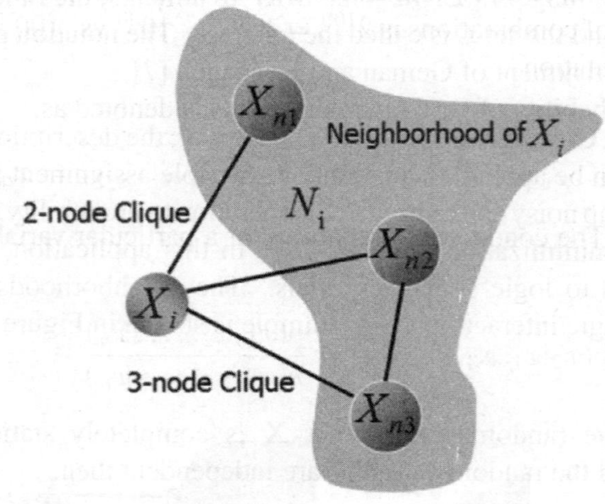


*Figure 5.1.* A MRF neighborhood and example cliques

entire set of nodes, $\mathcal{C}$. The use of the energy concept relates to the historical origins of the MRF model in physics.

The probability of a particular label assignment is given by,

$$p(x_1, x_2, \ldots, x_n) = \frac{1}{Z} e^{\frac{-1}{k_b T} U_c(x_{c1}, x_{c2} \ldots x_{cm})} \tag{5.8}$$

Equation (5.8) is called the Gibbs formulation. The fact that a general MRF is equivalent to the Gibbs form was established by the Hammersley and Clifford theorem [9]. For a given clique, $c$, $U_c(x_{c1}, x_{c2} \ldots x_{cm})$ is defined on the set of $m$ random variables (nodes) in the clique. The term $k_b T$ can be interpreted as thermal energy from the physical point of view, but in the calculations below it is merely treated as a constant in proportion to the clique energy that controls the sharpness of the Gibbs probability distribution. The term $Z$ is called the *partition function* and is a constant required to normalize the probability function to $[0, 1]$. It is the sum over all possible label assignments of the exponential term in the numerator.

The great power of the Gibbs representation is that the problem of finding a global site label assignment with maximum probability can be decomposed into that of minimizing clique energies. In most practical problems, the neighborhoods are small and the cliques involve only a few nodes. This computation is immensely smaller then that entailed by considering all the possible global site label assignments. There are $|\mathcal{L}|^n$ assignment combinations for the full set of variables, while only on the order of $n \times (|\mathcal{L}|^{m_{max}})$ using the Gibbs formulation. In this case, $m_{max}$ is the size of the largest clique. For example, if there are

100 sites, each with two possible labels, but the largest clique has three nodes, then the full set of combinations is $2^{100} = 1.267 \times 10^{30}$ vs. $100 \times 8 = 800$ for the Gibbs computation.

**Mapping Logic Circuits onto the MRF.** So far, the description of the MRF is such that it can be applied to any random variable assignment problem. The goal here is to map noisy and faulty circuit operation to probability maximization (clique energy minimization) on the MRF. In this application, the nodes or sites correspond to logic signal terminals. The neighborhoods of the MRF correspond to logic interactions. An example is shown in Figure 5.2.

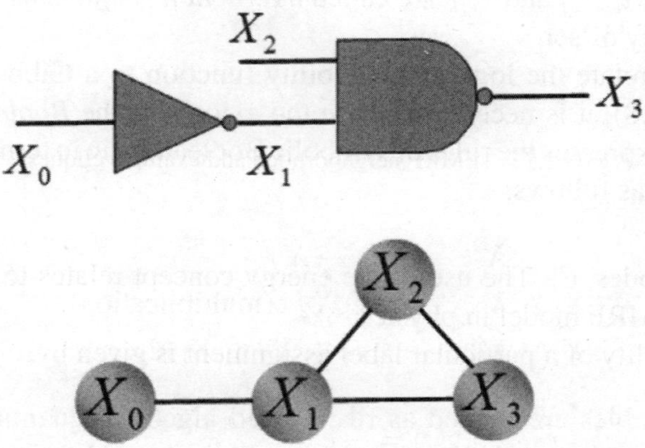


*Figure 5.2.* An example mapping from a logic circuit onto an MRF. The graph edges indicate neighborhood relations.

The input and output of the inverter are considered to be statistically dependent as indicated by the graph edge between the two nodes. The graph edge does not explicitly represent causality but just that there is a joint probability relationship between $X_0$ and $X_1$, i.e., $p(x_0, x_1)$. That is, one doesn't think of $X_0$ *causing* $X_1$, but instead their joint assignments must be maximally probable. Thus in the case of an isolated inverter with logic states taken from $\{0, 1\}$, there are two equally probable assignments $(X_0 = 0, X_1 = 1)$ and $(X_0 = 1, X_1 = 0)$. For the NAND gate, the graph structure indicates that cliques up to size three are required to represent the statistical dependence among the three logic terminals.

**Applying MRF in Nanoscale Architecture Design.** The MRF is a completely general computational framework and in principle any type of computation could be mapped onto the model. In order to concretely illustrate the operation of the model, we will use combinatorial logic as an example. The *programming* of the MRF is straightforward in this case, and will permit some analysis of the fault tolerance of the architecture.

Combinatorial logic can be implemented using a simple, yet powerful, form for the clique energy, called the *auto-model*. For cliques up to order three, the energy function is given by:

$$\begin{aligned} U_c &= \kappa + \Sigma_{i \in \mathcal{C}_0} \alpha_i x_i + \Sigma_{i,j \in \mathcal{C}_1} \beta_{ij} x_i x_j + \\ & \quad \Sigma_{i,j,k \in \mathcal{C}_2} \gamma_{ijk} x_i x_j x_k. \end{aligned} \tag{5.9}$$

The constants, $\alpha_i, \beta_{ij}$ and $\gamma_{ijk}$ are called *interaction coefficients*. The constant $\kappa$ acts an energy offset.

In order to relate the logic compatibility function to a Gibbs energy form in Equation (5.8), it is necessary to use the axioms of the *Boolean ring*. The Boolean ring expresses the rules of symbolic Boolean logic in terms of algebraic manipulations as follows:

$$\begin{aligned} X' &\rightarrow (1 - X) \\ X_1 \wedge X_2 &\rightarrow X_1 X_2 \quad \text{(multiplication)} \\ X_1 \vee X_2 &\rightarrow X_1 + X_2 + X_1 X_2. \end{aligned}$$

The logic variables are treated as real valued algebraic quantities and logic operations are transformed into arithmetic operations. Additionally, it is desired that valid input/output states of computational logic should have lower clique energies than invalid states so as to maximize the probability of being in a correct (i.e., valid) state as expressed in Equation (5.8). Thus, the clique energy expression is obtained by a negative sum over minterms from the valid states,

$$U_c = -\sum_i f_i(x_0, x_1, \ldots, x_n),$$

where $f_i = 1$, and the minterms are transformed using the Boolean ring rewrite rules. Note that this form exploits the simplification that cross-products of minterms vanish. For instance, the Boolean ring conversion for the minterm $(x_0, x_1, x_2) = 000$ is,

$$\begin{aligned} x_0' \wedge x_1' \wedge x_2' &= (1 - x_0)(1 - x_1)(1 - x_2) = (1 - x_0 - x_1 + x_0 x_1)(1 - x_2) \\ &= 1 - x_0 - x_1 - x_2 + x_0 x_1 + x_0 x_2 + x_1 x_2 - x_0 x_1 x_2. \end{aligned}$$

Case Study I: Exclusive-OR Gate

The effect of structure-based errors, or errors on the coefficients in the clique energy, is illustrated using an XOR example. There are three nodes in the

| $i$ | $x_0$ | $x_1$ | $x_2$ | $f$ |
|---|---|---|---|---|
| 0 | 0 | 0 | 0 | 1 |
| 1 | 0 | 0 | 1 | 0 |
| 2 | 0 | 1 | 0 | 0 |
| 3 | 0 | 1 | 1 | 1 |
| 4 | 1 | 0 | 0 | 0 |
| 5 | 1 | 0 | 1 | 1 |
| 6 | 1 | 1 | 0 | 1 |
| 7 | 1 | 1 | 1 | 0 |

*Figure 5.3.* The logic compatibility function for an exclusive-or gate with all possible states.

network: the inputs $x_0, x_1$, and the output $x_2$ of the gate. Successful operation of the gate is designated by the compatibility function, $f(x_0, x_1, x_2)$ as shown in Figure 5.3. Here we list all possible states (valid states with $f = 1$ and invalid state with $f = 0$) because our approach *adapts* to errors and we make no assumption about the occurrence of errors. For the exclusive-or example, by summing over the valid states based on the Boolean ring axiom, $000 = (1 - x_0)(1 - x_1)(1 - x_2), 011 = (1 - x_0)x_1x_2, 101 = x_0(1 - x_1)x_2$, and $110 = x_0x_1(1 - x_2)$, we can compute the clique energy as follows:

$$
\begin{aligned}
U_c &= -1 - (1 - x_0)(1 - x_1)x_2 - (1 - x_0)x_1(1 - x_2) - \\
&\quad x_0(1 - x_1)(1 - x_2) - x_0x_1x_2 \\
&= -1 + x_0 + x_1 + x_2 - 2x_0x_1 - 2x_0x_2 - \\
&\quad 2x_1x_2 + 4x_0x_1x_2.
\end{aligned} \tag{5.10}
$$

If we take the structural errors into consideration in our design, the clique energy in Equation(5.9) can be rewritten as:

$$
\begin{aligned}
U_c &= \kappa + Ax_0 + Bx_1 + Cx_2 - 2Dx_0x_1 - \\
&\quad 2Ex_0x_2 - 2Fx_1x_2 + 4Gx_0x_1x_2.
\end{aligned} \tag{5.11}
$$

where $\kappa$ is a constant, and the nominal *weight* values for the coefficients are: $A = B = C = D = E = F = G = 1$, as derived above for the error free case. In the modified equation above, the energy coefficients have been replaced by variables to indicate that their values can deviate from the ideal setting due to failures. The variables $A, B, C$ stand for the first-order clique energy coefficients, and $D, E, F$ are second coefficients. The third order coefficient, $G$, constrains the values of all the lower order coefficients as will be shown shortly. In the nanoarchitecture being described here, the coefficient values are determined by a set of gate connections, so coefficient error is caused by structure-based failure. For successful operation of the logic, it is necessary that the energy of correct logic state configurations always be *less* than invalid state configurations.

LEMMA 5.1 *For any combinational logic, the energy of a correct logic state is always less than that of an invalid state by a constant.*

*Proof.* For example, in a simple exclusive-or design shown in Figure 5.3, the clique energy is

$$U_c = -1 + x_0 + x_1 + x_2 - 2x_0x_1 - 2x_0x_2 - 2x_1x_2 + 4x_0x_1x_2.$$

By substituting the invalid and valid states into this energy equation, we get that the energy for valid states is always '$-1$' while that of invalid states is always '0'. The energy difference is a constant (in this case, it is one).

The reason is embedded in the definition of clique energy:

$$U_c(x_0, x_1, x_2) = -\sum_i f_i(x_0, x_1, x_2).$$

For a valid state of any logic, the summation of valid states, $f_i$, is always one. Or, clique energy $U_c$ is always $U_c = -1$. On the other hand, for any invalid state, the summation of valid states is always zero or $U_c = 0$. Therefore, the energy of a valid logic state is always less than an invalid state by a *constant*.□

For our example, the set of inequalities that must hold is given in Table 5.1. Here we relate a valid state to all possible invalid states. For example, for valid state $(x_0, x_1, x_2) = 000$ the clique energy in Equation (5.11) must evaluate to a lower energy state than all possible invalid states. These inequalities can be solved using a proposed algorithm similar to Gaussian elimination where a variable that appears with opposite signs in two equations can be eliminated. Applying this procedure to the inequalities in Table 5.1 the following constraints on the clique coefficients are obtained:

| | | | |
|---|---|---|---|
| 2G>D | 2F>C | 2E>A | 2D>B |
| 2G>F | 2F>B | 2E>C | 2D>A |
| 2G>E | | | |

The constraints should be viewed as being driven by coefficient $G$ which can take on any positive value. A selected value for $G > 0$ then determines bounds on coefficients, $D, E, F$ in terms of $(2G > D, 2G > F, 2G > E)$. They in turn bound $A, B, C$. The bounds are linear, and so the constraints form a *polytope* in the space of energy coefficients. This concept is illustrated in Figure 5.4 where a projection onto the $D, A, B$ subspace is depicted. In general, the polytope will

| | | | |
|---|---|---|---|
| 000 | $\kappa$ | < | $\kappa$ +C |
| | $\kappa$ | < | $\kappa$ +B |
| | $\kappa$ | < | $\kappa$ +A |
| | $\kappa$ | < | $\kappa$ +A+B+C-2D-2E-2F+4G |
| 011 | $\kappa$+B+C-2F | < | $\kappa$ +C |
| | $\kappa$+B+C-2F | < | $\kappa$ +B |
| | $\kappa$+B+C-2F | < | $\kappa$ +A |
| | $\kappa$+B+C-2F | < | $\kappa$ +A+B+C-2D-2E-2F+4G |
| 101 | $\kappa$+A+C-2E | < | $\kappa$ +C |
| | $\kappa$+A+C-2E | < | $\kappa$ +B |
| | $\kappa$+A+C-2E | < | $\kappa$ +A |
| | $\kappa$+A+C-2E | < | $\kappa$ +A+B+C-2D-2E-2F+4G |
| 110 | $\kappa$+A+B-2D | < | $\kappa$ +C |
| | $\kappa$+A+B-2D | < | $\kappa$ +B |
| | $\kappa$+A+B-2D | < | $\kappa$ +A |
| | $\kappa$+A+B-2D | < | $\kappa$ +A+B+C-2D-2E-2F+4G |

*Table 5.1.* The inequalities that must hold among the energy coefficients for successful gate operation.

be a cone whose cross-section increases linearly with the highest order clique coefficient. The nominal values for the coefficients of Figure 5.4 are $A_0 = D_0$ and $B_0 = D_0$.

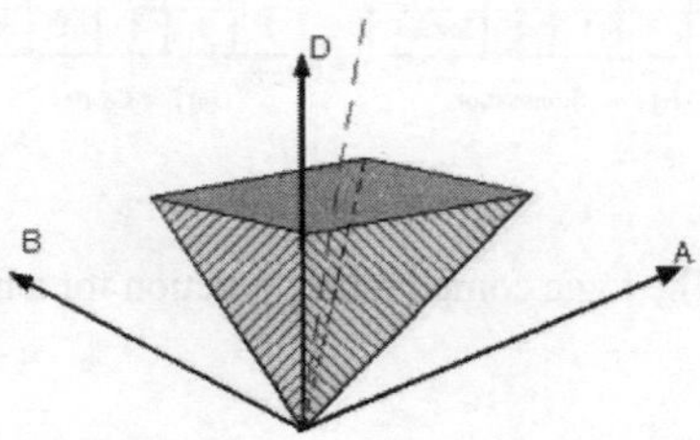


*Figure 5.4.* The constraints on the clique coefficients form a cone in the space of coefficient values for $2D > B$ and $2D > A$. The dotted line indicates the nominal coefficient values.

We assume a fixed error rate in the connections leads to a coefficient error proportional to its value. For example, for some error rate $\alpha$, if coefficient $D$ deviates from its nominal value by $D' = D_0 \pm \epsilon$, then $\epsilon = \alpha D_0$. The inequality relating $2D > A$ requires that,

$$\begin{aligned} 2D_0(1 \pm \alpha) &> D_0(1 \pm \alpha) \quad \text{or} \\ 2 * (1 - \alpha) &> (1 + \alpha), \end{aligned}$$

when the worst case condition is used. Thus, $\alpha$ can be as large as 1/3 without causing a failure of the inequality. The constraint on the $D$ coefficient also permits $\alpha < 1/3$. Similar conditions arise from considering the remaining constraints. Thus for the XOR circuit, up to one third of the connections can be bad and the correct logic state will still be achieved.

Case Study II: Half Adder

The effect of errors on the coefficients in the clique energy is illustrated using another more complicated example — a half-adder. There are four nodes in the network: the inputs $x_0, x_1$, the sum, $x_2$, and the carry bit, $x_3$, of the gate (here we don't consider the carry from the previous stage.). The successful operation of the gate is designated by the compatibility function, $f(x_0, x_1, x_2, x_3)$ as shown in Figure 5.5.

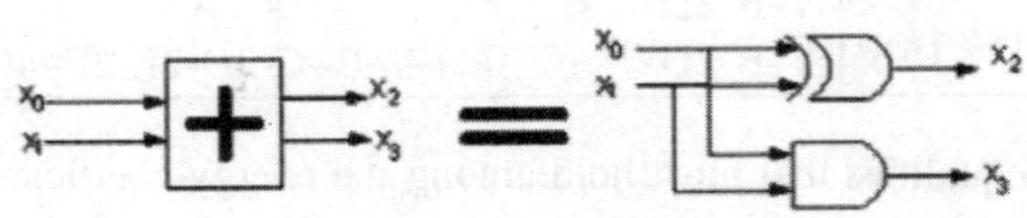


| $i$ | $x_0$ | $x_1$ | $x_2$ | State |
|---|---|---|---|---|
| 0 | 0 | 0 | 0 | Valid |
| 1 | 0 | 0 | 1 | Invalid |
| 2 | 0 | 1 | 0 | Invalid |
| 3 | 0 | 1 | 1 | Valid |
| 4 | 1 | 0 | 0 | Invalid |
| 5 | 1 | 0 | 1 | Valid |
| 6 | 1 | 1 | 0 | Valid |
| 7 | 1 | 1 | 1 | Invalid |

(a) For Summation

| $i$ | $x_0$ | $x_1$ | $x_3$ | State |
|---|---|---|---|---|
| 0 | 0 | 0 | 0 | Valid |
| 1 | 0 | 0 | 1 | Invalid |
| 2 | 0 | 1 | 0 | Valid |
| 3 | 0 | 1 | 1 | Invalid |
| 4 | 1 | 0 | 0 | Valid |
| 5 | 1 | 0 | 1 | Invalid |
| 6 | 1 | 1 | 0 | Invalid |
| 7 | 1 | 1 | 1 | Valid |

(b) For Carrier

*Figure 5.5.* The logic compatibility function for a half-adder gate.

The clique energy for summation is the same as the exclusive-or case in Equation (5.9) while the clique energy for the carry bit is:

$$\begin{aligned} U_c &= -(1-x_0)(1-x_1)(1-x_3) - \\ &\quad (1-x_0)x_1(1-x_3) - x_0(1-x_1)(1-x_3) - x_0x_1x_3 \\ &= -1 + x_3 + x_0x_1 - 2x_0x_1x_3. \end{aligned} \tag{5.12}$$

The reason why we want to separately compute the clique energy for summation and carry is that the summation ($x_2$) and carry ($x_3$) are independent outputs. Their results only depend on inputs $x_0$ and $x_1$. Based on such a design, we can drastically reduce the computational complexity of mixing both $x_2$ and $x_3$ into clique energy computation. The clique energy can then be expressed as follows:

$$\begin{aligned} U_c &= -1-(1-x_0)(1-x_1)(1-x_2)(1-x_3) - \\ &\quad (1-x_0)x_1x_2(1-x_3) - x_0(1-x_1)x_2(1-x_3) - \\ &\quad x_0x_1(1-x_2)x_3 \\ &= -1+x_0+x_1+x_2+x_3-x_0x_1-2x_0x_2-x_0x_3- \\ &\quad 2x_1x_2-x_1x_3-x_2x_3+3x_0x_1x_2+2x_0x_2x_3+ \\ &\quad 2x_1x_2x_3-2x_0x_1x_2x_3. \end{aligned} \tag{5.13}$$

If we take device errors into consideration in our design, the clique energy for the summation portion is the same as in Equation (5.11) while the clique energy for the carry portion is:

$$U_c = \kappa + Hx_3 + Dx_0x_1 - 2Ix_0x_1x_3. \tag{5.14}$$

Here we assume the same error coefficient $D$ for connection $x_0x_1$.

By combining constraints on the clique coefficients for both summation and carry cases, we can obtain the following results:

| | | | | |
|---|---|---|---|---|
| 2I>D | 2G>D | 2F>C | 2E>A | 2D>B |
| 2I>H | 2G>F | 2F>B | 2E>C | 2D>A |
| | 2G>E | | | |

From these two case studies, we observe that complex logic can be decomposed into simple designs by exploiting properties embedded in a circuit. In general, the highest order clique coefficient can be increased until the lowest order coefficient has sufficient connection redundancy to be guaranteed to attain the average error rate. This policy guards against catastrophic failure, where a few bad connections affect a large percentage of the coefficient values. The conical structure of the constraint surface (shown, for instance, in Figure 5.4) insures that this strategy is always possible.

It should be noted that the failure tolerance depends on the particular clique energy function and follows from the form of the inequalities that arise from correct logic operation. Work is underway to construct an algorithm for computing the inequalities for an arbitrary logic function and in turn to analyze the fault tolerance of the circuit.

## State Probability Computation

The use of a probabilistic approach has its advantages because the process is inherently fault tolerant to signal-based errors. The behavior of simple inverter and NAND gates will be used to illustrate this aspect of the Markov random network approach.

Case Study I: Inverter

By following the automodel computation in Section 5.2, we can get the Gibbs distribution for an inverter as:

$$p(x_0, x_1) = \frac{1}{Z} e^{-\frac{1}{k_bT}(2x_0x_1 - x_0 - x_1)}, \tag{5.15}$$

where $x_0$ is the input and $x_1$ is the output of an inverter. Here $U_c = 2x_0x_1 - x_0 - x_1$ is the clique energy or auto-model of an inverter. The partition function $Z$ normalizes the expression as required for a probability. Suppose the input, $x_0$, takes on values from $\{0, 1\}$. The dependence on the input $x_0$ can be marginalized away by summing over its possible values, i.e,

$$\begin{aligned} p(x_1) &= \frac{1}{Z} \sum_{x_0=\{0,1\}} e^{-\frac{1}{k_bT}(2x_0x_1 - x_0 - x_1)} \\ &= \frac{e^{\frac{x_1}{k_bT}} + e^{\frac{(1-x_1)}{k_bT}}}{2(1 + e^{\frac{1}{k_bT}})} \end{aligned} \tag{5.16}$$

In the marginalization it is assumed that the input to the inverter is equally likely to be a zero or one and that the inverter has exact clique energy weights. These assumptions are somewhat idealized, since in practice the inverter will have variable clique coefficients and the input will range over a continuous set of values near zero or one according to the distribution of signal noise and device error.

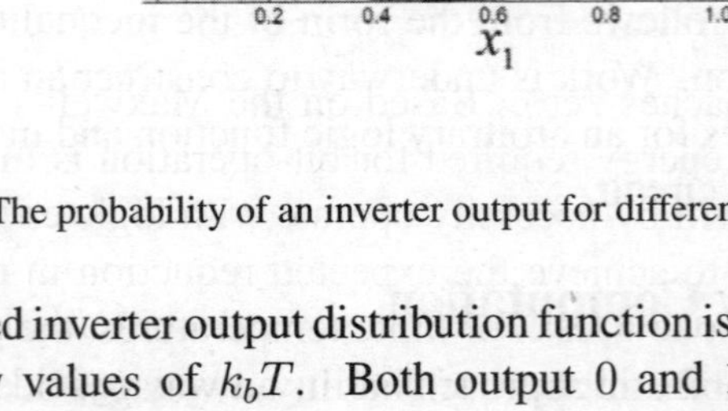


*Figure 5.6.* The probability of an inverter output for different values of $k_bT$.

The marginalized inverter output distribution function is plotted in Figure 5.6 for various energy values of $k_bT$. Both output 0 and 1 are equally likely, but note that the most likely outputs are 0 or 1 with the likelihood of any intermediate values becoming vanishingly small as $k_bT \rightarrow 0$. This behavior is characteristic of Markov random network processing. As long as the energy balance is favorable to the correct logic state, decreasing $k_bT$ will lock in the valid configurations.

Most applications of the MRF model treat $k_bT$ as a variable that can be manipulated in solving for the maximum probability state. For the purposes of the following examples, $k_bT$ is expressed in normalized units relative to the logic state energy. That is, thermal energy is expressed as a fraction of unit logic state energy. For example, the value $k_bT = 0.1$ means that the unit logic energy is ten times the thermal energy, so at room temperature, the unit logic energy in some physical realizations of the Markov network would be 0.026 electron-volts.

In actual operation of a logic circuit, the input states would not be equally likely but would have higher probability of being in a given state, as required by deterministic behavior. For example, suppose the input to the inverter has $p(1) = 0.7, p(0) = 0.3$ then the Gibbs distribution of Figure 5.6 is as shown in Figure 5.7. As the computing entropy increases, this probability margin

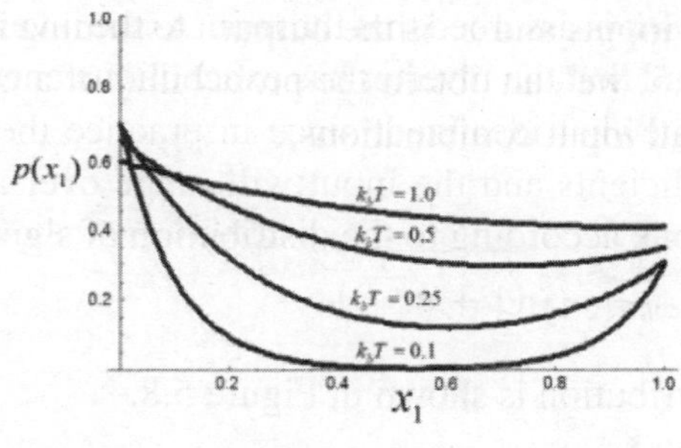


*Figure 5.7.* The probability of an inverter output for different values of $k_bT$ when the input is one, with probability 0.7.

asymptotically approaches zero. Based on the Maxwell's demon discussion in [3], the minimum energy required for bit-operation is $\ln 2\ k_bT$. Similarly, a nano logic device will by necessity operate with logic energies within a few tens of $k_bT$ in order to achieve the expected reduction in power afforded by the small scale of nanodevices. For finite temperatures, the policy of choosing the output state with the highest probability always yields the correct logic operation. However, it can be expected that errors will result if $|p(x) - 0.5|$ is small, since any physical realization of the Markov network will have significant fluctuation of the logic levels.

Case Study II: NAND Gate

In order to consider the error behavior of more complex circuits, it is necessary to describe the processing of logic signals through the Markov random network. This process is carried out by the chaining of conditional probabilities. The probability of logic variables can be determined by summing (marginalizing) over the set of possible values of clique neighborhood states except for the variable in question. What remains is the probability for the single variable. This probability can be propagated to the next node in the network and used for the next summation. An example of this basic algorithm has already been given in the case of Equation (5.15).

The Gibbs joint and conditional probability distributions for a NAND gate are given by,

$$
\begin{aligned}
p(x_a, x_b, x_c) &= \frac{1}{Z} e^{-\frac{1}{k_b T}(2x_a x_b x_c - x_a x_b - x_c)} \\
p(x_c | x_a, x_b) &= \frac{1}{(e^{\frac{x_a x_b}{k_b T}} + e^{\frac{1 - x_a x_b}{k_b T}})} e^{-\frac{1}{k_b T}(2x_a x_b x_c - x_a x_b - x_c)},
\end{aligned}
$$

where $x_a, x_b$ are the inputs and $x_c$ is the output. Assuming independent inputs, $p(x_a) = p(x_b) = 0.5$, we can obtain the probability of a one at the output by marginalizing over all input combinations,

$$
p(x_c) = \sum_{x_a, x_b \in \{0,1\}} \frac{1}{(1 + e^{\frac{1}{k_b T}})} e^{-\frac{1}{k_b T}(2x_a x_b x_c - x_a x_b - x_c)}.
$$

This probability distribution is shown in Figure 5.8.

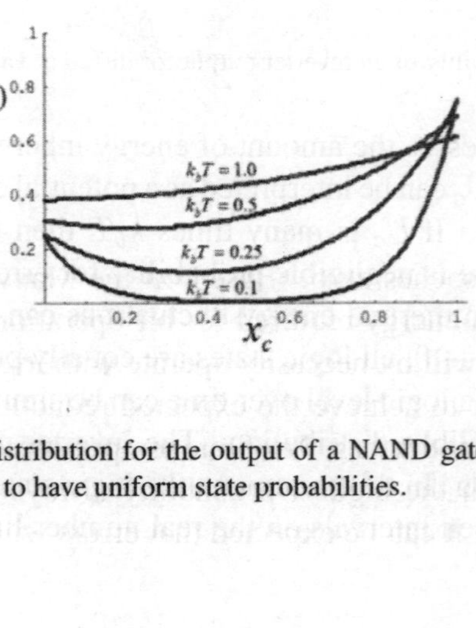


*Figure 5.8.* The marginal distribution for the output of a NAND gate for different values of $k_b T$. The inputs are assumed to have uniform state probabilities.

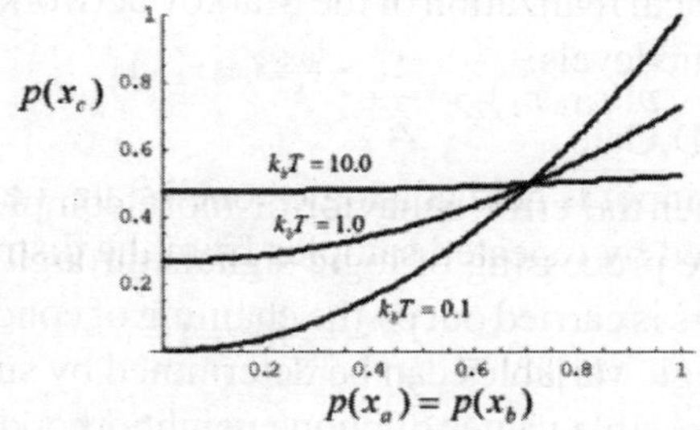


*Figure 5.9.* The probability, $p(x_c = 0)$, for a zero output state for a NAND gate as a function of the input state probabilities, $p(x_a) = p(x_b)$.

Note that the NAND gate is *asymmetrical* in its probability distribution and for a uniform distribution of inputs, the probability of a one output state is three times that of a zero state. This should be expected, since only one input combination produces a zero output. However, this asymmetry is detrimental to logic processing as shown in Figure 5.9. Note that it is necessary to have $p(x_a) = p(x_b) > 0.7$ in order to achieve any logic margin. This margin is reduced at higher input probabilities as the entropy increases. This result shows that logic structures should be as symmetrical as possible in order to operate close to the thermal limit of $\ln 2\ k_bT$.

## 5.3 Design for Signal-based Errors

One of the goals for the MRF-based architecture is to be able to operate much closer to the thermal limit of computation than with conventional logic circuits. The effect of thermal noise on the behavior of the logic signals can be expressed in the MRF model through the Gibbs distribution defined earlier,

$$p(x) = \frac{1}{Z} e^{-\frac{U_c}{k_b T}}.$$

The term $k_bT$ expresses the amount of energy inherent in thermal excitations. The clique energy $U_c$ can be interpreted as a potential energy well that maintains a given logic state. If $U_c$ is many times $k_bT$ then the logic states are well-delineated and there is negligible probability for errors. On the other hand, if $U_c$ is near $k_bT$ then thermal energy fluctuations can easily cause logic errors. In the limit of $\frac{U_c}{k_bT} \to 0$, all logic states are equally probable.

The variation of a logic level over time can be simulated by drawing random samples from the Gibbs distribution. The inverter model described earlier is taken for illustration. In this treatment, the logic states are nominally, $\{-1, 1\}$ in order to have open intervals on the real number line. With this mapping of states, the joint probability of the input, $x_0$ and the output $x_1$ is,

$$p(x_0, x_1) = \frac{1}{Z} e^{-\frac{0.5}{k_b T}(x_0 x_1 - 1)}.$$

Suppose that the input is held at a logic "one" state, i.e., $x_0 = 1$. The output can then be simulated by repeated samples from the distribution,

$$p(x_1) = \frac{1}{Z} e^{-\frac{1}{k_b T}(x_1 - 1)}. \tag{5.17}$$

A series of samples is shown in Figure 5.10 for two different ratios of logic energy to thermal energy. The decision as to the output logic state can be taken

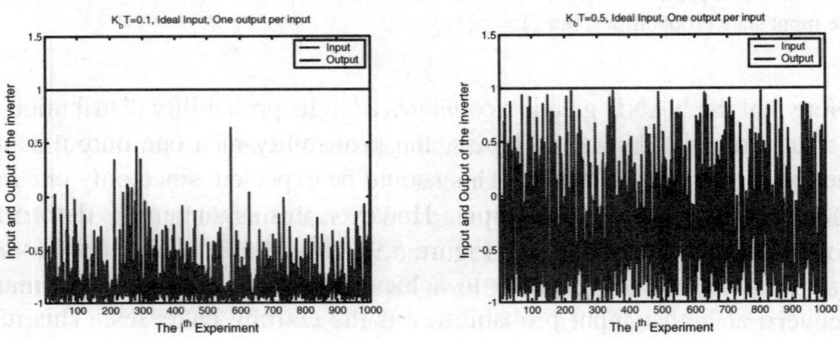


(a) (b)

*Figure 5.10.* A series of samples from $\frac{1}{Z} e^{-\frac{1}{k_b T}(x_1 - 1)}$, (a) with $\frac{1}{k_b T} = 10$, (b) with $\frac{1}{k_b T} = 2$.

as $x_1 > 0$. Obviously, many errors will result even at $\frac{1}{k_b T} = 10$. The plot b) corresponds to the ultimate limit of computation in the presence of thermal fluctuations, $U_c = 2k_b T$. This limit is established from entropy arguments in a study of logic decisions in the presence of random thermal fluctuations [3].

There are two possible approaches to reducing the probability of error:

1 increase the logic energy ratio;

2 exploit temporal redundancy by averaging a series of output samples.

The first approach is certainly feasible since current CMOS logic operates at energies many orders of magnitude above $k_b T$. For example, for a junction capacitance of 1 $pF$ and a logic level of one volt, the logic energy ratio is approximated by,

$$\frac{U_c}{k_b T} = \frac{CV^2}{2\, k_b T} = 1.2 \times 10^8.$$

This ratio can be reduced by many orders of magnitude and still have highly stable MRF logic states. It should be emphasized that operation below room temperature is not being advocated here. As will be shown in the following experiments, it is possible to reduce the logic energy ratio by a million times and still have a negligible probability of logic error.

In the next section considers the second approach, temporal redundancy, to eliminate errors. These studies will provide insight into logic operation at or near the thermal limit.

## Error Tolerance Through Temporal Redundancy

In following experiments a logic decision is based on an ensemble average of samples, rather than a single observation of the output state. That is, a logic signal, $x_i$, is determined to be "one" if,

$$(\sum_{k=-m}^{k=-1} x_i) > 0$$

Here it is assumed that $m$ samples in the past are averaged to produce the logic state decision at the the current moment.

The effect of averaging can be illustrated by sampling the output distribution of a two-inverter chain. The input to the first inverter is held at logic "one" and thus its output distribution follows Equation 5.17. The samples from the first inverter output are used as inputs to the second inverter. Each sample is held fixed while the final inverter output is sampled to produce the signal to be studied. This arrangement insures that the input to the second inverter has the statistical distribution that would be present in a logic circuit implemented using the MRF model. The temperature dependence of the effect of ensemble averaging will also be consistent with the Gibbs distribution by using the first inverter as a MRF "signal generator".

Figure 5.11 shows the joint effect of the number of samples in the ensemble average and temperature. In this figure, the ratio of $\frac{U_c}{k_bT}$ is expressed as the reciprocal, $\frac{k_bT}{U_c}$ with $U_c = 1$. So here, low values of $\frac{k_bT}{U_c}$ indicate more stable logic states. A given effective temperature ratio can be also be achieved by adjusting the logic switching energy with $k_bT$ held fixed, which is what would be done in practice.

It is also interesting to consider the effect of long cascaded logic paths. Intuitively it is expected that error rate will increase at each stage due to the increased randomness of the signal. A set of experiments were carried out to simulate the output of a long chain of inverters. The result is shown in Figure 5.12. As expected, the error rate rises dramatically with the number of stages for operation at $\frac{U_c}{k_bT} = 2$, the theoretical limit of computation. In actual practice, the logic energy would more on the order of $\frac{U_c}{k_bT} = 100$ and so a reasonable number of logic stages can be implemented with negligible error rate.

| sampling | Inve=2, MU=1 (Deterministic), 1_in_1_out | | | |
|---|---|---|---|---|
| N | $K_bT$=0.1 | $K_bT$=0.25 | $K_bT$=0.5 | $K_bT$=1 |
| 1 | 3.3 | 26.3 | 42.9 | 46.5 |
| 2 | 1 | 20.7 | 40.8 | 44.9 |
| 5 | 0 | 11.7 | 35.2 | 44.5 |
| 10 | 0 | 4.9 | 28.9 | 43.7 |
| 20 | 0 | 0.5 | 21.5 | 41.9 |
| 30 | 0 | 0.2 | 15.5 | 40.6 |
| 40 | 0 | 0 | 14.7 | 39.1 |
| 50 | 0 | 0 | 11.4 | 38.2 |
| 100 | 0 | 0 | 4.3 | 33.1 |
| 125 | 0 | 0 | 2.7 | 31.6 |
| 150 | 0 | 0 | 1.8 | 30.8 |
| 200 | 0 | 0 | 0.3 | 30.1 |

(a)

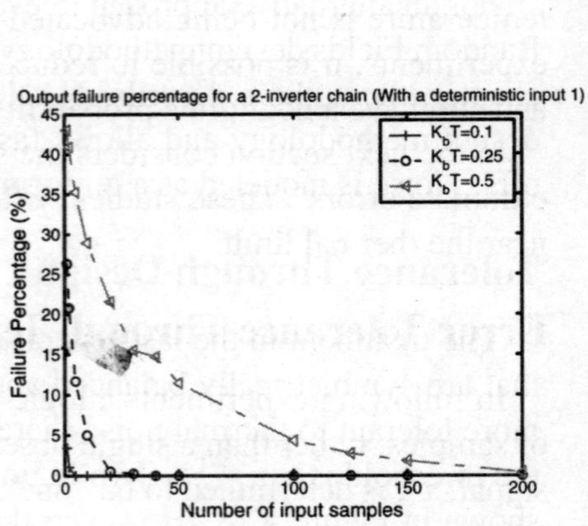


(b)

*Figure 5.11.* The effect of ensemble averaging for a chain of two inverters. The input to the first inverter is fixed at ideal logic "one". The output of this first inverter follows the Gibbs distribution. (a) A table showing percentage of failure vs. the number of samples in the ensemble average of the final output. Blank table entries indicate the failure rate was zero. (b) A graph showing the same information from the table in (a).

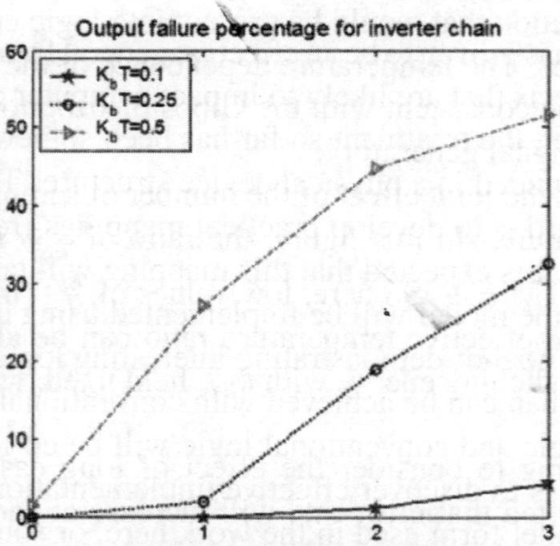


*Figure 5.12.* The effect of ensemble averaging for a chain of inverters. inverters. The input to the first inverter is fixed at ideal logic "one". The output of the final inverter is averaged over a 20 samples.

For an alternative approach to evaluating signal errors based on the Markov Random Fields design methodology, the reader is referred to Chapter 6. In this chapter, the authors describe NANOLAB, which automates the probabilistic design methodology and allows fast analysis of reliability/redundancy trade-offs. Noise is modeled as a uniform or Gaussian distribution.

## Tolerance Through Design

The treatment in the last several sections has demonstrated that logic states that are symmetrically balanced about the decision threshold are going to be more tolerant to thermal noise (here the logic states are taken from $\{0, 1\}$ and the threshold is $x_i > \frac{1}{2}$). Two logic circuits with equivalent logic function are shown in Figure 5.13. However, the probability distribution of logic states is not symmetrical as shown in the figure. The marginal distribution of the output state, $p(x_6)$ is shown in Figure 5.14. Circuit b) has a perfectly symmetrical output distribution and thus can be expected to have significantly better failure tolerance than circuit a), even though they have the same logic function.

This example demonstrates that the Markov framework can be used to characterize circuit configurations for the best thermal noise reliability. However, it should be noted that this treatment assumes perfect device operation. A complete analysis would include both simulation of device structural errors and logic signal errors.

## 5.4 Future Directions

The MRF model has proven to be an effective general framework for studying the fundamental effects that are likely to impact computer circuit design at the nano-scale. However, the treatment so far has been somewhat abstract in that the model is not grounded in a physical device structure. The highest research priority going forward is to develop practical mappings from the MRF model to physical devices. It is expected that this mapping will take several forms.

In the near term, the model will be implemented using conventional silicon technology with the goal of demonstrating interesting logic operation at much lower power levels than can be achieved with conventional circuits. It is envisioned that MRF logic and conventional logic will be co-located on the same chip. The key issue is to discover effective implementations of clique energy, either of the automodel form used in the work here, or some new form that is more aligned with silicon device physics.

In any case, the maximization of logic state probability will occur as a natural physical process of energy minimization. It is envisioned that the entire logic circuit will be continuously optimizing in the presence of random thermal excitations. The output to macro circuits will have negligible error probability through the use of structural and temporal redundancy.

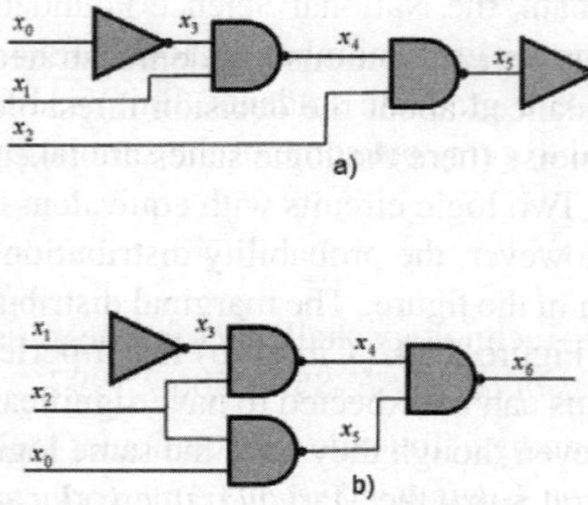


*Figure 5.13.* a) The logic function $x_6 = x_2 \wedge (x_0 \vee x_1')$. b) An equivalent logic circuit.

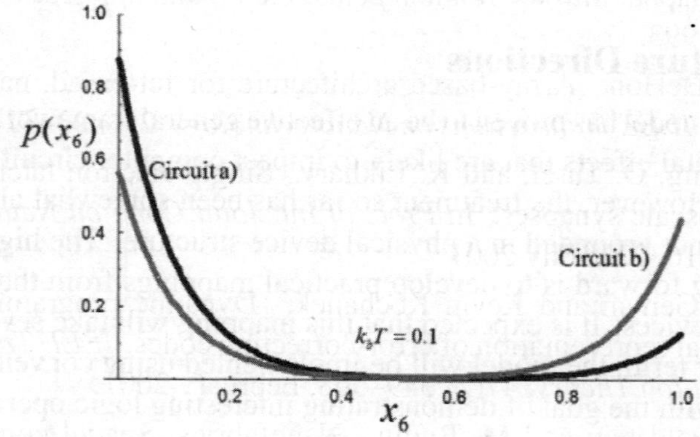


*Figure 5.14.* The marginal output distributions for the circuits in Figure 5.13. Note that circuit a) is highly asymmetrical while circuit b) is perfectly symmetrical.

In the longer term, the mapping will be extended to non-silicon devices such as chemical gates and carbon nanotubes. Again the computation will take the form of energy minimization and continuous adaptation to structural and signal errors.

## 5.5 Acknowledgments

The authors wish to thank the National Science Foundation for providing support for this project under grant number CCR-0304284. In addition, we also would like to thank our graduate students, Bai Yu, Kundan Nepal, and Yingda Chen, who helped to carry out some of the simulations.

## References

[1] S. Asai and Y. Wada. Technology challenges for inegration near and below 0.1 $\mu m$. *Proceedings of the IEEE*, 85(4):506–, 1997.

[2] P. Beckett and A. Jennings. Towards nanocomputer architecture. In *Asia-Pacific Conference on Computer Systems Architecture*, volume 6, pages 141–150, 2002.

[3] Charles H. Bennett. The thermodynamics of computation—a review. *Intl. Journal of Theoretical Physics*, 21(12):905–940, 1982.

[4] R. Chellappa. *Markov random fields: theory and applications*. Academic Press, 1993.

[5] Andre DeHon. Array-based architecture for fet-based, nanoscale electronics. *IEEE Transactions on Nanotechnology*, 2(1):23–32, March 2003.

[6] S. Folling, O. Turel, and K. Likharv. Single-electron latching switches as nanoscale synapses. In *Proc. of Int. Joint Conf. on Neural Networks*, pages 216–221, July 2001.

[7] Stuart. Geman and Kevin Kochanek. Dynamic programming and the graphical representation of error-correcting codes. *IEEE Transactions on Information Theory*, 47(2):549–568, February 2001.

[8] S. C. Goldstein and M. Budiu. Nanofabrics: Spatial computing using molecular electronics. In *The 28th Annual International Symposium on Computer Architecture*, pages 178–189, 2001.

[9] J. Hammersley and P. Clifford. Markov fields on finite graphs and lattices. Technical report, University of California, Berkeley, 1968.

[10] Jie Han and Pieter Jonker. A system architecture solution for unreliable nanoelectronic devices. *IEEE Transactions on Nanotechnology*, 1(4):201–208, December 2002.

[11] S. Z. Li. *Markov Random Field Modeling in Computer Vision*. Springer, 1995.

[12] Samuel Luckenbull. Building bayesian netorks with analog subthreshold cmos circuits. http://zoo.cs.yale.edu/classes/cs490/01-02b/luckenbill.samuel.sbl23/.

[13] R. R. Schulz and R. L. Stevenson. A Bayesian approach to image expansion for improved definition. *IEEE Transactions on Image Processing*, 3(3):233–242, 1994.

[14] Y. Taur, D. Buchanan, and *et. al*. CMOS scaling into the nanometer regime. *Proceedings of the IEEE*, 85(4):486–, 1997.

[15] J. von Neumann. *Probabilistic Logic and the Synthesis of Reliable Organisms from Unreliable Components*. Automata Studies. Princeton University Press, 1956.

# Chapter 6

# TOOLS AND TECHNIQUES FOR EVALUATING RELIABILITY TRADE-OFFS FOR NANO-ARCHITECTURES

Debayan Bhaduri
*FERMAT Lab*
*Center for Embedded System for Critical Applications*
*Virginia Tech, Blacksburg, VA-24061 U.S.A*
dbhaduri@vt.edu

Sandeep K. Shukla
*FERMAT Lab*
*Center for Embedded System for Critical Applications*
*Virginia Tech, Blacksburg, VA-24061 U.S.A*
shukla@vt.edu

**Abstract** Nano-computing in the form of quantum, molecular and other computing models is proliferating as we scale down to nano-meter fabrication technologies. According to many experts, it is expected that nano-scale devices and interconnections will introduce unprecedented level of defects in the substrates and architectural designs need to accommodate the uncertainty inherent at such scales. This consideration motivates the search for new architectural paradigms based on redundancy based defect-tolerant designs. However, redundancy is not always a solution to the reliability problem, and often too much or too little redundancy may cause lack of reliability. The key challenge is in determining the granularity at which defect tolerance is designed, and the level of redundancy to achieve a specific level of reliability. Various forms of redundancy such as NAND multiplexing, Triple Modular Redundancy (TMR), Cascaded Triple Modular Redundancy (CTMR) have been considered in the fault-tolerance literature. Also, redundancy has been applied at different levels of granularity, such as gate level, logic block level, logic function level, unit level etc. Analytical probabilistic models to evaluate such reliability-redundancy trade-offs are error prone and cumbersome. In this chapter, we discuss different analytical and automation methodologies that can evaluate the reliability measures of combinational logic blocks, and can be used to analyze trade-offs between reliability and redundancy for different architec-



*S.K. Shukla and R.I. Bahar (eds.), Nano, Quantum and Molecular Computing*, 157-211.

tural configurations. We also illustrate the effectiveness of our reliability analysis tools pointing out certain anomalies which are counter intuitive and can be obtained easily by designers through automation, thereby providing better insight into defect-tolerant design decisions. We foresee that these tools will help furthering research and pedagogical interests in this area, expedite the reliability analysis process and enhance the accuracy of establishing reliability-redundancy trade-off points.

**Keywords:** Nanotechnology, Gibbs distribution, TMR, CTMR, reliability, entropy, PRISM, probabilistic model checking, interconnect noise, modeling, defect-tolerant architecture, granularity, Gaussian

## 6.1 Introduction

New technologies for building nanometer-scale devices are expected to provide the means for constructing much denser logic and thinner wires. These technologies provide a mechanism for the construction of a useful Avogadro computer [23] that makes efficient use of extremely large number (Avogadro number is in the order of $10^{23}$) of small devices computing in parallel. But the economic fabrication of complete circuits at the nanometer level with devices computing in parallel remains challenging because of the difficulty of connecting nanodevices to one another. Also, experts predict that these devices will have high defect density due to their minuscule dimension, quantum physical effects, reduced noise margins, system energy levels reaching thermal limits of computation, manufacturing defects, aging and many other factors. In the near future we will be unable to manufacture large, defect-free integrated circuits. Thus, designing reliable system architectures that can work around these problems at run-time becomes important. General computer architectures till date have been based on principles that differentiate between memory and processing and rely on communication over buses. Nanoelectronics promises to change these basic principles. Processing will be cheap and copious, interconnection expensive and prevalent. This will tend to move computer architecture in the direction of locally connected, redundant and reconfigurable hardware meshes that merge processing and memory. At the same time, due to fundamental limitations at the nanoscale, micro-architects will be presented with new design challenges. For instance, the methodology of using global interconnections and assuming error-free computing may no longer be possible. Due to the small feature size, there will be a large number of nanodevices at a designer's disposal. This will lead to redundancy based defect-tolerant architectures, and thus some conventional techniques such as Triple Modular Redundancy (TMR), Cascaded Triple Modular Redundancy (CTMR) and multistage iterations of these may be implemented to obtain high reliability. However, too much redundancy does not necessarily lead to higher reliability, since the defects affect the redundant parts as well. As a result, in-depth analysis is required to find the optimal redundancy level for each specific architecture. [43] and [6] discuss the key challenges in determining reliability-redundancy trade-off points. These are as follows:

- The arbitrary augmentation of unreliable devices could result in the decrease of the reliability of an architecture.
- For each specific architecture and a given failure distribution of devices, once an optimal redundancy level is reached, any increase or decrease in the number of devices may lead to less reliable computation.
- Redundancy may be applied at different levels of granularity, such as gate level, logic block level, functional unit level etc.
- Determining the correct granularity level for a specific Boolean network is crucial in such trade-off analysis.

Analytical methodologies [31, 32, 42] have been proposed in the past for computing reliability-redundancy trade-offs of classical defect-tolerant architectures. These techniques become complicated and cumbersome for complex combinational logic. This motivates the need for automating reliability analysis of systems. [44] proposes a probabilistic model checking based automation approach, that found a flaw in the the analytical approach of [31]. We have developed automation methodologies to evaluate reliability measures of different redundant architectural configurations for arbitrary Boolean networks.

In this chapter, we describe a MATLAB based tool (code named **NANOLAB** since it is based on MATLAB) [5, 7] and a probabilistic model checking based tool named **NANOPRISM** [6]. One difference between our automation techniques and the standard analytical approaches is that we evaluate the reliability of specific cases as opposed to considering the general framework, and hence are not necessarily restricted by the analytical bounds of reliability.

## Brief Introduction to Our Tools

- Conventional digital signals are bimodal, meaning that the logic low or high are defined as discrete voltage/current levels. Due to non-silicon manufacturing technologies and device miniaturization in the nanotechnology era, the notion of being a binary zero or one will change. A probabilistic design methodology based on Markov Random Fields (MRF) [40] proposed in [2] (details in Chapter 5) introduces a new information encoding and computation scheme where signals are interpreted to be logic low or high over a continuous energy distribution. The inputs and outputs of the gates in a combinational block are realized as nodes of a Markov network and the logic function for each gate is interpreted by a Gibbs distribution [48] based transformation. This computational scheme also exploits the fact that maximizing the probability of energy levels at each node of the network is equivalent to minimizing the entropy of Gibbs energy distribution that depends on neighboring nodes. The probability of a logic variable can be calculated by marginalizing over the set of possible states of the neighboring logic variables and propagated to the next node in a Boolean network by using Belief Propagation [36]. **NANOLAB** automates this probabilistic design methodology by computing energy distribution and entropy at the primary/intermediate outputs and interconnects of Boolean networks, and by implementing Belief Propagation. The logic compatibility functions (similar to truth table) [2] for the different component gates of the Boolean network and the energy distribution at the primary inputs are specified to the tool. We have also augmented the capability to model uniform and Gaussian noise at the primary inputs and interconnects of combinational blocks and analyze such systems in terms of entropy at the outputs. Such modeling features in NANOLAB will expedite and enhance the analysis of reliability measures for different defect tolerant architectures.

- **NANOPRISM** is a probabilistic model checking based tool that applies probabilistic model checking techniques to calculate the likelihood of occurrence of transient defects in the devices and interconnections of nano architectures. NANOPRISM is based on the conventional Boolean model of computation and can automatically evaluate reliability at different redundancy and granularity levels, and most importantly show the trade offs and saturation points. By saturation point we mean the granularity based redundancy vs. reliability reaches a plateau meaning that there cannot be any more improvements in reliability by increasing redundancy or granularity levels. It consists of libraries built on PRISM [37, 49] (probabilistic model checker) for different redundancy based defect-tolerant architectural configurations. These libraries also support modeling of redundancy at different levels of granularity, such as gate level, logic block level, logic function level, unit level etc. Arbitrary Boolean networks are given as inputs to these libraries and reliability measures of these circuits are evaluated. This methodology also allows accurate measurements of variations in reliability on slight changes in the behavior of the system's components, for example the change in reliability as the probability of gate failure varies.

## Features of these Tools

Here we enlist some major advantages of our tools:

- Our tools allow fast analysis of reliability-redundancy trade-offs for alternative defect tolerant architectures. In the case of NANOLAB, Figures 6.9 and 6.10 which evaluate the computational entropy (measure of the failure probability) at the outputs of two different circuits, show that a particular fault tolerance technique CTMR, requires different depths of iteration for optimum circuit reliability. Whereas, the plots obtained in Figure 6.19 (a) from NANOPRISM indicate that for a 1st order CTMR configuration of a logic circuit (given in Figure 6.8) with logic block and gate level granularity, there is a degradation of reliability of computation at higher probabilities of gate failure. Such analysis can help system designers quickly decide the redundancy scheme required for defect tolerance while building the designs targeted for nanoscale substrates.

- The idea of a new model of computation in [2] for uncertainty based computation is extended in the direction of reliability evaluation and automated by NANOLAB. Previously [2, 16] only show the viability of a MRF based computation scheme and how it can be mapped to nano devices such as Y CNTs [54]. These do not extend the model of computation in the direction of reliability-redundancy trade-off analysis.

- NANOLAB can also compute reliability of systems in the presence of signal noise at interconnects and inputs. Noise is modeled as uniform or Gaussian distribution or combinations of both. This models practical situations the circuits are subjected to during normal operation. Analysis of reliability measures for redundant architectural configurations of these logic circuits when exposed to such real world noise distributions makes our methodology more effective.

- NANOPRISM is a tool that automates the evaluation of redundancy vs. reliability vs. granularity trade-offs. In [44] a probabilistic model checking based CAD framework is used to evaluate reliability measures of a specific redundancy technique namely NAND multiplexing, and only redundancy vs. reliability trade-offs are computed. NANOPRISM uses such a framework too, but we have developed a library of generic redundancy based defect tolerant architectures where different Boolean networks can be plugged in and analyzed. NANOPRISM analyzes optimal redundancy and granularity levels for specific logic networks and expedites the process of finding the correct trade-offs.

- We are able to illustrate some anomalous counter intuitive trade-off points that would not be possible to observe without significant and extensive theoretical analysis, and the automation makes it easier to analyze these critical parameters.

- From our literature search, we found that the results on reliability measures for different defect tolerant architectures were mostly analytical [53, 31, 42, 27] and none considered *granularity* and *entropy* as parameters. However, for complex network of gates, such analytical results may be error prone. As a result, we believe that scalable automation methodologies to quickly evaluate these figures of merits is crucial for practical use by engineers.

## Organization

We begin with an introduction to the basic concepts of defect-tolerant computing, the probabilistic model of computation from [2] and how to use such a computational scheme to model noise. We also discuss about granularity and measure of redundancy, and introduce probabilistic model checking. We discuss some interesting analytical reliability computation models and hybrid defect-tolerant architectures in Section 3. Section 4 describes NANOLAB and our approach to analyze reliability of systems with the model of computation proposed by Bahar et al. A detailed example along with a code snippet is used to elucidate this methodology. In Section 5 we report reliability measures of different logic networks computed by NANOLAB. Section 6 focuses on explaining how we use the NANOPRISM framework to model defect-tolerant architectures such as TMR, CTMR and their iterations for single gates and logic blocks respectively. Reliability-redundancy trade-off points for specific combinational circuits and interesting anomalies computed by NANOPRISM are presented in Section 7. Finally, Section 8 concludes the chapter and summarizes plans to extend these automation methodologies in the future.

## 6.2 Background

Nanotechnology currently involves various technologies that exploit quantum mechanical effects, molecular bindings, inherent properties of materials at atomic scale, weak forces, van der waal forces etc. Some of these emerging technologies are single electron transistors [17], nanowire transistors [19], quantum dots [56], resonant tunneling devices [12], negative differential resistors (NDR) [14], chemically assembled electronic nanocomputers (CAENs) [18, 46], reconfigurable switches [15, 18] and molecular switches [24]. In all the cases at least one dimension of the fabricated devices are of the order of a few nanometers. Such miniaturization leads to high density of devices on system on chips (*SOCs*). It has been shown in [58] that it is possible to build a trillion ($10^{12}$) devices in a square centimeter. But there is going to be a very high degree of failures due to (i) manufacturing defects, (ii) transient faults resulting from thermal perturbations and reduced noise tolerance at reduced voltage and current levels (less amiable operating environments), and (iii) faults due to aging, etc. Defect tolerant architectures are possible solutions to this problem specifically by using redundant devices and functional units. The importance of evaluating these defect-tolerant architectures is many-fold. Usually there are many alternatives available in terms of configuration and parameters. Hence, the perfect choice for a specific Boolean network is a matter of major analysis.

## Defect Tolerant Architectures

Formally, a *defect-tolerant architecture* is one which uses techniques to mitigate the effects of defects in the devices that make up the architecture, and guarantees a given level of reliability. There are a number of different canonical defect-tolerant computing schemes most of which are based on redundancy. The techniques which we look at are highly generic and concerned with resource redundancy and can be easily extended to nano architectures. Some of these canonical techniques are *Triple modular Redundancy(TMR), Cascaded Triple Modular Redundancy(CTMR)* and multi-stage iterations of these [42].

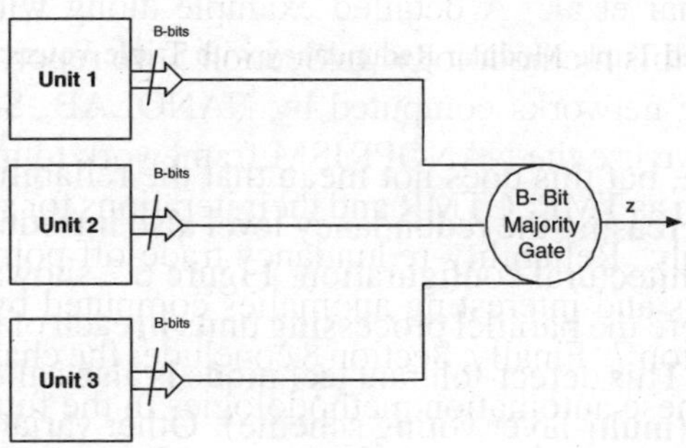


*Figure 6.1.* Generic Triple Modular Redundancy Configuration

Triple Modular Redundancy as shown in Figure 6.1 entails three similar units working in parallel, and comparison of their outputs with a majority voting logic [42]. The units could be gates, logic blocks, logic functions or functional units. The TMR provides a functionality similar to one of the three parallel units but provides a better probability of working. The tradeoff is that instead of $n$ devices, $3n$ devices and a majority gate are needed in this configuration. The R-fold modular redundancy is a generalization of the TMR configuration where R units work in parallel instead of 3 and $R \in \{3,5,7,9.....\}$.

Cascaded Triple Modular Redundancy is similar to TMR, wherein the units working in parallel are TMR units combined with a majority gate. This configuration forms a CTMR of the first order and therefore TMR can be considered to be 0th order CTMR. Higher orders of CTMR are obtained by multi-stage

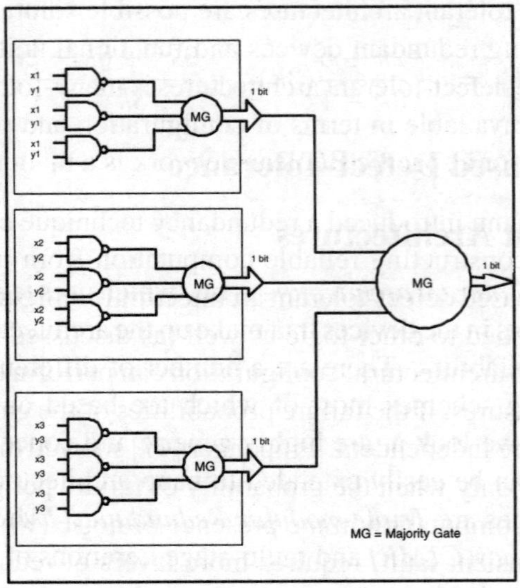


*Figure 6.2.* Cascaded Triple Modular Redundancy with Triple Voters: Multi-layer Voting

iterations of these, but this does not mean that the reliability of a system goes up due to this. Increasing the redundancy level also introduces more unreliable devices in the architectural configuration. Figure 6.2 shows a 1st order CTMR configuration where the parallel processing units in each of the three TMR units are NAND gates. This defect-tolerant technique is also called a Cascaded TMR with triple voters (multi-layer voting scheme). Other variations such as CTMR with triple voters (smaller granularity of redundancy) are discussed in details in Chapter 2.

Let n be the order of the CTMR configuration. The majority logic requires four gates i.e. if a, b and c are inputs, the logic operation is $(a\wedge b)\vee(b\wedge c)\vee(c\wedge a)$. If $F_{n-1}$ is the total number of devices for a (n-1)th order CTMR, then we can say:

$$F_n = 3F_{n-1} + 5 \tag{6.1}$$

For a single gate, the total number of redundant gates in a nth order CTMR configuration (where $n \in \{0,1,2....\}$) is:

$$F_n = \frac{21}{2}\cdot 3^n - \frac{5}{2} \tag{6.2}$$

If the total number of gates in a logic circuit is *k*, then the total number of devices in any nth order configuration will be:

$$F_n = 3 \cdot \frac{2k+5}{3k-1}(3k)^n + \frac{9k^2+6k-20}{3k-1} \tag{6.3}$$

## Multiplexing Based Defect-Tolerance

In 1952, von Neumann introduced a redundancy technique called NAND multiplexing [57] for constructing reliable computation from unreliable devices. This multiplexing based defect-tolerant architectural configuration is more general and can be applied to other logic as well (as discussed in Chapter 2). He showed that such an architectural configuration can perform computations with high reliability measures, if the failure probabilities of the gates are sufficiently small and failures are independent. Pippenger [47] showed that von Neumann's construction works only when the probability of failure per gate is strictly less than $1/2$, and that computation in the presence of noise (which can be seen as the presence of transient fault) requires more layers of redundancy. [21] also showed that a logarithmic redundancy is necessary for some Boolean function computations, and is sufficient for all Boolean functions. In [42], NAND multiplexing was compared to other techniques of fault-tolerance and theoretical calculations showed that the redundancy level must be quite high to obtain acceptable levels of reliability.

The multiplexing based defect-tolerance scheme replaces a single logic device by a multiplexed unit with $N$ copies of every input and output of the device. The $N$ devices in the multiplexing unit process the copies of the inputs in parallel to give $N$ outputs. Each element of the output set will be identical and equal to the original output of the logic device, if all the copies of the inputs and devices are reliable. However, if the inputs and devices are in error, the outputs will not be identical. To tackle such an error-prone scenario, some critical level $\Delta \in (0, 0.5)$ is defined. The output of the multiplexing unit is considered stimulated (taking logical value `true`) if at least $(1-\Delta)\cdot N$ of the outputs are stimulated, and non-stimulated (taking logical value `false`) if no more than $\Delta \cdot N$ outputs are stimulated. If the number of stimulated outputs is in the interval $(\Delta \cdot N, (1-\Delta)\cdot N)$, then the output is undecided, and hence a malfunction will occur. The basic design of a von Neumann multiplexing technique consists of two stages: the *executive stage* which performs the basic function of the processing unit to be replaced, and the *restorative stage* which reduces the degradation in the executive stage caused by erroneous inputs and faulty devices. As shown in Figure 6.3, NAND multiplexing is an architectural configuration wherein a single NAND gate is the processing unit of a multiplexing based defect-tolerant system. The unit $U$ performs random permutation of the input signals i.e. each signal from the first input bundle is randomly paired with a signal from the second input bundle to form the input pair of one of the duplicated NAND gates.

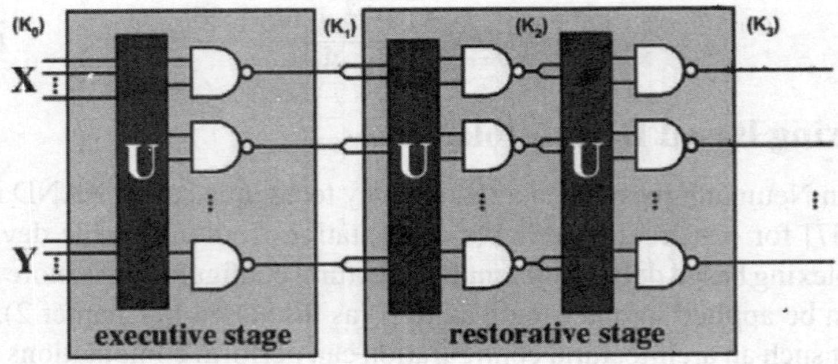


*Figure 6.3.* A NAND multiplexing unit from [44]

Creating computing systems built from devices other than silicon-based transistors is a research area of major interest. A number of alternatives to silicon VLSI have been proposed, including techniques based on molecular electronics, quantum mechanics, and biological processes. One such nanostructure proposed by Lent et al [39, 38] is quantum-dot cellular automata (QCA) that employs arrays of coupled quantum dots to implement Boolean networks. Due to the miniscule size of the quantum dots, extremely high packing densities are possible in CA based architectures. Also, the other advantages are simplified interconnections, and extremely low power-delay product. The fundamental QCA logic device is a three-input majority logic gate consisting of an arrangement of five standard quantum cells: a central logic cell, three inputs and an output cell. A majority gate can be programmed to act as an OR gate or an AND gate by fixing any of the three inputs as a program line [55]. This indicates that majority gates are going to play an important role in the future architectural configurations.

This motivates us to consider multiplexing when the logic device is a single majority gate. We therefore replace the inputs and output of the gate with $N$ copies and duplicate the majority gate $N$ times in the executive stage, as shown in Figure 6.4. Again, the unit $U$ represents a *random permutation* of the input signals, that is, for each input set to the $N$ copies of the majority gate, three input signals are randomly chosen from the three separate input bundles respectively. Figure 6.4 also shows the restorative stage which is made using the same technique as the executive stage, duplicating the outputs of this stage to use as inputs to the restorative stage. Note that, applying this approach only once will invert the result, therefore two steps are required. To give a more effective restoration mechanism this stage can be iterated [57].

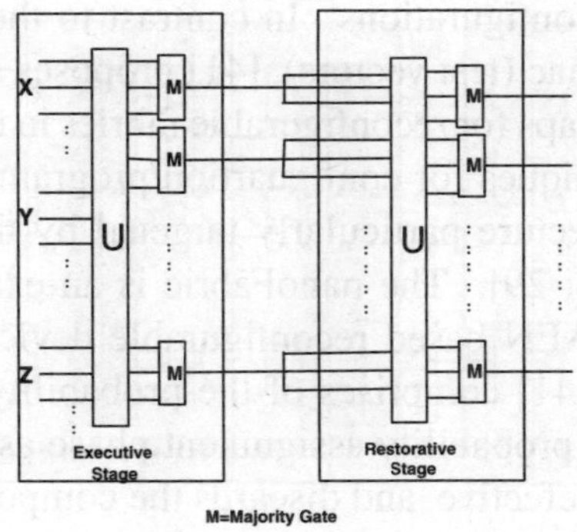


*Figure 6.4.* A majority multiplexing unit from [9]

## Defect-Tolerance through Reconfiguration

A computer architecture that can be configured or programmed after fabrication to implement desired computations is said to be *reconfigurable*. Reconfigurable fabrics such as Field-Programmable Gate Arrays (FPGAs) are composed of programmable logic elements and interconnects, and these can be programmed, or configured, to implement any circuit. Such reconfigurable architectures may mitigate manufacturing defects that will be rampant at the nano substrates. The key idea behind defect tolerance in FPGAs is that faulty components are detected during testing and excluded during reconfiguration. It is expected [41] that reconfigurable fabrics made from next generation manufacturing techniques (CAEN-based technologies where molecular junctions can be made which hold their own state) will go through a post-fabrication testing phase during which these fabrics will be configured for self-diagnosis. Testing for error-prone devices will not incur either an *area* or a *delay* penalty because the test circuits placed on the fabric during this self-diagnosis phase will utilize resources that will be available later for normal fabric operation (unlike BIST structures). The defect map obtained from this test phase will contain locations of all the defects, and this map can be used by place-and-route tools to layout circuits on the fabric that avoid the defective devices. Thus, the built-in defect-tolerance in such a reconfigurable digital system will ease the requirements on the manufacturing process.

Teramac [34] is such an extremely defect-tolerant reconfigurable machine built in HP laboratories. The basic components in Teramac are programmable switches (memory) and redundant interconnections. The high communication bandwidth is critical for both parallel computation and defect tolerance. With about 10% of the logic cells and 3% of the total resources defective, Teramac could still operate 100 times faster than a high-end single processor workstation for some of its configurations. In contrast to the static defect discovery process used in Teramac (test vectors), [41] proposes scalable testing methods that generate defect maps for reconfigurable fabrics in two phases, and dynamic place-and-route techniques for configuartion/programming of the fabric. The reconfigurable architecture particularly targeted by the methodology in [41] is the nanoFabric [28, 29]. The nanoFabric is an example of a possible implementation of a CAEN based reconfigurable device. The post-fabrication testing suggested in [41] comprises of the probability assignment and defect location phases. The probability assignment phase assigns each component a probability of being defective, and discards the components which have a high probability. Thus, this phase identifies and eliminates a large fraction of the defective components. The remaining components are now likely to have a small enough defect rate and can be tested in the defect location phase using simple test-circuit configurations. The authors also point out the non-adaptiveness of the test-circuit generation. This implies that the results of previous tests are not used to generate new circuits (details in Chapter 3). The Cell Matrix architecture [23] (details in Chapter 4) also supports dynamic defect identification and elimination. The Cell Matrix is a fine-grained reconfigurable fabric composed of simple, homogeneous cells and nearest-neighbor interconnect. The cells are programmable, gate-level processors where a small set of properties holds true for cell structure, function, and intercellular communication. This set of cellular properties provides ways to directly program cells to bring about useful computations and data processing [22]. The homogeneity of this architecture makes it inherently fault-tolerant. Like Teramac, the Cell Matrix can handle large manufacturing defect rates (permanent faults), and provide high reliability of computation. However, the Cell Matrix architecture also provides defect-tolerance to transient errors (caused due to less amiable operating conditions) due to the inherent self-analyzing, self-modifying, and dynamic local processing capabilities of the architectural configuration. Also, the embryonics architecture [20] is a potential paradigm for future nanoscale computation systems. The objective of developing defect-tolerant and ultra-large integrated circuits capable of self-repair and self-replication makes this architecture viable for future nanoelectronic system design.

## Markov Random Field based Model of Computation

The advent of non-silicon manufacturing technologies and device miniaturization will introduce uncertainty in logic computation. Newer models of computations have to be designed to incorporate the strong probabilistic notions inherent in nanodevices. Such a probabilistic design methodology has been proposed in [2] (details in Chapter 5) that adapts to errors by maximizing the

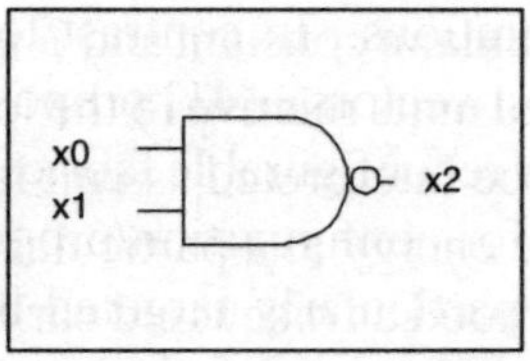


*Figure 6.5.* A NAND gate

probability of being in valid computational states. This model of computation introduces a new information encoding and computation scheme, where signals are interpreted to be logic low or high over a continuous energy distribution. The basis for this architectural approach is the Markov random network which is based on Markov random fields (MRF) [40]. The Markov random field is defined as a finite set of random variables, $\Lambda = \{\lambda_1, \lambda_2, ......, \lambda_k\}$. Each variable $\lambda_i$ has a neighborhood, $N_i$, which are variables in $\{\Lambda - \lambda_i\}$.

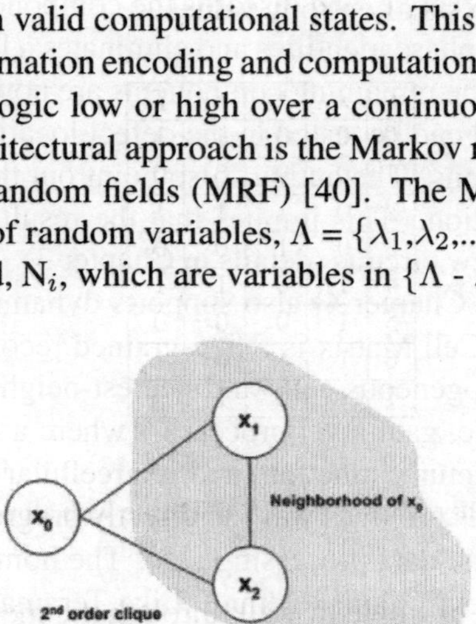


*Figure 6.6.* The neighborhood of the input of a NAND gate depicted as a network node

The energy distribution of a given variable depends only on a (typically small) neighborhood of other variables that is called a clique. These variables may represent states of nodes in a Boolean network and we might be able to consider effects of noise and other uncertainty factors on a node by considering the conditional probabilities of energy values with respect to its clique. Due to the Hammersley-Clifford theorem [4],

$$P(\lambda_i | \{\Lambda - \lambda_i\}) = \frac{1}{Z} e^{-\frac{1}{KT} \sum_{c \in C} U_c(\lambda)} \tag{6.4}$$

The conditional probability in Equation 6.4 is Gibbs distribution. Z is the normalizing constant and for a given node $i$, C is the set of cliques affecting the node $i$. $U_c$ is the clique energy function and depends only on the neighborhood of the node whose energy state probability is being calculated. $KT$ is the thermal energy ($K$ is the Boltzmann constant and $T$ is the temperature in Kelvin) and is expressed in normalized units relative to the logic energy (clique energy). For example, $KT = 0.1$ can be interpreted as unit logic energy being ten times the thermal energy. The logic energy at a particular node of a Markov network depends only on its neighborhood as discussed earlier. Also, the logic margins of the nodes in a Boolean network decrease at higher values of $KT$ and become significant at lower values. The logic margin in this case is the difference between the probabilities of occurrence of a logic low and a logic high which if high leads to a higher probability of correct computation. This formulation also allows correct analysis of entropy values.

| i | $x_0$ | $x_1$ | $x_2$ | $F$ |
|---|---|---|---|---|
| 0 | 0 | 0 | 1 | 1 |
| 1 | 0 | 0 | 0 | 0 |
| 2 | 0 | 1 | 1 | 1 |
| 3 | 0 | 1 | 0 | 0 |
| 4 | 1 | 0 | 1 | 1 |
| 5 | 1 | 0 | 0 | 0 |
| 6 | 1 | 1 | 0 | 1 |
| 7 | 1 | 1 | 1 | 0 |

*Table 6.1.* Logic compatibility function for a NAND gate with all possibilities

Let us take a specific example to walk through the methodology in [2]. For a two input NAND gate as shown in Figure 6.5, there are three nodes in the assumed logic network: the inputs $x_0$ and $x_1$, and the output $x_2$. Figure 6.6 represents $x_0$ as a network node, and shows its neighborhood. The energy state of this node depends on its neighboring nodes. The edges in Figure 6.6 depict the conditional probabilities with respect to the other input $x_1$ and the output $x_2$ (nodes in the same clique). The operation of the gate is designated by the logic compatibility function $F(x_0,x_1,x_2)$ shown as a truth table in Table 6.1. $F=1$ when $x_2 = (x_0 \wedge x_1)\prime$ (valid logic operations). Such a function takes all valid and invalid logic operation scenarios into account so as to represent an energy based transformation similar to the NAND logic. The axioms of the Boolean ring [13] are used to relate $F$ to a Gibb's energy form. Also, the valid input/output states should have lower clique energies than invalid states to maximize the probability of a valid energy state at the nodes. Thus the clique energy (logic energy) is the summation over the minterms of the valid states and is calculated as :

$$U(x_0, x_1, x_2) = -\sum_i F_i(x_0, x_1, x_2) \tag{6.5}$$

where $F_i = 1$ ($i$ is the index for each row of Table 6.1). This clique energy definition reinforces that the energy of invalid logic state is greater than valid state. As shown in [16], for a valid state, the summation of valid states ($F_i = 1$) is '-1' and for any invalid state, this summation value is '0'. The clique energy for the NAND gate is:

$$U(x_0, x_1, x_2) = -x_2 + 2x_0x_1x_2 - x_0x_1 \tag{6.6}$$

The probability of the different energy configurations of $x_2$ is:

$$p(x_2) = \frac{1}{Z} \sum_{x_0 \in \{0,1\}} p(x_0) \sum_{x_1 \in \{0,1\}} p(x_1)\, e^{-\frac{1}{KT}U(x_0,x_1,x_2)} \tag{6.7}$$

As according to Equation 6.7, the probability of different energy states at $x_2$ is calculated as a function of $x_2$ after marginalizing over the distributions of $x_0$ and $x_1$. The probability of the energy state configurations at the outputs of any logic gate can be calculated by the methodology above. Given the input probability distributions and logic compatibility functions, using Belief Propagation algorithm [36] it is also possible to calculate entropy and energy distributions at different nodes of the network. Thus, [2] gives a probabilistic model of computation which we exploit to compute reliability-redundancy trade-offs for different nanoscale architectures.

## Modeling Noise at Inputs and Interconnects

The probabilistic non-discrete computing scheme described above can be extended to incorporate the impact caused by continuous noise at the inputs and the interconnects of combinational circuits. Let us take the same NAND gate example to illustrate this. For the inputs being logic low or high, the Gaussian noise distribution below zero will be filtered out because negative values are invalid [16]. To make the Gaussian noise symmetrically distributed about the inputs of the gate, the coordinate system has to be shifted such that $x = 0 \rightarrow x' = -1$, and $x = 1 \rightarrow x' = 1$. Thus, $x_0$, $x_1$ and $x_2$ can be rewritten as :

$$x'_0 = 2(x_0 - \tfrac{1}{2}),\ x'_1 = 2(x_1 - \tfrac{1}{2}),\ x'_2 = 2(x_2 - \tfrac{1}{2})$$

The clique energy in Equation 6.6 for the NAND gate is modified as follows:

$$U(x'_0,x'_1,x'_2) = -\tfrac{1}{2} - \tfrac{1}{4}x'_2 + \tfrac{1}{4}x'_0x'_1 + \tfrac{1}{4}x'_1x'_2 + \tfrac{1}{4}x'_0x'_1x'_2$$

We have modeled noise as a *Gaussian process* with mean $\mu$ and variance $\sigma^2$. The probability distribution of $x_2'$ being in different energy configurations $\in$ {-1.0, -0.9, -0.8,...., 0.1, 0.2, 0.3,...., 1.0} is:

$$p(x_2') = \frac{1}{Z}\int_{-1}^{1} \sum_{x_1' \in \{-1,1\}} e^{-\frac{U}{KT}} \left( \frac{e^{-(x_0'-\mu)^2/2\sigma^2}}{\sqrt{2\pi}\sigma} \right) dx_0 \cdot p(x_1') \quad (6.8)$$

The energy distribution at $x_2'$, if uniform distribution is used to model signal noise is given by:

$$p(x_2') = \frac{1}{Z}\int_{-1}^{1} \sum_{x_1' \in \{-1,1\}} e^{-\frac{U}{KT}} dx_0 \cdot p(x_1') \quad (6.9)$$

Equation 6.8 or 6.9 is used to compute the probability of $x_2$ at different energy states, marginalizing over the distributions of $x_0$ (ranges between $-1$ and 1) and $x_1$ [11]. The energy distribution at the output of any logic gate can be calculated in the presence of noisy inputs and interconnects by the methodology above. For more details refer to Chapter 5.

## Granularity and Measure of Redundancy

Redundancy based defect-tolerance can be implemented for Boolean networks at different levels of granularity. For a specific logic circuit, all the gates could be replicated as a particular CTMR configuration and the overall architecture can be some other CTMR configuration. For example, each gate in the circuit could be a $k$th order CTMR configuration and the overall logic block could be a $n$th order configuration where $k \neq n$. Redundancy can thus be applied at different levels of granularity, such as gate level, logic block level, logic function level etc. Later in this chapter, we discuss a few experiments that show us that reliability is often dependent on the granularity level at which redundancy is injected in a logic network. NANOPRISM indicates the correct level of granularity beyond which the reliability measures for a system do not improve.

## Probabilistic Model Checking and PRISM

Probability is the measure of level of uncertainty or randomness. It is widely used to design and analyze software and hardware systems characterized by incomplete information, and unknown and unpredictable outcomes. *Probabilistic model checking* is a range of techniques for calculating the likelihood of the occurrence of certain events during the execution of unreliable or unpredictable systems. The system is usually specified as a state transition system, with probability values attached to the transitions. A probabilistic model checker applies algorithmic techniques to analyze the state space and calculate performance measures.

NANOPRISM consists of libraries built on PRISM [37, 49], a probabilistic model checker developed at the University of Birmingham. PRISM supports analysis of three types of probabilistic models: discrete-time Markov chains (DTMCs), continuous-time Markov chains (CTMCs) and Markov decision processes (MDPs). We use DTMCs to develop libraries for generic defect-tolerant architectural configurations. This model of computation specifies the probability of transitions between states, such that the probabilities of performing a transition from any given state sums up to 1. DTMCs are suitable for conventional digital circuits and the fault models considered. The fault models are manufacturing defects in the gates and transient errors that can occur at any point of time in a Boolean network.

The PRISM description language is a high level language based on guarded commands. The basic components of the language are *modules* and *variables*. A system is constructed as a number of modules which can interact with each other by means of the standard process algebraic operations [50]. A module contains a number of variables which represents its state. Its behavior is given by a set of guarded commands of the form:

[] <guard> → <command>;

The guard is a predicate over the variables of the system and the command describes a transition which the module can make if the guard is true (using primed variables to denote the next values of variables). If a transition is probabilistic, then the command is specified as:

<prob> : <command> + ··· + <prob> : <command>

The properties of a DTMC model can be verified by PCTL model checking techniques. PCTL [33] is a probabilistic temporal logic, an extension of CTL. It can be used to express properties such as "termination will eventually occur with probability at least 0.98". Our work is based on modeling defect tolerant architectures as state machines with probabilistic assumptions about the occurrence of defects in the devices and interconnections, and then use Markov analysis and probabilistic model checking techniques to evaluate reliability measures.

## 6.3 Analytical Approaches for Reliability Analysis

In this section we survey some theoretical models that have been used to analyze reliability-redundancy trade-off points of some classical defect-tolerant architectures. Several techniques have been suggested in the past to mitigate the effects of both faults and defects in designs. One of the techniques discussed exhaustively is the multiplexing technique initiated by von Neumann [57], and is based on massive duplication of imperfect devices and randomized error-prone interconnects. Han and Jonker [31] extend this technique to a rather low degree of redundancy, and elucidate the stochastic markov [45] nature of the system. They also investigate a system architecture based on NAND multiplexing for SETs, where transient errors are introduced due to random background charges. Stochastic analysis of the chains of stages (Figure 6.3) is performed to evaluate optimal redundancy factors for systems, and bounds on the failure probability per gate that can be tolerated for specific reliability and redundancy levels.

In current digital systems, memory and caches use the bulk of available logic devices. The processor is made from a number of functional units, each of which can be separated into function blocks. These function blocks are composed of many logic circuits cascaded together, however there are specific cascading bounds to avoid timing problems (hazard). [31] assumes that such a function block can be made entirely by $n$ stages of $N$ parallel NAND gates, and various trade-offs can be computed by statistical analysis. In a design with unreliable logic devices, the upper bound is that we must replace each logic gate with $n \cdot N$ unreliable and smaller gates.

For the NAND multiplexing system shown in Figure 6.3, let $X$ be the set of lines in the first input bundle that are stimulated. Consequently, $(N - X)$ lines are not stimulated (logic low). $Y$ and $Z$ are also corresponding sets for the second input and the output bundles. [31] also assumes a constant probability of gate failure $\varepsilon$, and that the faulty gates invert their outputs (von Neumann fault). If the sets $X, Y$ and $Z$ have $(\bar{x}, \bar{y}, \bar{z})$ elements respectively, then these are the relative excitation levels of the two input bundles and of the output bundle, respectively. The stochastic distribution of $\bar{z}$ has to be determined in terms of the given input distributions ($\bar{x}$ and $\bar{y}$).

In [57], von Neumann concluded that for extremely large $N$, the stochastic variable $\bar{z}$ is approximately normally distributed. He also determined an upper bound of 0.0107 for the gate failure probability that can be tolerated. In other words, according to von Neumann, if the failure probability per gate is greater than this threshold, then the probability of the NAND multiplexing system failing will be larger than a fixed, positive lower bound, no matter how large a bundle size is used. Recently, it was shown that, if each NAND gate fails independently, the tolerable threshold probability of each gate will be 0.08856 [25]. [31] shows how this technique can be used with rather low degrees of redundancy rather than massive duplication of imperfect nanodevices.

Let us discuss this interesting analytical methodology in brief. For a single NAND gate in the multiplexing scheme (Figure 6.3), assume $\bar{x}N$ and $\bar{y}N$ are input lines stimulated in each input bundle respectively. If the error probabilities for the two input bundles of the multiplexing configuration are independent, the probability of each gate's output being logic high is $\bar{z} = 1 - \bar{x}\bar{y}$ (assuming error-free NAND operation). If the gate failure probability is $\varepsilon$, the probability of the output being stimulated is:

$$\bar{z} = (1 - \bar{x}\bar{y}) + \varepsilon(2\bar{x}\bar{y} - 1) \tag{6.10}$$

Equation 6.10 is valid only for the von Neumann fault model (erroneous gate inverts the correct output). The NAND multiplexing unit constitutes a Bernoulli sequence because the gates are assumed to function independently. Therefore, the probability distribution of the stimulated outputs can be expressed as a binomial distribution. The probability of exactly $k$ outputs being stimulated is:

$$P(k) = \binom{N}{k} \bar{z}^k (1 - \bar{z})^{N-k} \tag{6.11}$$

When $N$ is extremely large and $\bar{z}$ is extremely small, the probability distribution of exactly $k$ outputs being stimulated from the $N$ output lines of the NAND multiplexing stage can be approximated to a Poisson distribution. If both inputs of the NAND gates have high probability of being logic high, the stimulated outputs are then considered erroneous. The reliability of the system can then be computed as the probability of the number of stimulated outputs being below a threshold ($P(k \leq x)$). Since the number of stimulated outputs is a stochastic variable that is binomially distributed, the central limit (De Moivre-Laplace) theorem applies when N is extremely large and $0 < \bar{z} < 1$. In this case, Equation 6.11 can be rewritten as:

$$P(k \leq x) = \int_{-\infty}^{x} \frac{1}{\sqrt{2\pi}\sqrt{N\bar{z}(1-\bar{z})}} e^{-1/2(t - N\bar{z}/\sqrt{N\bar{z}(1-\bar{z})})^2} \tag{6.12}$$

Thus, [31] shows analytically that for the executive stage (Subsection 6.2.0) of a NAND multiplexing configuration, for smaller $N$, the probability of the number of stimulated outputs is theoretically a binomial distribution. Also, the authors elucidate that the probability could be approximated to a Gaussian distribution with mean $N\bar{z}$ and standard deviation $\sqrt{N\bar{z}(1-\bar{z})}$, when $N$ is extremely large and $0 < \bar{z} < 1$. The authors then go on to demonstrate how additional restorative stages improve fault-tolerance. The discussion above illustrates that the number of stimulated outputs of each NAND multiplexing stage is a stochastic variable and its state space is $A = [0, 1, 2, \dots. N - 1, N]$. This stochastic variable is denoted as $\bar{\xi}_n$, where $n$ is the index of the multiplexing stage. Thus, the evolution of $\bar{\xi}_n$ in the multiplexing configuration is a stochastic process, and with fixed bundle size and gate failure probability, the distribution of $\bar{\xi}_n \forall n$ (stages) depends on the number of stimulated inputs of the $n$th multiplexing unit (outputs of the $n - 1$ th unit). This can be expressed as follows:

$$P(\bar{\xi}_n \in A \mid \bar{\xi}_1 = k_1, \bar{\xi}_2 = k_2, ....\bar{\xi}_{n-1} = k_{n-1}) \\ = P(\bar{\xi}_n \in A \mid \bar{\xi}_{n-1} = k_{n-1}) \quad (6.13)$$

Equation 6.13 is the condition for a stochastic process to be a Markov process. As indicated in Figure 6.3, $k_1, k_2, ....k_{n-1}$ are the number of stimulated outputs of the stages represented by the indices respectively. The evolution of $\bar{\xi}_n$ in the NAND multiplexing system is therefore a Markov process, or a discrete Markov chain. The transition probability of a stochastic Markov chain indicates the conditional probability from one specified state to another. As elucidated in [31], the transition probability matrix $\Psi$ for each $\bar{\xi}_n$ is identical and independent of the multiplexing stage ($n$). Thus, it can be inferred that $\bar{\xi}_n$ evolves as a homogeneous Markov chain. Therefore, an initial probability distribution and a transition probability matrix are sufficient to get all output distributions. For a NAND multiplexing system with $n$ individual stages, the output distribution of the configuration is:

$$P_n = P_0 \Psi^n \quad (6.14)$$

[31] also points out the fact that when the number of multiplexing stages ($n$) is large, $\Psi^n$ approaches a constant matrix $\pi$, and each row of $\pi$ is identical. This indicates that as $n$ becomes extremely large, not only the transition probabilities in a NAND multiplexing system will get stable, but also the output distribution will become stable and independent of the number of multiplexing stages. Experiments indicate that this defect-tolerant technique requires a rather large amount of redundant components to give acceptable reliability levels. This makes NAND multiplexing inefficient for the protection against permanent faults, normally compensated by reconfiguration techniques. However, this architectural configuration may be a system solution for ultra large integration of highly unreliable nanometer-scale devices affected by dominant transient errors.

[32] reviews the NAND multiplexing and reconfiguration fault tolerant techniques, and presents a defect and fault tolerant architecture in which multiplexing (low degree of redundancy) is combined with a massively reconfigurable architecture. The system performance of this architecture is evaluated by studying its reliability, and this shows that the suggested architecture can tolerate a device error rate of up to $10^{-2}$. The architectural configuration is seen to be efficiently robust against both permanent and transient faults for an ultra-large integration of highly unreliable nanodevices. In Section 6.2.0, the defect-tolerant capabilities of different reconfigurable architectures are discussed, and it is seen that the Cell Matrix architecture can also provide defect-tolerance for both permanent and transient faults. For the NAND multiplexing configuration, Equation 6.11 can be used to compute the reliability of the system if the faulty devices are

independent and uniformly distributed. This scenario may be reasonable when the dominant faults are transient in nature. However, this binomial distribution model is not sufficient to describe the manufacturing defects and permanent faults. The device components are not statistically independent but rather correlated since defects tend to cluster on a chip [35]. Thus, Equation 6.11 is not appropriate for computing reliability measures of the multiplexing system. [32] indicates that such manufacturing defects can be modeled with a continuous probability distribution function $f(r)$ where $r$ is the component reliability. Thus, the new reliability evaluation formula is:

$$R(k) = \int_0^1 \binom{N}{k} z^k (1-z)^{N-k} f(r) dr \tag{6.15}$$

The success of the approach depends on finding appropriate parameters for Equation 6.15, and [32] follows Stapper's beta distribution model [30]. Han and Jonker also discuss about the analytical methodology to compute reliability of reconfigurable architectures. The basic logic circuit blocks in the processor arrays on a reconfigurable chip are referred to as processing elements (PEs), and these are sometimes associated with local memories. In very large chips, reliability can be enhanced by adding spare PEs to the architectural configuration. Instead of trying to achieve complete fault tolerance, most techniques aim at optimizing probability of survival, defined as the percentage of defects that can be eliminated by the reconfiguration approach. Reconfiguration approaches are categorized as local or global [26]. Global approaches usually involve far more complex reconfiguration algorithms than local solutions. [32] assumes that all PEs (also called modules) are identical, so that any spare module can be substituted for a defective one, provided sufficient interconnection paths exist in the cluster. If in an array there are $r$ spares out of $n$ identical modules, then at least $n - r$ must be error-free for proper operation. The reliability of the array is given by $R_n = \sum_{m=n-r}^{n} R_{mn}$, where $R_{mn}$ is the probability of exactly $m$ out of $n$ modules being fault free. Assuming the modules have the same reliability measure $R_0$, and are statistically independent, the probability distribution of the number of defect free modules $m$ can be modeled as a binomial distribution:

$$R_{mn} = \binom{n}{m} R_0^m (1 - R_0)^{n-m} \tag{6.16}$$

Again, these defective modules in an array are not uniformly distributed but rather correlated, and Stapper's model is used to improve the reliability calculation of correlated modules [30]. The authors compute the reliability measures of a hierarchical approach that uses NAND multiplexing at the lowest level and then uses redundancy (spares) for reconfiguration at three additional implementation levels. The authors show that for an individual gate failure probability of

$10^{-2}$ and with $10^{12}$ devices on the chip, this architectural approach achieves a chip-level reliability of 99.8% with a redundancy factor of less than 10.

[42] also uses theoretical models to compare the relative reliability measures of R-fold modular redundancy, NAND multiplexing and reconfiguration for different device failure rates. It is shown that for a chip with $10^{12}$ devices, and with a reliability level that specifies that the chip must work with 90% probability, RMR is the least effective followed by NAND multiplexing and reconfiguration providing the best computational reliability. Table 1.1 in Chapter 2 provides these comparison data in details. It is also indicated in [42] that for individual device failure rates of 0.01 to 0.1, the redundancy factors are very large ($10^3$ to $10^5$). The authors also derive the failure rate of a chip with $N$ devices [27] for a RMR defect-tolerant architectural configuration (explained in Subsection 6.2.0) and this can be expressed as:

$$P_{fail}^{chip} = \frac{N}{RN_c + mB}[C(N_c p_f)^{(R+1)/2} + mBp_f] \tag{6.17}$$

In equation 6.17, $C = \begin{pmatrix} R \\ \frac{R-1}{2} \end{pmatrix}$ is the binomial factor, $p_f$ is the probability of an individual device failing, $N_c$ is the total number of devices in one of the $R$ units working in parallel, $N = RN_c$ is the total number of devices and imperfect majority gates have $B$ outputs and $mB$ devices. The probability that an i-th order RMR configuration (cascaded redundancy) works is as follows:

$$P_w^{(i)} = (1 - p_{fail})^{mB}[P_w^{(i-1)^3} + 3P_w^{(i-1)^2}(1 - P_w^{(i-1)})] \tag{6.18}$$

where the majority gate contains $mB$ imperfect devices and $P_w = 1 - P_{fail}$. [42] also show that CTMR is not advantageous when the redundant units have small number of devices and the majority logic is also made from the same devices as the units. For complex Boolean networks, most of these analytical methodologies do not scale well, and are error-prone and cumbersome. These limitations motivate the need for automating reliability-redundancy trade-off analysis methodologies.

## 6.4 NANOLAB: A MATLAB Based Tool

In this section, we discuss how information theoretic entropy coupled with thermal entropy can be used as a metric for reliability evaluation [7]. We also present the automation methodology for the NANOLAB tool with a detailed example and code snippet.

### Reliability, Entropy and Model of Computation

[2] not only provides a different non-discrete model of computation, in fact, it relates information theoretic entropy and thermal entropy of computation

in a way so as to connect reliability to entropy. It has been shown that the thermodynamic limit of computation is $KT \ln 2$ [3]. What the thermodynamic limit of computation means is that the minimum entropy loss due to irreversible computation amounts to thermal energy that is proportional to this value. If we consider energy levels close to these thermal limits, the reliability of computation is likely to be affected, and if we can keep our systems far from the temperature values that might bring the systems close to this amount of entropy loss, the reliability is likely to improve. The model of computation in [2] considers thermal perturbations, discrete errors and continuous signal noise as sources of errors. The idea is to use a Gibbs distribution based technique to characterize the logic computations by Boolean gates and represent logic networks as MRFs, and maximize probability of being in valid energy configurations at the outputs.

## Automation Methodology

NANOLAB consists of a library of functions implemented in MATLAB [1]. The library consists of functions based on the probabilistic non discrete model of computation discussed earlier (details in [2]), and can handle discrete energy distributions at the inputs and interconnects of any specified architectural configuration. We have also developed libraries that can compute energy distribution at the outputs given continuous distributions at the inputs and interconnects, introducing the notion of signal noise. Therefore, this tool supports the modeling of both discrete and continuous energy distributions.

The library functions work for any generic one, two and three input logic gates and can be extended to handle n-input logic gates. The inputs of these gates are assumed to be independent of each other. These functions are also parameterized and take in as inputs the logic compatibility function (Table 6.1) and the initial energy distribution for the inputs of a gate. If the input distribution is discrete, the energy distribution at the output of a gate is computed according to Equation 6.7, by marginalizing over the set of possible values of the nodes that belong to the same clique. In the case of generic gates these nodes are their inputs. These probabilities are returned as vectors by these functions and indicate the probability of the output node being at different energy levels between 0 and 1. These probabilities are also calculated over different values of $KT$ so as to analyze thermal effects on the node. The Belief Propagation algorithm [36] is used to propagate these probability values to the next node of the network to perform the next marginalization process. The tool can also calculate entropy values at different nodes of the logic network. It also verifies that for each logical component of a Boolean network, the valid states have an energy level less than the invalid states as shown theoretically in [2].

Our tool also consists of a library of functions that can model noise either as an uniform or gaussian distributions or combinations of these, depending on the user specifications. The probability of the energy levels at the output of a gate is calculated by the similar marginalizing technique but using Equation 6.8 or 6.9 or both depending on the characterization of signal noise. Similar to the library functions discussed previously, the signal library returns output energy distributions as vectors but the energy levels are between -1 and 1 due to rescaling of the logic level for reasons discussed in Section 6.2. Entropy values at the primary outputs of Boolean networks are also computed for different thermal energy levels. Arbitrary Boolean networks in any redundancy based defect-tolerant architectural configuration can be analyzed by writing simple MATLAB scripts that use the NANOLAB library functions. Also, generic fault tolerant architectures like TMR, CTMR are being converted into library functions such that these can be utilized in larger Boolean networks where more than one kind of defect-tolerant scheme may be used for higher reliability of computation.

## Detailed Example

We now discuss a detailed example to indicate the power of our methodology. Figure 6.2 shows a CTMR configuration with three TMR blocks working in parallel with a majority gate logic. The code listing shown in Figure 6.7 is a MATLAB script that uses NANOLAB functions and Belief Propagation algorithm to evaluate the probability of the energy configurations at the output of the CTMR.

The probability distributions for $x_1, y_1, x_2, y_2, x_3$ and $y_3$ for the NAND gates in Figure 6.2 are specified as vectors. These vectors specify the probability of the inputs being a logic low or high (discrete). The input probability distributions for all the TMR blocks are the same in this case but these can be varied by having separate input vectors for each TMR block.

| z=0.0 | z=0.2 | z=0.5 | z=0.8 | z=1.0 |
|---|---|---|---|---|
| 0.809 | 0.109 | 0.006 | 0.025 | 0.190 |
| 0.798 | 0.365 | 0.132 | 0.116 | 0.201 |
| 0.736 | 0.512 | 0.324 | 0.256 | 0.263 |
| 0.643 | 0.547 | 0.443 | 0.379 | 0.356 |

*Table 6.2.* Probability of the output $z$ of a logic gate being at different energy levels for values of $KT \in \{0.1, 0.25, 0.5, 1.0\}$

```
no_of_blocks = 3; % number of TMR blocks
prob_input1 = [0.1 0.9]; % prob distbn of input1 of NAND gate
prob_input2 = [0.1 0.9]; % prob distbn of input2 of NAND gate
BT_Values = [0.10.25 0.5 1.0];% different kbT values

for TMR_block = 1:no_of_blocks
   counter = 1;
   % energy_2_input_gates_function is a NANOLAB function and takes in as parameters
   % the logic compatibility function, input prob distbns and the kbT values.
   % The output gives the state configurations of the primary output of the logic
   % gate at different kbT values.

   prob1 = energy_2_input_gates_function(input1,prob_input1,prob_input2,BT_Values);
   prob2 = energy_2_input_gates_function(input1,prob_input1,prob_input2,BT_Values);
   prob3 = energy_2_input_gates_function(input1,prob_input1,prob_input2,BT_Values);
   [a,b] = size(prob1);

   % req_pb1, req_pb2 , req_pb3 are vectors which contain probabilities
   % of the output being a 0 or a 1 for a particular kbT value for Belief
   % Propagation.
   for i = 1:a
        req_pb1 = [prob1(i,1) prob1(i,b)];
        req_pb2 = [prob2(i,1) prob2(i,b)];
        req_pb3 = [prob3(i,1) prob3(i,b)];
        BT = BT_Values(i);

        % energy_3_input_gates_function is a part of NANOLAB and takes in input
        % and output parameters similar to the previous 2 input gates function.
        t_p = energy_3_input_gates_function(input2,req_pb1,req_pb2,req_pb3,BT_Values(i));
        prob(TMR_block,counter) = t_p(1,1);
        counter = counter + 1;
        prob(TMR_block,counter) = t_p(1,b);
        counter = counter + 1;
   end
end
```

*Figure 6.7.* Script for 1st order CTMR with discrete input distribution

The NANOLAB functions return vectors similar to the one shown in Table 6.2. These indicate the probability of the output of a logic network being at specified energy levels for different $KT$ values. In the CTMR configuration, *for each TMR block*, the energy configurations at the outputs of each of the three NAND gates are obtained from the function for a two input gate. Then these probabilities are used as discrete input probability distributions to the function for a three input gate. This computes the energy distribution at the output of the majority logic gate. Similarly, the probabilities of the final output of the CTMR is calculated. It can be seen that our methodology provides a very convenient way of evaluating this CTMR configuration and only requires minor modifications to be extended to any i-th order CTMR. Additionally, it is desired that valid input/output states should have lower clique energies than invalid states [2]. We have checked for the conformance to this property for the different clique energy functions which are generated by NANOLAB.

Similarly, MATLAB scripts can be written to model signal noise, and evaluate energy distribution and entropy at the output of the CTMR shown in Figure 6.2. The NANOLAB library functions can be used to interject unform or gaussian noise or combinations of these. Our tool provides a wide range of modeling options to system architects, and expedites reliability analysis of defect tolerant architectural configurations for combinational logic circuits.

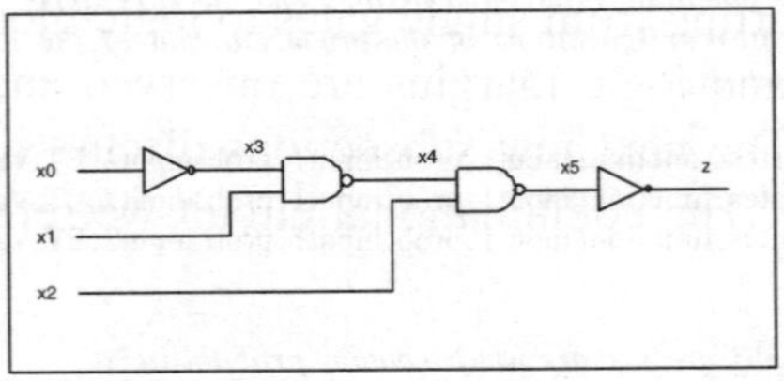


*Figure 6.8.* A Boolean network having the logic function z = $x_2 \wedge (x_0 \vee x_1\prime)$

## Shortcomings of NANOLAB

NANOLAB libraries can be used to inject random noise at the inputs and interconnects of logic circuits. Signal noise can be modeled as discrete and continuous distributions. We have experimented with different such combinations at the primary inputs of a Boolean network as well as interjected such noise distribution at the interconnects of a logic circuit. While trying to specify two noise spikes at the inputs of a NAND gate as independent gaussian distributions, the expression for the energy distribution at the output of the NAND gate turns out to be as follows [11]:

$$p(output) = \frac{1}{Z}\int_{-1}^{1}\int_{-1}^{1} e^{-\frac{U}{k_bT}} \frac{e^{-(x_0-\mu_0)^2/2\sigma_0^2}}{\sqrt{2\pi}\sigma_0} \frac{e^{-(x_1-\mu_1)^2/2\sigma_1^2}}{\sqrt{2\pi}\sigma_1} dx_0 dx_1 \tag{6.19}$$

The two gaussian processes have mean $\mu_0$ and $\mu_1$ which may or may not be equal, and variance $\sigma_0$ and $\sigma_1$. Equation 6.19 doesnot evaluate to an explicit integral and probability values for the output being at different energy levels cannot be computed. To tackle this scenario, we are developing a procedure in MATLAB to approximate the multiplication of two gaussian distributions. Note that specifying noise as a bivariate gaussian leads to an explicit integral as the probability density function of a bivariate gaussian leads to a less complex integrand. Also, the accuracy of the probabilities computed by our libraries and the Belief Propagation algorithm [36] are limited by the floating point accuracy of MATLAB.

## 6.5 Reliability Analysis of Boolean Networks with NANOLAB

We analyze different defect-tolerant architectural configurations of arbitrary Boolean networks starting from single gates to logic blocks with NANOLAB. The entropy values and logic margins are observed and these determine interesting facts [5]. The next few subsections discuss in details the different experimental results. The combinational circuit we refer to is shown in Figure 6.8.

### Reliability and Entropy Measures of a NAND Gate

Figure 6.9 shows the entropy curves at the outputs of a single NAND gate and different orders of a NAND CTMR configuration at different $KT$ values. The output probability distribution of a NAND gate is asymmetrical. This should be expected since only one input combination produces a logic low at the output. Figure 6.9 (a) indicates that the entropy is lower when the inputs of the single NAND gate are uniformly distributed. At higher $KT$ values, in both the uniform and non-uniform probability distribution cases, the entropy increases, resulting in the logic margins (probability of being in valid energy configurations) being reduced. This indicates that the logic margins in both scenarios of distribution almost approach zero as thermal energy levels increase. When thermal energy becomes equal to logic energy ($KT = 1$), the entropy values in the case of the uniform distribution is lower implying that the logic margins are better than when the inputs have a higher probability of being logic high.

Figure 6.9 (b) shows the entropy curves for different orders of a NAND CTMR configuration at different $KT$ values. The inputs in this case are uniformly distributed. It can be observed that the 0th order CTMR (TMR) has higher entropy value i.e. less logic margin than the 1st order CTMR at lower thermal energy levels and in both cases the entropy values almost converge at $KT = 0.5$. At this point (at close proximities of the thermodynamic limit of computation) logic margins at the output of the CTMR configurations become insignificant and there is total disorder, and computation becomes unpredictable. The interesting plot is for the 2nd order CTMR. The entropy shoots up at a $KT$ value of 0.25 indicating that the 2nd order CTMR degrades the reliability of computation. This aptly shows that every architectural configuration has reliability/redundancy trade-off points and even if redundancy levels are increased beyond these, the reliability for the architecture either remains the same or worsens.

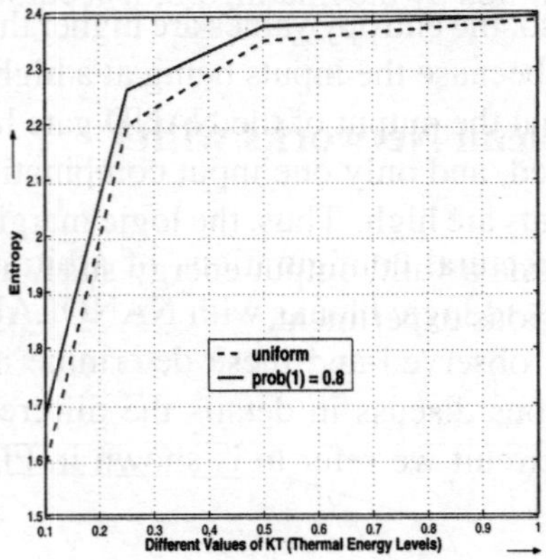


(a) Entropy of a NAND gate for different input distributions

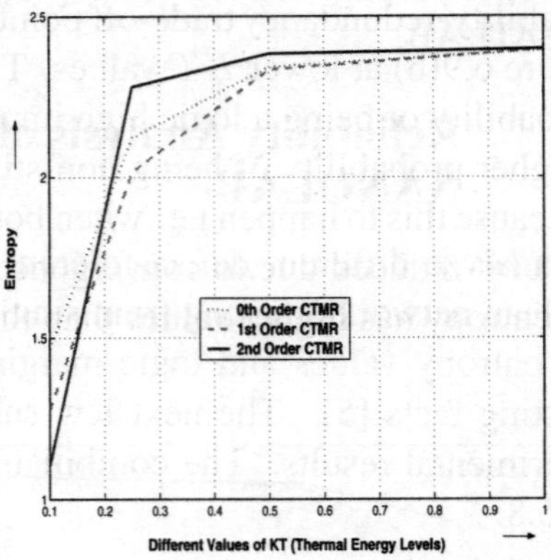


(b) Entropy of a CTMR NAND configuration when inputs are uniformly distributed

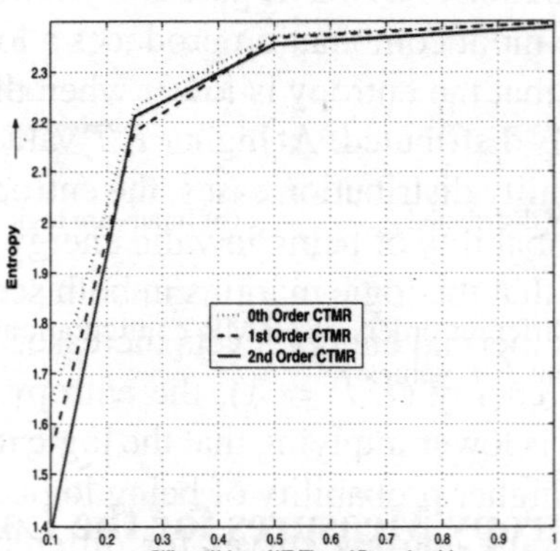


(c) Entropy of a CTMR NAND configuration when inputs are logic high with 0.8 probability

*Figure 6.9.* Entropies for the output of a Single NAND gate and CTMR configurations of the NAND gate at different $KT$ values

Figure 6.9 (c) indicates the entropy values when the inputs are non-uniformly distributed with a probability of 0.8 to be logic high. In this case, it is observed that the 2nd order CTMR degrades reliability of the system, but does better than the 0th order CTMR configuration. This result demonstrates that varying energy distributions at the inputs of a Boolean network may result in different reliability-redundancy trade-off points. Also, the entropy values are higher than Figure 6.9(b) at lower $KT$ values. This is because the inputs being at a higher probability of being a logic high implies that the output of the NAND gate has a higher probability of being non-stimulated, and only one input combination can cause this to happen i.e. when both inputs are high. Thus, the logic margins are a bit reduced due to convergence towards a valid output energy state, and the entropy has higher values than the previous experiment.

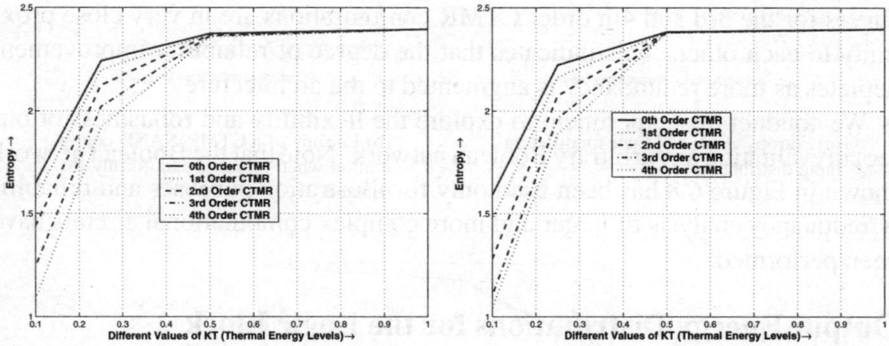


(a) Inputs are uniformly distributed (b) Inputs are logic high with 0.8 probability

*Figure 6.10.* Entropy for different orders of CTMR configurations for the Boolean network given in Figure 6.8 at different $KT$ values

# Reliability and Entropy Measures for the Logic Block

Figure 6.10 shows the entropy curves at the outputs of the CTMR configurations for the logic block. Figure 6.10 (a) shows the entropy when the input distribution is uniform. It can be observed that as the CTMR orders are increased, the entropy decreases (logic margin increases) at lower $KT$ values, and the entropy converges at $KT = 0.5$ for all the CTMR orders. This indicates that as the system approaches the thermal limit of computation ($KT \ln 2$) [3], increase in redundancy level does not improve the logic margins. The reliability measures remain the same. But at lower thermal energy levels, increasing the redundancy level (orders of the CTMR) results in improvement of reliability. This is because the 4th order CTMR has entropy values less than the other lower orders at thermal energy levels below 0.5. Whereas, in the NAND CTMR configuration, the 2nd order CTMR has higher entropy at lower $KT$ values indicating degradation in reliability of computation. This experiment illustrates that each specific Boolean network has an *unique* reliability-redundancy trade-off point.

It may also be inferred that the redundancy level is proportional to the device density of a logic network.

Figure 6.10 (b) indicates the entropy values when the inputs have a non-uniform probability distribution i.e the probability of being logic high is 0.8. The same facts are observed in this case as in Figure 6.10 (a). But, the entropy is higher than the previous experiment at lower thermal energy levels. This is because the inputs being at a higher probability of being logic high means that the output of the logic block will be stimulated and only a few input combinations can cause this to happen. Thus, the logic margins are slightly reduced as compared to when the inputs are uniformly distributed. Also, the entropy curves for the 3rd and 4th order CTMR configurations are in very close proximity to each other. This indicates that the degree of reliability improvement depletes as more redundancy is augmented to the architecture.

We conduct this experiment to explore the flexibility and robustness of our tool in evaluating any arbitrary Boolean network. Note that the Boolean network shown in Figure 6.8 has been used only for illustration purposes and reliability/redunancy analysis of larger and more complex combinational circuits have been performed.

## Output Energy Distributions for the Logic Block

NANOLAB has another perspective to it: the probability of being in different energy configurations at the primary outputs of a Boolean network can also be computed. Reliability measures of logic circuits can also be analyzed from these probability distributions. Figure 6.11 shows the energy distributions at the outputs of the different CTMR orders applied to the logic circuit when the inputs have uniform energy distribution. Note that the probability values are based on bin sizes of 0.1. As the order of the CTMR is increased, it can be seen that the logic margins for the output (*z*) at $KT$ values of 0.1, 0.25 and 0.5 keep on increasing. Due to the asymmetrical nature of the logic network, the probability of *z* (*p(z)*) being at energy level zero is almost always higher than being at one and hence such plots are obtained. It is also observed that at a $KT$ value of one, the logic margin for any order of the CTMR configuration becomes considerably small (output energy distribution becomes almost uniform), and remains the same even with an increase of redundancy resulting in unreliable computation. Comparing the different orders of CTMR in Figure 6.11, we infer that for lower thermal energy levels, the probability of being in a valid energy configuration increases as more redundancy is added to the architecture. But this increase in probability slows down as higher orders of CTMR are reached. This can be understood as follows: the logic margin of the system reaches a saturation point after which reliability can no longer be improved. The experimental results for single NAND gate CTMR configurations show similar plots with the difference

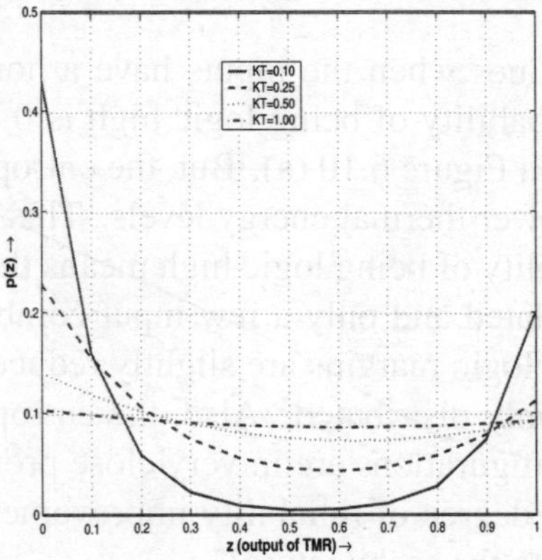


(a) The output distributions of a TMR configuration

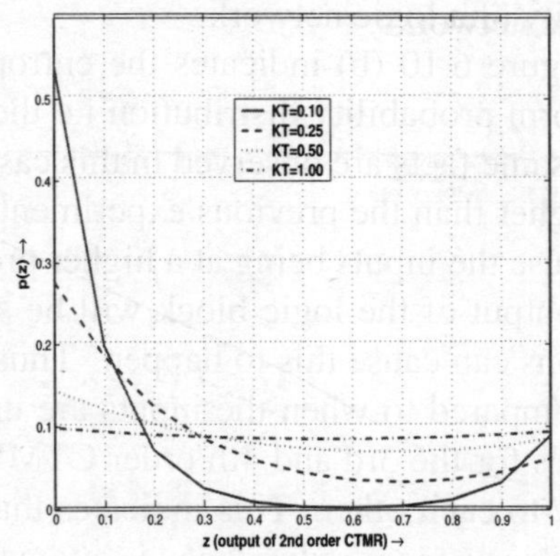


(b) The output distributions of a 2nd Order CTMR configuration

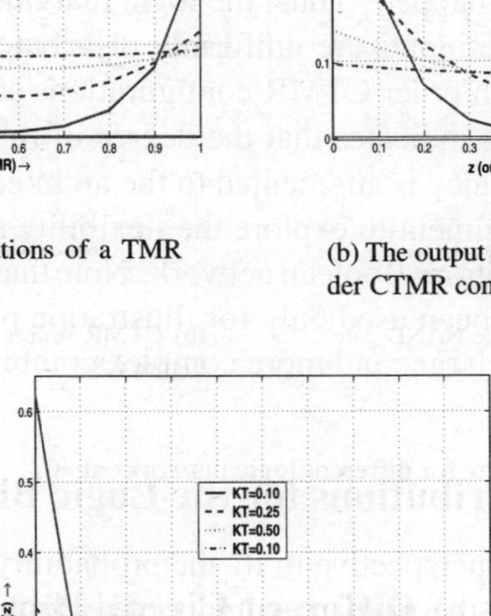


(c) The output distributions of a 4th Order CTMR configuration

*Figure 6.11.* Energy distributions at the output of the Boolean network for different orders of CTMR configuration at different $KT$ values. Inputs are uniformly distributed

that the output of the 2nd order CTMR configuration is non-stimulated (valid state) with a higher probability at $KT = 0.1$. This can be attributed to the intuitive fact that if a unit being duplicated has a lesser number of devices, a stable logic margin is reached with lower redundancy levels than a unit which has higher number of components. Such observations give a clear notion of the optimal redundancy points of different architectural configurations for specific logic networks.

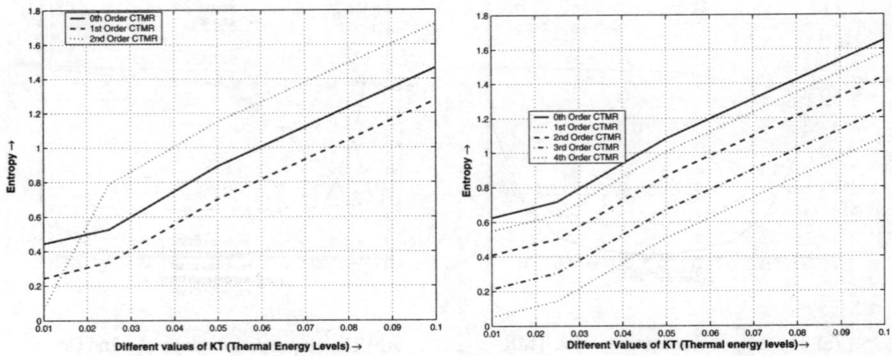


(a) CTMR orders of single NAND gate

(b) CTMR orders of logic circuit shown in Figure 6.8

*Figure 6.12.* Entropy for different logic networks at $KT$ values $\in [0.01, 0.1]$

## Reliability Analysis of Different Circuits for Smaller KT values

Reliability analysis of real systems at thermal energy levels below 0.1 are of practical importance. Figure 6.12 presents computational entropy values for a single NAND gate and the Boolean network. These results are similar to the ones indicated by Figures 6.9 and 6.10. It is interesting to note that in this case, there is a linear increase in the entropy plots for both the circuits with increase in $KT$ values. The previous experimental results had a convergence of the entropy values at around $KT = 0.5$ and remained steady for all thermal energy levels beyond that point. Similar reliability/redundancy trade-off points are observed in both the logic networks. The 2nd order CTMR configuration for the NAND gate does worse than the lower orders, whereas increasing the redundancy level for the logic block yields better reliability.

## Reliability Analysis in the Presence of Noisy Inputs and Interconnects

We have also analyzed reliability of different logic networks in the presence of random noise at the inputs and interconnects. Entropy and energy distributions at the outputs are computed automatically and reliability/redundancy trade-off points are inferred. Logic margins in the presence of signal noise is the difference between the probabilities of occurrence of -1 (logic low) and 1 (logic high) due to rescaling the energy levels as discussed in Section 6.2. The Boolean network we consider here is the one given in Figure 6.8.

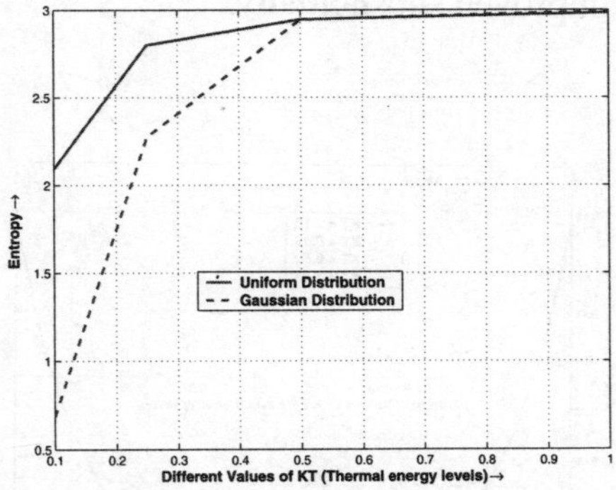


(a) Entropy at the output of a NAND gate

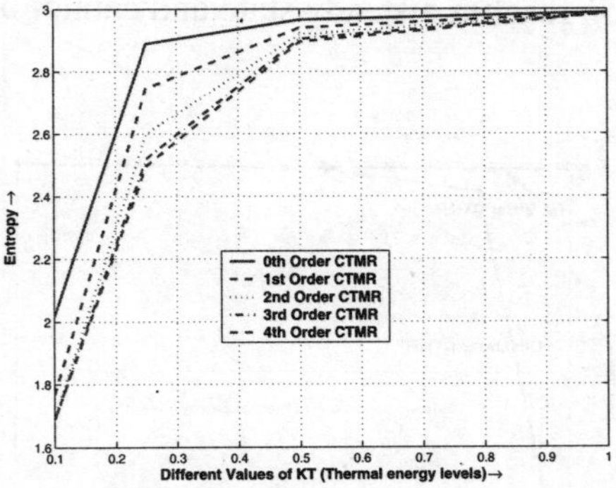


(b) Entropy for different CTMR orders of NAND gate

*Figure 6.13.* Entropy at the outputs of different NAND gate configurations in the presence of noisy inputs and interconnects

**Reliability and Entropy Measures of a NAND Gate:** Figure 6.13 shows the entropy values at the outputs of different NAND gate configurations in the presence of noisy inputs and interconnects. The plots in Figure 6.13 (a) are obtained when (i) one of the inputs of the gate has equal probability of being at logic low or high and the other is subjected to a gaussian noise spike centered around 1 with a variance of 2 and (ii) when the gaussian noise signal is changed to an uniformly distributed one ranging from $\{-1, 1\}$. The plots show that the entropy has lower values in the presence of gaussian noise as compared to uniformly distributed noise at lower $KT$ values. This is because a gaussian noise spike centered around logic high alleviates the probability of the output being in a valid energy state and decreases the degree of randomness the system is in (entropy). Normally gaussian noise spikes have mean at invalid energy levels, and thus prevents the system from converging to a valid logic state. Such noise signals have also been modeled for complex logic circuits.

The entropy curves for different CTMR configurations of a NAND gate are indicated in Figure 6.13 (b). The entropy values for the NAND gate are plotted till the 4th order CTMR for different thermal energy levels. The inputs to the logic block and the interconnects are subjected to both uniformly distributed and gaussian noise. We have assumed two gaussian noise spikes, one centered around 1 with variance of 2 and the other centered around 0.5 with variance of

0.2. It can be observed that as redundancy is increased by adding more CTMR orders, the entropy decreases (logic margin and hence reliability increases) at lower $KT$ values. However, the rate of improvement in reliability decreases as 3rd order CTMR is reached. This result emphasizes that beyond a certain redundancy level, the system's reliability for a given defect-tolerant configuration reaches a steady state and cannot be improved substantially.

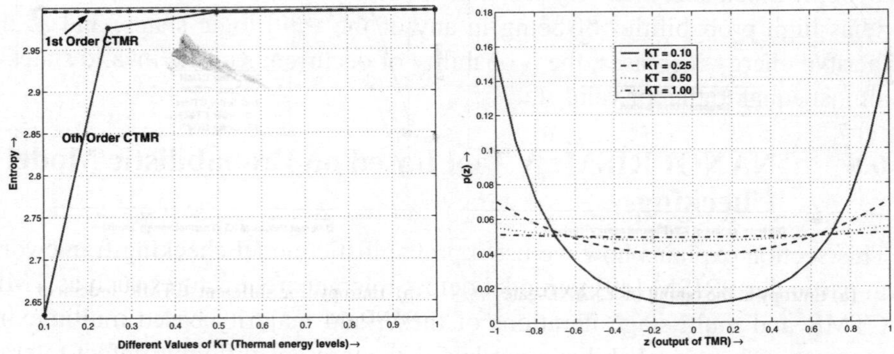


(a) Entropy for different CTMR orders of the logic block

(b) Output distribution for the TMR of the Boolean network

*Figure 6.14.* Entropy and Energy distribution at the outputs of different CTMR configurations of the logic block in the presence of noisy inputs and interconnects

**Entropy and Energy Distribution for the Boolean Network:** The entropy and energy distribution at the output of different CTMR configurations of the logic block are presented in Figure 6.14. The entropy values for the different CTMR orders of the Boolean network are given in Figure 6.14 (a). The experimental setup is as follows: the inputs and interconnects are subjected to both uniformly distributed and gaussian noise. Two gaussian noise signals are considered, with equal means at 0.5 energy level and variances 2 and 0.2 respectively. It can be inferred from the results that TMR (0th order CTMR) has low entropy values at $KT$ values less than 0.5. The higher CTMR orders have very high entropy irrespective of the thermal energy. We have only indicated entropy for the 1st order CTMR because higher orders do not improve the reliability. Considering this result, we can state that due to a very noisy and error-prone environment, the augmentation of any level of redundancy beyond the TMR configuration causes a steep degradation of reliability. Comparing these results with Figure 6.13(b), we surmise that reliability/redundancy trade-off points are not only dependent on specific logic networks, these vary with the normal operation scenarios (such as random noise) of the circuits.

The energy distribution at the output of the TMR configuration for the circuit at different thermal energy levels is presented in Figure 6.14 (b). As discussed earlier, these probability values can also indicate reliability measures. Noise characterizations at the inputs and interconnects are similar to the previous experiment. Due to the asymmetrical nature of the logic network, the probability of $z$ being at energy level zero is almost always higher than being at one when the inputs have equal probability of being logic low or high. But due to the uncertainty interjected in the form of noise, the logic margins at the output of the logic block becomes very sensitive to thermal fluctuations. At $KT = 0.25$, $z$ has high probabilities of being in any of the valid logic states, and as the thermal energy increases, the probability of occurrence of the invalid states is almost equal to that of valid states.

## 6.6 NANOPRISM: A Tool Based on Probabilistic Model Checking

This section explains how we use a probabilistic model checking framework, in particular PRISM [49] to model defect-tolerant architectures such as TMR, CTMR and multi-stage iterations of these, and majority based multiplexing systems. These probabilistic models that represent different defect-tolerant architectural configurations are integrated to form a library that composes NANOPRISM [8, 10].

### Modeling Single Gate TMR, CTMR

In this subsection we explain the PRISM model construction of a single gate TMR, CTMR and multistage iterations of these. The first approach is directly modeling the systems as given in Figures 6.1 and 6.2. For each redundant unit and the majority voting logic, construct separate PRISM modules and combine these modules through synchronous parallel composition. However, such an approach leads to the well know state space explosion problem. At the same time, we observed that the actual values of the inputs and outputs of each logic block is not important, instead one needs to keep track of only the total number of stimulated (and non-stimulated) inputs and outputs. Furthermore, to compute these values, without having to store all the outputs of the units, we replace the set of replicated units working in *parallel* with ones working in *sequence*. This folds space into time, or in other words reuse the same logic unit over time rather than making redundancy over space. This approach does not influence the performance of the system since each unit works independently and the probability of each gate failing is also independent.

The different orders of CTMR configurations are built incrementally from the models of the previous iterations. In this case too, two approaches seem to emerge. One of the approaches is incorporating PRISM modules of the previous CTMR iteration as the redundant functional units for the current order and adding a majority voting logic. This causes the model to grow exponentially as the higher orders of configuration are reached. The other approach is to use already calculated probability values (probability of being a logic low or high) at the output of the last CTMR configuration as the output probability distributions of the three redundant functional units of the current order of a CTMR configuration.

```
prob p_err = 0.1; //probability that gate has error
prob p_in=0.9; //probability an input is logic high

const R=3; //number of redundant processing units const
const R_limit=1;

module TMRNAND

    x : bool;
    y : bool;
    s : [0..3] init 0; // local state
    z : [0..R] init 0; // number of outputs that are stimulated
    z_output : [0..1] init 0; // output of majority logic
    c : [0..4] init 0;     // count of the redundant unit being processed

    [] s=0 & c>R-> (s'=0);

    // processed all redundant units
    [] s=0 & c=R-> (s'=3)&(c'=c+1);

    // initial choice of x and y
    [] s=0 & c<R-> p_in : (x'=1)&(s'=1)&('c=c+1) + (1-p_in) : (x'=0)&(s'=1)&('c=c+1);
    [] s=1-> p_in : (y'=1)&(s'=2) + (1-p_in) : (y'=0)&(s'=2);

    // NAND operation
    [] s=2-> p_err : (z'=z+(x&y))&(s'=0) + (1-p_err) : (z'=z+(!(x&y)))&(s'=0);

    // majority logic
    [] s=3 & z>=0 & z<=R_limit -> (s'=0) & (z_output'=0);
    [] s=3 & z>R_limit & z<=R-> (s'=0) & (z_output'=1);

endmodule
```

*Figure 6.15.* PRISM description of the TMR configuration of a single NAND gate

The DTMC model of the TMR configuration of a NAND gate is shown in Figure 6.15 for illustration purposes. We assume in this case that the inputs $X$ and $Y$ have identical probability distribution (probability of being logic high is 0.9), and the failure (inverted output) probability of NAND gates is 0.1. However, the input probability distributions and failure distribution of the NAND gates can be changed easily by modification of the constants given at the start of the description. The probabilistic state machine for this DTMC model built by PRISM has 115 states and 182 transitions. Also, model checking is performed to compute the probability distribution of the TMR configuration's output being in an invalid state for different gate failure probabilities. Furthermore, since PRISM can also represent non-deterministic behavior, one can set upper and lower bounds on the probability of gate failure and then obtain (best and worst case) reliability characteristics for the system under these bounds. As discussed before, the CTMR configuration uses three TMR logic units and majority voter. The probability distribution obtained for the TMR block of a single NAND gate can be used directly in the CTMR configuration, thus reducing the state space.

## Modeling Block Level TMR, CTMR

We explain the PRISM model construction for different orders of block level CTMR configurations. Each unit as shown in Figure 6.1 is no longer composed of a single gate, but contains a logic circuit such as the one shown in Figure 6.8. As discussed earlier, we replace the set of replicated units working in *parallel* with ones working in *sequence*, thus folding space into time. The DTMC model for the TMR configuration of the logic block in Figure 6.8 has 1407 states with this approach of modeling. The models for different orders of CTMR configurations are built using similar techniques that are used for single gate CTMR.

We have also incorporated different levels of granularity in these redundancy based architectural configurations. Let us walk through an example to explain this concept clearly. For the logic circuit under discussion, we model the different levels of CTMR wherein each redundant unit contains the logic block itself. We consider this redundancy at the logic block granular level. Next , for each redundant logic block, we further replicate each gate so as to form different orders of CTMR configurations at the gate level of granularity. Our experimental results indicate the intuitive fact that CTMR configurations at different levels of granularity and lower gate failure probabilities give better reliability than the conventional flat architectural configuration.

## Model Construction of Majority Multiplexing System

Our attention to the importance of majority gates in the nanotechnology context was drawn by [52, 51], where analytical results for similar trade-off evaluations were presented. However, since these evaluations involve complex combinatorial arguments, and conditional probability computations, analytical results often lead to mistakes. This motivates our work of automating the detailed reliability analysis of von Neumann multiplexing architecture for majority gates by building a generic multiplexing library for NANOPRISM.

In this subsection, we explain the PRISM model of a majority gate multiplexing configuration. The first approach is directly modeling the system as shown in Figure 6.4. A PRISM module is constructed for each multiplexing stage comprising $N$ majority gates and these modules are combined through synchronous parallel composition. However, such a model construction scheme leads to the well know state space explosion problem. Due to the observations stated earlier, a model construction technique similar to the TMR and CTMR architectural configurations is adopted. The set of $N$ majority gates working in *parallel* is replaced by $N$ majority gates working in *sequence*. The same methodology is applied to the multiplexing stages of the system so as to reuse the same module for each of the stages while keeping a record of the outputs from the previous stage.

The unit $U$ in Figure 6.4 performs random permutation. Consider the case when $k$ outputs from the previous stage are stimulated for some $0 < k < N$. Since there are $k$ stimulated outputs, the next stage will have $k$ of the inputs stimulated if $U$ performs random permutation. Therefore, the probability of either all or none of inputs being stimulated is 0. This implies that each of the majority gates in a stage are dependent on one other, for example, if one majority gate has a stimulated input, then the probability of another having the same input stimulated decreases. It is difficult to calculate the reliability of a system by means of analytical techniques for such a scenario. To change the number of restorative stages, bundle size, input probabilities or probability of the majority gates failing requires only modification of parameters given at the start of the model description. Since PRISM can also represent non-deterministic behavior, one can set upper and lower bounds on the probability of gate failure and then obtain best and worst case reliability characteristics for the system under these bounds.

Figure 6.16 shows the DTMC model of the von Neumann majority multiplexing system. We assume in this case that the inputs $X$ and $Y$ have identical probability distribution (probability of being logic high is 0.9), and the input $Z$ can be non-stimulated with the same probability value. The failure (inverted output) probability of the majority gates is 0.01. However, these parameters can be modified by changing the value of the constants (*const*) given at the start of the model . The probabilistic state machine built by PRISM for the multiplexing configuration shown in Figure 6.16 has 1227109 states and 1846887 transitions. This indicates the complexity of analyzing such architectural configurations. The models for the different multiplexing configurations are verified for PCTL [33] properties such as "probability of having less than 10% errors at the primary output".

## 6.7 Reliability Analysis of Logic Circuits with NANOPRISM

NANOPRISM has been applied for reliability analysis of different combinational circuits starting from single gates to logic blocks. The output error distributions for different granularity levels and failure distributions of the gates

```
const N = 20;  // number of inputs in each bundle
const M = 3;   // number of restorative stages equals (M-1)
const low_lim = 1;  // the higher limit for logic low at majority gate output
const high_lim = 3;  // the higher limit for logic high at majority gate

prob p_err = 0.01;  // probability that majority gate has von Neumann fault
prob p1_in=0.9;  // probability one of the inputs is logic high
prob p2_in=0.1;  // probability one of the inputs is logic high

module majority_multiplex

    u : [1..M]; //current stage
    c : [0..N]; // counter (number of copies of the majority gate done)
    s : [0..5]; // local state

    nx : [0..N]; // number of stimulated X inputs (outputs of previous stage)
    ny : [0..N]; // number of stimulated Y inputs (outputs of previous stage)
    nz : [0..N]; // number of stimulated Z inputs (outputs of previous stage}

    x : [0..1]; // value of first input
    y : [0..1]; // value of second input
    z : [0..1]; // value of third input
    k : [0..3]; // value of addendum of all inputs

    out : [0..N];  // number of stimulated outputs

    [] s=0 & c<N -> (s'=1); // next majority gate of current stage
    // move on to next stage
    [] s=0 & c=N & u<M -> (s'=1)&(nx'=out)&(ny'=out)&(nz'=out)&(u'=u+1)&(c'=0);
    // finished (reset variables)
    [] s=0 & c=N & u=M -> (s'=0)&(nx'=0)&(ny'=0)&(nz'=0)&(x'=0)&(y'=0)&(z'=0)&(k'=0);
    // choose x (initially random)
    [] s=1 & u=1 -> p1_in : (x'=1)&(s'=2) + (1-p1_in) : (x'=0)&(s'=2);
    // permute
    [] s=1 & u>1 ->nx/(N-c) : (x'=1)&(s'=2)&(nx'=nx-1) + (1-(nx/(N-c))) : (x'=0)&(s'=2);
    // choose y (initially random)
    [] s=2 & u=1 -> p1_in : (y'=1)&(s'=3) + (1-p1_in) : (y'=0)&(s'=3);
    // permute
    [] s=2 & u>1 ->ny/(N-c) : (y'=1)&(s'=3)&(ny'=ny-1) + (1-(ny/(N-c))) : (y'=0)&(s'=3);
    // choose z (initially random)
    [] s=3 & u=1 -> p2_in : (z'=1)&(s'=4) + (1-p2_in) : (z'=0)&(s'=4);
    // permute
    [] s=3 & u>1 ->nz/(N-c) : (z'=1)&(s'=4)&(nz'=nz-1) + (1-(nz/(N-c))) : (z'=0)&(s'=4);
    // add values for the inputs to check for majority
    [] s=4 -> (k'=x+y+z) & (s'=5);
    // decide majority logic low or high
    [] s=5 & k>=0 & k<=low_lim -> (1-p_err) : (out'=out+0)&(s'=0)&(c'=c+1)&(k'=0)
                                  + p_err : (out'=out+1)&(s'=0)&(c'=c+1)&(k'=0);
    [] s=5 & k>low_lim & k<=high_lim -> (1-p_err) : (out'=out+1)&(s'=0)&(c'=c+1)&(k'=0)
                                        + p_err : (out'=out+0)&(s'=0)&(c'=c+1)&(k'=0);
endmodule
```

*Figure 6.16.* PRISM description of the von Neumann majority multiplexing system

are observed, and these demonstrate certain anomalous and counter-intuitive facts. The next subsections discuss in details the different experimental results [8]. Figure 6.8 shows the Boolean network used for the illustration of the effectiveness of NANOPRISM in evaluating arbitrary logic circuits.

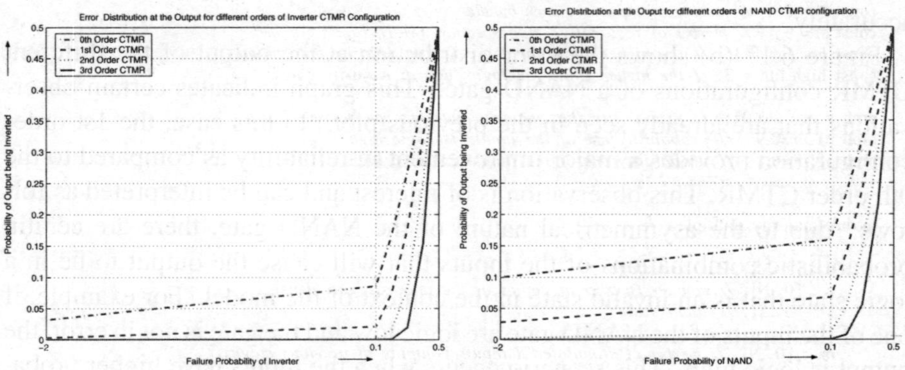


(a) Inverter CTMR Configurations

(b) NAND CTMR Configurations

*Figure 6.17.* Error Distribution at the Outputs for the different Orders of CTMR Configurations of Single Gates

## Reliability Measures of Single Gate CTMR Configurations

Figure 6.17 shows the error distribution at the output of CTMR configurations for the Inverter and NAND gates. The input distribution in both the cases is assumed to be logic high with 0.9 probability. The probability of the gates being in error $\in \{10^{-2}, 0.1, 0.2, ......, 0.5\}$. Figure 6.17 (a) shows the probability of the output being in error (inverted with respect to the expected output) for the inverter CTMR configurations upto order three. It is observed that as the order increases, the probability of having the output in error goes down. But at lower gate error probabilities, it is clearly seen that the 2nd order and 3rd order configurations provide the same reliability measures. Thus increasing redundancy beyond this point does not make sense. At higher gate failure rates, there is a difference between the reliability obtained by the 2nd and 3rd order CTMR architectures. We extended the number of orders further and observed that at the 5th CTMR order when gate failure distributions are high, the difference in reliability measures obtained from this iteration and the previous one reduces but never becomes equivalent. This is intuitive because at higher device failure rates, introducing more redundancy causes more error prone computation and at same point the reliability can not be improved anymore and may sometimes degrade as we will observe shortly. Also, for each CTMR configuration, at lower gate failure distributions a plateau is reached i.e. lowering the device failure rate does not lower the probability of an erroneous output. This means that a redundancy based architecture cannot provide a reliability level lower than a certain "point", and NANOPRISM can analyze such reliability measures accurately.

Figure 6.17 (b) shows the error distribution at the output of the different CTMR configurations of a NAND gate. This graph indicates certain observations that are already seen in the previous plot. In this case, the 1st order configuration provides a major improvement in reliability as compared to the 0th order CTMR. This observation is of interest and can be interpreted as follows: due to the asymmetrical nature of the NAND gate, there are ceratin probabilistic combinations of the inputs that will cause the output to be in a logic state that is an invalid state in the context of the model. For example, if one of the inputs of the NAND gate are logic low and the gate is not in error, the output is logic high. This scenario occurs when the inputs have higher probability of being 1 and a logic high is expected at the output if there is an error thus mildly elevating the probability of errors at the output.Also, the 2nd and 3rd order CTMR configurations provide almost the same reliability measures for lower gate failure rates.

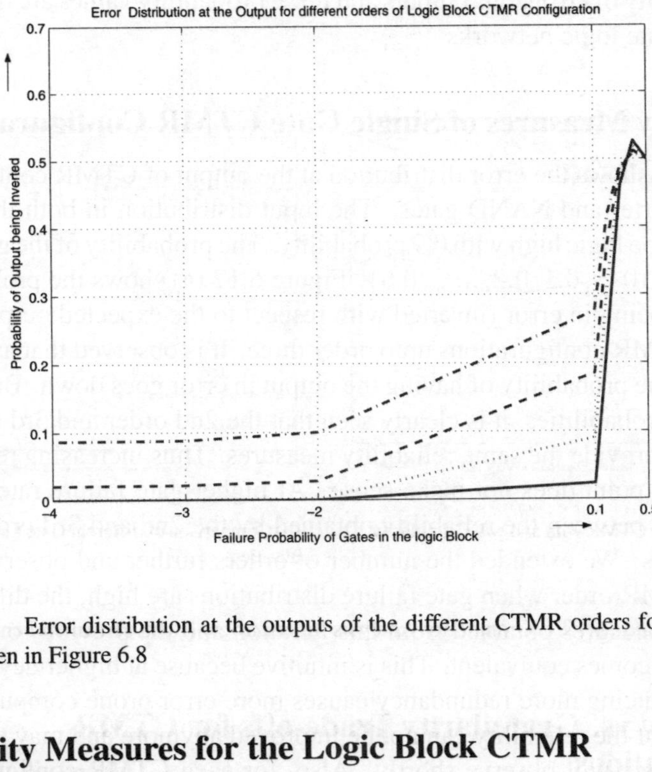


*Figure 6.18.* Error distribution at the outputs of the different CTMR orders for the Boolean network given in Figure 6.8

## Reliability Measures for the Logic Block CTMR Configurations

We also analyze the combinational circuit given in Figure 6.8. This is to illustrate that our tool can be used to compute reliability measures of any arbitrary logic circuit. Figure 6.18 shows the output error distributions for different orders of CTMR configurations for the aforesaid logic network. The probability of the inputs being stimulated (logic high) is assumed to be 0.9. Also, the component inverter and NAND gates are assumed to have same failure distribution. These failure probability values $\in \{10^{-4}, 10^{-3}, 10^{-2}, 0.1, 0.2, ......, 0.5\}$. The plots observed in Figure 6.18 are similar to the previous experiment. It can be seen that as the CTMR orders increase, the probability of having the output in error goes down. Also, at lower gate error probabilities, it is clearly seen that the 2nd and 3rd order configurations provide the same reliability measures. Interestingly, in this case, at higher gate failure rates, the rate of improvement in the reliability of the system slows down steadily as the redundancy (in terms of CTMR orders) is increased. This is due to the augmentation of unreliable devices to the architecture. Note that this degree of slow down is higher than what is observed in the case of single gate logic networks. This important observation can be interpreted as follows: for Boolean networks with higher number of unreliable component gates, increasing redundancy factor increases the probability of erroneous outputs and these probability values are higher than for single gate logic networks.

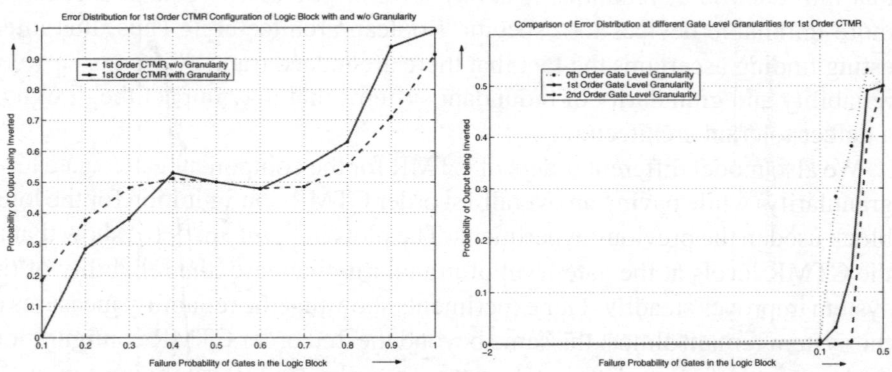


(a) 1st Order CTMR with and w/o Granularity

(b) Comparison of different Gate Level Granularities

*Figure 6.19.* Granularity Analysis for the Logic Block shown earlier

## Reliability vs. Granularity Trade-offs for CTMR Configurations

We discuss the impact of applying reliability at different granularity levels for specific Boolean networks. We determine interesting facts and anomalies while experimenting with this form of redundancy based architectures. Figure 6.19 shows error distributions at the outputs of 1st order CTMR configurations with and without granularity. For illustration purposes, the granularity considered here is at the gate level but our generic libraries can handle granularity based redundancy at logic block level, logic function level, unit level etc. In Figure 6.19 (a), we compare the reliability measures of two different defect-tolerant architectural configurations of the logic circuit shown in Figure 6.8. One of these is a 1st order CTMR configuration with redundancy at the logic block level (only the logic circuit is duplicated), whereas the other has gate level and logic block level redundancy and we call this a granularity based architectural configuration. In this experiment, the logic circuit is configured as a 1st order CTMR and the component gates are individually configured as TMR configurations. It is observed that at lower device failure rates, the reliability of the system for the granularity based architecture is better. This is intuitive because introduction of more redundant devices causes better reliability. But counter-intuitive phenomena is observed at higher gate failure rate distributions. At the 0.4 failure probability mark, the probability of an erroneous output for the granular architecture becomes more than the one without granularity. This shows that for the CTMR configuration with gate level granularity, there is a degradation of the reliability of the system. The reason for this anomaly is that introduction of redundancy at any level of granularity entails addition of more unreliable devices to a specific Boolean architecture. Thus, this interesting finding ascertains the fact that there are indeed trade-off points between reliability and granularity of redundant systems that may impact the design of a defect-tolerant architecture.

We also model different orders of CTMR for the component gates (gate level granularity) while having an overall 1st order CTMR configuration for the logic block used in the previous experiment. The plots in Figure 6.19 (b) show that as the CTMR levels at the gate level of granularity increase, the reliability of the system improves steadily. Our experiments show that the reliability measures of the system remain almost the same beyond the 2nd order CTMR configuration. At lower gate failure distributions, the 1st order CTMR configuration at the gate level can be considered to be the optimal redundancy level of the overall architecture as increasing the level of redundancy further at the gate level does not yield much improvement in the overall reliability.

## 6.8 Reliability Evaluation of Multiplexing Based Majority Systems

In this section we study the reliability measures of multiplexing based majority systems [9] both when the I/O bundles are of size 10 and 20. These bundle sizes are only for illustration purposes and we have investigated the performance of these systems for larger bundle sizes. In all the experiments reported in this paper, we assume that the inputs $X$, $Y$ and $Z$ are identical (this is often true in circuits containing similar devices) and that two of the inputs have high probability of being logic high (0.9) while the third input has a 0.9 probability of being a logic low. Thus the circuit's correct output should be stimulated. Also, it is assumed that the gate failure is a von Neumann fault, i.e. when a gate fails, the value of its output is inverted.

The PRISM model in Figure 6.16 is verified for properties such as the probability of having $k$ non-stimulated outputs which in terms of Figure 6.16 corresponds to the probability of reaching the state where out $= N - k$, u = M and c = N, for $k = 0, \ldots, N$ where $N$ is the I/O bundle size. This is the computation of the error distribution at the output of the majority gate architectural configuration. Hence any measure of reliability can be calculated from these results. PRISM can be also be used to directly compute other reliability measures such as, the probability of errors being less than than 10% and the expected number of incorrect outputs of the system. Our analysis of reliability for the majority multiplexing system using NANOPRISM concentrates on the effects of the failure probabilities of the majority gates and the number of restorative stages. The results we present show:

- the error distribution at the output for different gate failure probabilities (Figure 6.20).
- the error distribution at the output for different gate failure probabilities when additional restorative stages are augmented (Figure 6.21).
- reliability measures in terms of the probability that at the most 10% of the outputs are erroneous, as the probability of gates failure varies (Figure 6.22).
- reliability almost reaches steady state for small gate failure probabilities and can be improved marginally once a certain number of restorative stages have been augmented to the architecture. (Figure 6.22).
- the maximum probability of gate failure allowed for the system to function reliably by comparing the probability that at most 10% of the outputs are incorrect and the expected percentage of incorrect outputs for different numbers of restorative stages (Figures 6.23 and 6.24).

## Error Distribution at the Outputs of Different Configurations

We consider a majority multiplexing system as shown in Figure 6.4, where the I/O bundle size equals 10 and 20. We first investigate the effect of varying the failure probabilities of the majority gates on the reliability of the system. Figure 6.20 shows the error distribution at the output of the system in the cases

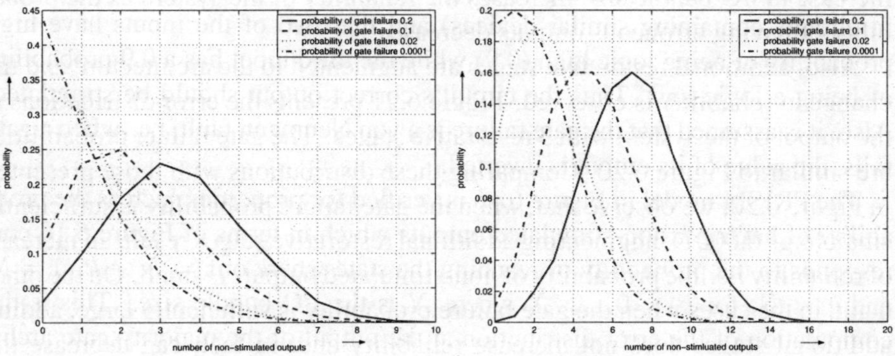


(a) I/O bundle size equals 10 (b) I/O bundle size equals 20

*Figure 6.20.* Error distributions at the output with 1 restorative stage under different gate failure rates and I/O bundle sizes

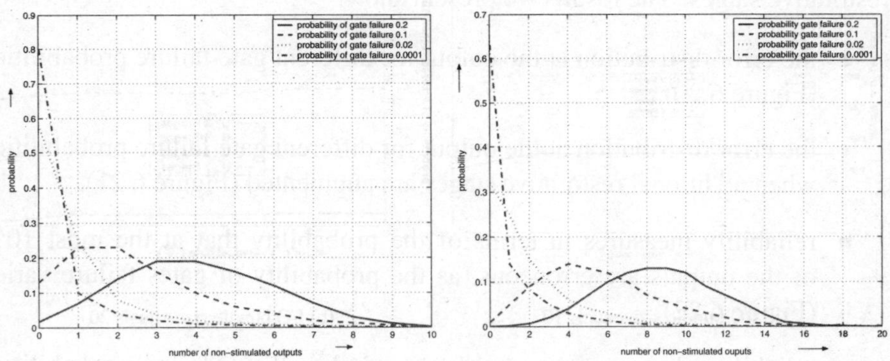


(a) I/O bundle size equals 10 (b) I/O bundle size equals 20

*Figure 6.21.* Error distributions at the output with 3 restorative stages under different gate failure rates and I/O bundle sizes

when the probability of gate failure equals 0.2, 0.1, 0.02 and 0.0001. Two of the inputs of the majority gate are stimulated when working correctly, and the correct output of the majority gate should be logic high. Hence, the more outputs that are non-stimulated, the lesser the reliability of the system.

As expected, the distributions given in Figure 6.20 show that, as the probability of a gate failure decreases, the reliability of the multiplexing system increases, i.e. the probability of the system returning incorrect results diminishes. If Figure 6.20 (a) and (b) are compared, it can be determined that the increase in I/O bundle size increases the reliability of the system as the probability of having erroneous outputs decreases.

Also, additional restorative stages are augmented to the architecture and the change in reliability is observed. Figure 6.21 presents the error distributions at the output of the system with 3 restorative stages. The gate failure probabilities are similar to Figure 6.20. Comparing these distributions with those presented in Figure 6.20, we observe that, when the gate failure probability is sufficiently small (e.g. 0.0001), augmenting additional restorative stages results in increase of reliability i.e. the probability of non-stimulated outputs is small. On the other hand, in the cases when the gate failure probability is sufficiently large, adding additional stages does not increase reliability and, in fact, can decrease the reliability of the system (compare the distributions when the failure probability equals 0.2 for each bundle size).

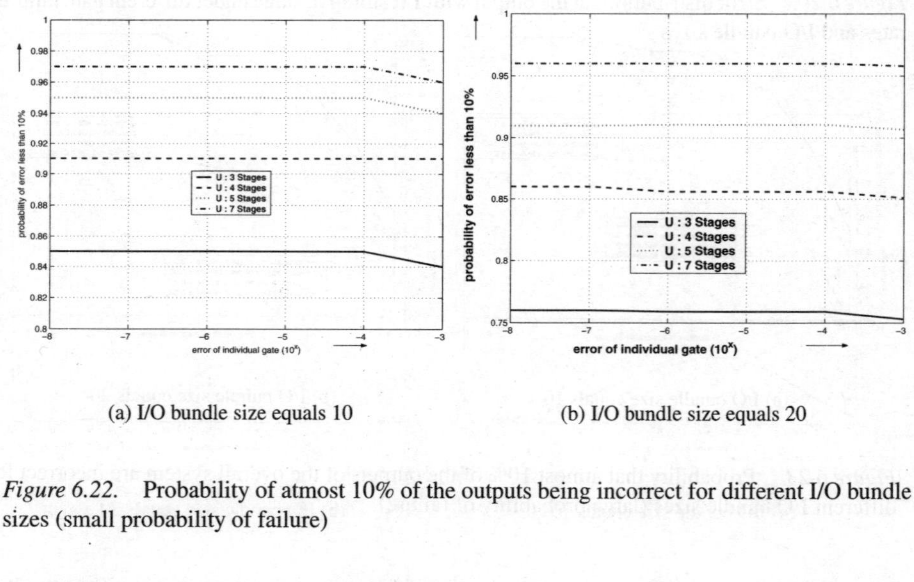


(a) I/O bundle size equals 10

(b) I/O bundle size equals 20

*Figure 6.22.* Probability of atmost 10% of the outputs being incorrect for different I/O bundle sizes (small probability of failure)

## Small Gate Failure Probabilities

In Figure 6.22, we have plotted the probability that less than 10% of the outputs are incorrect against small gate failure probabilities for multiplexing stages of 3, 4, 6 and 7. These plots show that for small gate failure probabilities, the reliability of the multiplexing system almost reaches a steady state. Comparing Figure 6.22(a) and (b), it can be inferred that increasing the bundle size results in improvement of the reliability of the system. We have also experimented with higher I/O bundle sizes such as 40 and 45, and the results from these multiplexing configurations show that the rate of increase in reliability of the system decreases as the bundle size increases.

Also, these results demonstrate that increasing the number of stages can greatly enhance the reliability of the system. However, the degree of improvement slows down as more restorative stages are added. Moreover, there exists an optimal redundancy level (in terms of number of restorative stages) beyond which the reliability improvement is marginal. For example, let us consider the plots presented in Figure 6.22 that indicate probability of erroneous outputs being less than 10% when the number of restorative stages equals 6 and 7. These show that the probability values increase marginally for different gate failure probabilities as compared to when the architecture has lower number of restorative stages. We should also mention that this result corresponds to the observation made in [31] that, as the number of stages increases, the output distribution of the system will eventually become stable and independent of the number of stages.

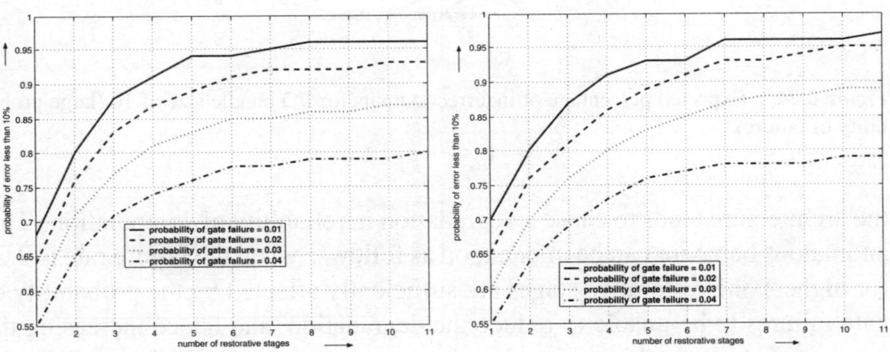


(a) I/O bundle size equals 10

(b) I/O bundle size equals 20

*Figure 6.23.* Probability that atmost 10% of the outputs of the overall system are incorrect for different I/O bundle sizes (large probability of failure)

## Large Gate Failure Probabilities

In this subsection, we consider the case when the probability of gate failure of the majority gates becomes too large for the multiplexing system to function reliably. Figure 6.23 plots the probability that less than 10% of the outputs are incorrect against large gate failure probabilities (between 0.01 and 0.04) for different number of multiplexing stages $\in \{1, 2, \cdots, 10, 11\}$. As can be seen from the results, when the probability of gate failure is equal to 0.04 increasing the I/O bundle size does not improve the reliability of the system. Comparing the same plots in Figure 6.23 (a) and (b), it can be observed that augmenting additional restorative stages and increasing the I/O bundle size of

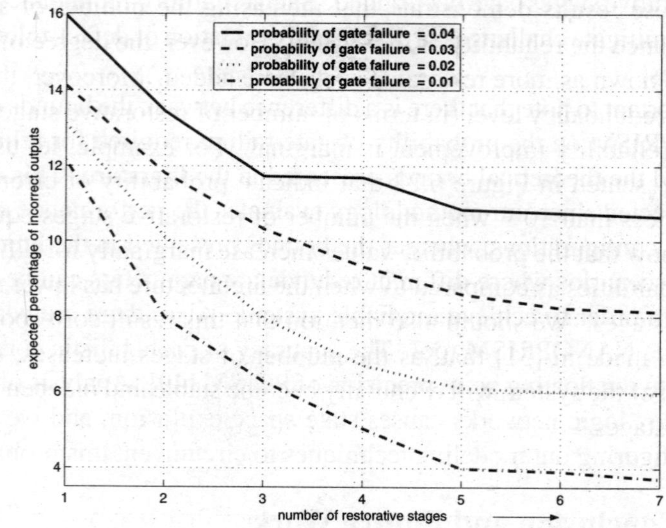


*Figure 6.24.* Expected percentage of incorrect outputs for I/O bundle size of 10 (large probability of failure)

the architecture tends to cause a degradation in reliability of computation. This anomalous behavior can be understood as follows: when the failure rate is 0.04 (or higher), the restorative stages are sufficiently affected by the probability of gate failures to be unable to reduce the degradation, and hence increasing the number of stages in this case makes the system more amenable to error.

We can infer from these observations that for gate probabilities of 0.04 and higher, increasing bundle size or addition of more restorative stages cannot make the system more reliable and may even cause degradation. On the other hand, in the case when the gate failure probability is less than 0.04, the results demonstrate that the system can be made reliable once a sufficient number of restorative stages have been added.

In Figure 6.24, we have plotted the expected percentage of incorrect inputs when I/O bundle size equals 10. The gate failure probabilities and restorative stages are configured similar to Figure 6.23. Figure 6.24 and Figure 6.23 determine similar intuitive results but have different perspectives to reliability evaluation of the system. By similar facts, we mean that for gate failure probabilities of 0.04 and higher, increasing the number of restorative stages cannot improve reliability measures. In fact, when the multiplexing system is configured to have certain specific number of restorative stages, the reliability may decrease as compared to a system with less redundancy (less number of multiplexing stages). It can also be observed that for all gate failure probabilities, the reliability of the architecture reaches a steady state once a sufficient number of restorative stages have been added. Such reliability-redundancy trade-off points may mitigate challenges in design and evaluation of defect-tolerant nano-architectures.

It is important to note that there is a difference between the bounds computed by NANOPRISM on the probability of gate failure required for reliable computation and the theoretical bounds presented in the literature. This difference is to be expected since our methodology evaluates the performance of systems under fixed configurations, whereas the bounds presented in the literature correspond to scenarios where different redundancy parameters can be increased arbitrarily in order to achieve a reliable system. Also, there are certain limitations of the NANOPRISM tool. The accuracy of probabilistic computations are limited to the floating point precision of PRISM [49]. Analysis of large and very complex logic networks causes state space explosion, and we are in the process of figuring out modeling techniques to circumvent this problem.

## 6.9 Conclusion and Future Work

In this chapter, we discuss a few analytical methodologies and propose two automation methodologies that can expedite reliability evaluation of defect-tolerant architectures for nanodevices. We have developed tools that can determine optimal redundancy and granularity levels of any specific logic network, given the failure distribution of the devices and the desired reliability of the system. These tools can also be used to inject signal noise at inputs and interconnects and create practical situations the circuits are subjected to during normal operation.

We have developed NANOLAB, a tool that is based on a non-discrete probabilistic design methodology proposed by Bahar et al. This computational scheme has two aspects. First, the gates are assumed to be defect free, and the model of computation based on Markov Random Fields correlates information theoretic entropy and the thermal entropy of computation. Since at the reduced voltage level in nano-architecture this issue can become significant, reliability may suffer when the computation is carried out close to the thermal limit of computation. However, we show that by considering various defect-tolerant architectural techniques such as TMR, CTMR and multi-stage iterations of these, the effects of carrying out computation within close thermal limits can be reduced substantially. Second, signal noise can be injected at the inputs and interconnects of a logic network. We have introduced the effects of signal noise in our automation methodology. Noise is modeled using Gaussian and uniform probability distributions, and this goes beyond the thermal aspects as described above. NANOLAB automatically computes reliability measures of systems in terms of energy distributions and entropy in the presence of discrete input distributions and random noise. This tool consists of MATLAB based libraries for generic logic gates that can calculate the probability of the outputs being in different energy levels, and Belief Propagation algorithm that can compute such distributions at the primary outputs of arbitrary Boolean networks (given as inputs to the tool). It also supports various interconnect noise models, and can be very useful in modeling transient faults and provide the designers a way to figure out the configuration that is best in terms of reduced entropy (which in turn means higher reliability). This is indeed an effective tool and technique for evaluating reliability and redundancy trade-offs in the presence of thermal perturbations and interconnect noise models. Our tool can also inject error probabilities at the gates and do similar trade-off calculations.

We have reported on the results obtained for single gate and logic block level architectural configurations and investigated the performance of the system in terms of entropy and energy distribution, as the number of CTMR orders vary. Using NANOLAB, we were able to compute the energy distributions at the outputs of systems, and hence construct a complete picture of the reliability of the systems under study. We chose to analyze TMR and CTMR for illustration purposes as these are canonical fault tolerant architectures standard in the literature, and since these enabled us to compare the results with others. However, as explained, this approach can equally be applied to alternative fault-tolerant architectures for nanotechnology.

This chapter also presents NANOPRISM, a reliability evaluation tool based on probabilistic model checking. This tool consists of libraries for different redundancy based defect-tolerant architectural configurations. We have analyzed reliability measures of specific logic circuits with this tool, and shown that for a given probability of gate failure, we can find the minimum level of redundancy and *granularity* that enables reliable computation. The NANOPRISM libraries support implementation of redundancy at different levels of granularity, such as gate level, logic block level, logic function level, unit level etc. We illustrate the proficiency of our methodology by modeling von Neumann multiplexing systems, and different orders of CTMR for arbitrary logic circuits. We also describe a methodology that reduces the effect of the well known state space explosion problem, and hence allows for the analysis of larger configurations. The experimental results from our tool show anomalous and counter-intuitive phenomena that may not be detected quickly by analytical methods.

NANOLAB applies an approach to reliability analysis that is based on Markov Random Fields as the probabilistic model of computation. Whereas, NANOPRISM applies probabilistic model checking techniques to calculate the likelihood of occurrence of probabilistic transient defects in devices and interconnections. Thus, there is an inherent difference in the model of computation between these two approaches. Although these two methodologies are different, they offer complementary alternatives to analytical methodologies and allow system architects to obtain sharp bounds and study anomalies for specific architectures. Future work includes extending these tools to support reliability analysis of sequential circuits. We are also in the process of building libraries for other frequently used fault tolerance schemes.

## 6.10 Acknowledgments

We acknowledge the support of NSF grant CCR-0340740.

## References

[1] Web Page: www.mathworks.com.

[2] R. I. Bahar, J. Mundy, and J. Chen, *A probability-based design methodology for nanoscale computation*, in *International Conference on Computer-Aided Design* (IEEE Press, San Jose, CA, November 2003).

[3] C. H. Bennett, 'The thermodynamics of computation–a review', *International Journal of Theoretical Physics* **21** (1982.), no. 905-940.

[4] J. Besag, 'Spatial interaction and the statistical analysis of lattice systems', *Journal of the Royal Statistical Society* **Series B** (1994), no. 36(3), 192–236.

[5] D. Bhaduri and S. K. Shukla, 'Nanolab: A tool for evaluating reliability of defect-tolerant nano architectures', *Tech. Report* (Fermat Lab, Virginia Tech, 2003). Available at http://fermat.ece.vt.edu/Publications/pubs/techrep/techrep0309.pdf.

[6] D. Bhaduri and S. K. Shukla, 'Nanoprism: A tool for evaluating granularity vs. reliability trade-offs in nano architectures', *Tech. Report* (Fermat Lab, Virginia Tech, 2003). Available at http://fermat.ece.vt.edu/Publications/pubs/techrep/techrep0318.pdf.

[7] D. Bhaduri and S. K. Shukla, *Nanolab: A tool for evaluating reliability of defect-tolerant nano architectures*, in *IEEE Computer Society Annual Symposium on VLSI* (IEEE Press, Lafayette, louisiana, February 2004). `http://fermat.ece.vt.edu`.

[8] D. Bhaduri and S. K. Shukla, *Nanoprism: A tool for evaluating granularity vs. reliability trade-offs in nano architectures*, in *GLSVLSI* (ACM, Boston, MA, April 2004).

[9] D. Bhaduri and S. K. Shukla, 'Reliability evaluation of multiplexing based defect-tolerant majority circuits', *IEEE Conference on Nanotechnology* (2004). Sent for Publication.

[10] D. Bhaduri and S. K. Shukla, *Tools and techniques for evaluating reliability of defect-tolerant nano architectures*, in *International Joint Conference on Neural Networks* (IEEE Press, Budapest, Hungary, July 2004).

[11] D. Bhaduri and S. Shukla, 'Reliability analysis in the presence of signal noise for nano architectures', *IEEE Conference on Nanotechnology* (2004). Sent for Publication.

[12] F. Buot, 'Mesoscopic physics and nanoelectronics: nanosicence and nanotechnology', *Physics Reports* (1993), 173–174.

[13] S. Burris, *Boolean algebra* (March 2001). Available at : `http://www.thoralf.uwaterloo.ca/htdocs/WWW/PDF/boolean.pdf`.

[14] J. Chen, M. Reed, and A. Rawlett, 'Observation of a large on-off ratio and negative differential resistance in an electronic molecular switch', *Science* **286** (1999), 1550–2.

[15] J. Chen, W. Wang, M. Reed, M. Rawlett, D. W. Price, and J. Tour, 'Room-temperature negative differential resistance in nanoscale molecular junctions', *Appl. Phys. Lett.* **1224** (2000), 77.

[16] J. Chen, J. Mundy, Y. Bai, S.-M. Chan, P. Petrica, and I. Bahar, *A probabilistic approach to nano-computing*, in *IEEE non-silicon computer workshop* (IEEE Press, San Diego, June 2003).

[17] R. Chen, A. Korotov, and K. Likharev, 'Single-electron transistor logic', *Apll. Phys. Lett.* (1996), 68:1954.

[18] C. Collier, E. Wong, M. Belohradsky, F. Raymo, J. Stoddart, P.J.Kuekes, R. Williams, and J. Heath, 'Electronically configurable molecular-based logic gates', *Science* **285** (1999), 391–3.

[19] Y. Cui and C. Lieber, 'Functional nanoscale electronic devices assembled using silicon nanowire building blocks', *Science* **291** (2001), 851–853.

[20] M. D, S. M, S. A, and T. G, 'Toward robust integrated circuits: the embryonics approach', in *IEEE*, **88**, 516–41.

[21] R. Dobrushin and E. Ortyukov, 'Upper bound on the redundancy of self-correcting arrangements of unreliable functional elements', *Problems of Information Transmission* **13** (1977), no. 3, 203–218.

[22] L. Durbeck, *Underlying future technology talk on computing: Scaling up, scaling down, and scaling back* (American Nuclear Society's International Meeting on Mathematical Methods for Nuclear Applications, September 2001). Available at `http://www.cellmatrix.com/entryway/products/pub/ANS_Abstract.html`.

[23] L. J. K. Durbek, *An approach to designing extremely large, extremely parallel systems*, in *Conference on High Speed Computing* (Oregon,, USA,, April 2001).

[24] J. C. Ellenbogen and J. C. Love, 'Architectures for molecular electronic computers: Logic structures and an adder designed from molecular electronic diodes', in *IEEE, 3* **88** (2000), 386–426.

[25] W. Evans and N. Pippenger, 'On the maximum tolerable noise for reliable computation by formulas', *IEEE Transactions on Information Theory* **44** (1998), no. 3, 1299–1305.

[26] D. F, S. M. G, and S. R, 'Fault-tolerance and reconfigurability issues in massively parallel architectures', in *Computer Architecture for Machine Perception* (IEEE Computer Society Press, Los Alamitos, CA, 1995), 340–9.

[27] M. Forshaw, K. Nikolic, and A. Sadek, 'Ec answers project (melari 28667)', *Third Year Report*. Available at http://ipga.phys.ucl.ac.uk/research/answers.

[28] S. C. Goldstein and M. Budiu, 'Nanofabrics: Spatial computing using molecular electronics', in *Annual International Symposium on Computer Architecture (ISCA)* (July 2001), 178–191.

[29] S. C. Goldstein and D. Rosewater, 'Digital logic using molecular electronics', in *IEEE International Solid-State Circuits Conference (ISSCC)* (San Francisco, CA, Feb 2002), 204–205.

[30] S. C. H, 'A new statistical approach for fault-tolerant vlsi systems', in *Int. Symp. on Fault-Tolerant Computing* (IEEE Computer Society Press, Los Alamitos, CA, 1992), 356–65.

[31] J. Han and P. Jonker, 'A system architecture solution for unreliable nanoelectronic devices', *IEEE Transactions on Nanotechnology* **1** (2002), 201–208.

[32] J. Han and P. Jonker, 'A defect- and fault-tolerant architecture for nanocomputers', *Nanotechnology* (2003), 224–230. IOP Publishing Ltd.

[33] H. Hansson and B. Jonsson, 'A logic for reasoning about time and probability', *Formal Aspects of Computing* **6** (1994), no. 5, 512–535.

[34] J. Heath, P. Kuekes, G. Snider, and R. Williams, 'A defect tolerant computer architecture: Opportunities for nanotechnology', *Science* **80** (1998), 1716–1721.

[35] K. I and K. Z, 'Defect tolerance in vlsi circuits: techniques and yield analysis', in *IEEE*, **86** (1998), 1819–36.

[36] J.Yedidia, W.Freeman, and Y.Weiss, *Understanding belief propagation and its generalizations*, in *Proc. International Joint Conference on AI* (2001). Distinguised Lecture.

[37] M. Kwiatkowska, G. Norman, and D. Parker, 'Prism: Probabilistic symbolic model checker', in *TOOLS 2002*, *LNCS* **2324** (Springer-Verlag, April 2002), 200–204.

[38] C. Lent, 'A device architecture for computing with quantum dots', in *Porceedings of the IEEE*, **541** (April 1997), 85.

[39] C. S. Lent, P. D. Tougaw, W. Porod, and G. H. Bernstein, 'Quantum cellular automata', *Nanotechnology* **4** (1993), 49–57.

[40] S. Li, *Markov random field modeling in computer verification* (Springer-Verlag, 1995).

[41] M. Mishra and S. C. Goldstein, *Defect tolerance at the end of the roadmap*, in *International Test Conference (ITC)* (Charlotte, NC, Sep 30-Oct 2 2003).

[42] K. Nikolic, A. Sadek, and M. Forshaw, 'Architectures for reliable computing with unreliable nanodevices', in *Proc. IEEE-NANO'01* (IEEE, 2001), 254–259.

[43] G. Norman, D. Parker, M. Kwiatkowska, and S. Shukla, 'Evaluating reliability of defect tolerant architecture for nanotechnology using probabilistic model checking', *Tech. Report* (Fermat Lab, Virignia Tech, 2003). Available at http://fermat.ece.vt.edu/Publications/pubs/techrep/techrep0314.pdf.

[44] G. Norman, D. Parker, M. Kwiatkowska, and S. Shukla, *Evaluating reliability of defect tolerant architecture for nanotechnology using probabilistic model checking*, in *IEEE VLSI Design Conference* (IEEE Press, Mumbai, India, January 2004).

[45] J. R. Norris, *Markov chains*, in *Statistical and Probabilistic Mathematics* (Cambridge University Press, October 1998).

[46] C. C. P, M. G, W. E. W, L. Y, B. K, S. J, R. F. M, S. J. F, and H. J. R, 'A [2] catenane-based solid state electronically reconfigurable switch', *Science* **280** (2000), 1172–5.

[47] N. Pippenger, 'Reliable computation by formulas in the presence of noise', *IEEE Transactions on Information Theory* **34** (1988), no. 2, 194–197.

[48] D. B. Pollard, *Hammersley-clifford theorem for markov random fields* (2004). Handouts, available at `http://www.stat.yale.edu/~pollard/251.spring04/Handouts/Hammersley-Clifford.pdf`.

[49] Web Page: `www.cs.bham.ac.uk/~dxp/prism/`.

[50] W. Roscoe, *Theory and practice of concurrency* (Prentice Hall, 1998).

[51] S. Roy, V. Beiu, and M. Sulieman, *Majority multiplexing: Economical redundant fault-tolerant designs for nano architectures*.

[52] S. Roy, V. Beiu, and M. Sulieman, *Reliability analysis of some nano architectures*.

[53] N. Saxena and E. McCluskey, 'Dependable adaptive computing systems', in *IEEE Systems, Man, and Cybernetics* (San Diego, CA, Oct 11-14 1998), 2172–2177. The Roar Project.

[54] E. J. Siochi, P. T. Lillehei, K. E. Wise, C. Park, and J. H. Rouse, 'Design and characterization of carbon nanotube nanocomposites', *Tech. Report* (NASA Langley Research Center, Hampton, VA, 2003). Available at http://techreports.larc.nasa.gov/ltrs/PDF/2003/mtg/NASA-2003-4ismn-ejs.pdf.

[55] G. L. Snider, A. O. Orlov, I. Amlani, G. H. B. adn Craig S. Lent, J. L. Merz, and W. Porod, 'Quatum-dot cellular automata: Line and majority logic gate', *Japanese Journal of Applied Physics* **38** (1999), no. 12B, 7227–7229.

[56] R. Turton, *The quantum dot: A journey into the future of microelectronics* (Oxford University Press, U.K, 1995).

[57] J. von Neumann, 'Probabilistic logics and synthesis of reliable organisms from unreliable components', *Automata Studies* (1956), 43–98.

[58] T. G. Y and E. J. C, 'Toward nanocomputers', *Science* **294** (2001), 1293–4.

# Chapter 7

# LAW OF LARGE NUMBERS SYSTEM DESIGN

André DeHon
*California Institute of Technology*
andre@acm.org

**Abstract** To date we have relied on the "Law of Large Numbers" below the device level to guarantee deterministic device behavior (*e.g.* dopant ratios, transition timing, electron state storage). However, at the nanoscale, we hope to build devices with small numbers of atoms or molecules (*e.g.* wires which are 3-10 atoms wide, diodes built from 1-10 molecules), and we hope to store state with small numbers of electrons (*e.g.* 10's). If we are to build devices at these scales, we will no longer be able to rely on the "Law of Large Numbers" **below** the device level. We must, instead, employ the "Law of Large Numbers" **above** the device level in order to obtain predictable behavior from atomic-scale phenomena which are statistical in nature. At the same time, the "Law of Large Numbers" can also help us by providing statistical differentiation at scales smaller than those we can pattern directly or economically using lithography. In this chapter, we examine various applications of the "Law of Large Numbers" above the device level to build reliable and controllable systems from nanoscale devices and processes that only have statistically predictable behavior.

## 7.1 Introduction

*How do we engineer systems when we can only statistically control behavior?*

As we approach the design of engineered systems at the atomic scale, we must confront the fact that the behaviors of individual atoms, molecules, and electrons is something we can control or predict only statistically. Quantum mechanics, Heisenberg uncertainty, and thermodynamics tell us that we can know the likely behavior of these devices, but not the absolute behavior of an individual element. How do we cope with this uncertainty in design?

In fact, we have coped with this uncertainty for decades. All of our engineered systems are built out of these same building blocks. The difference is one of scale. In the past, we were able to build devices from billions of atoms,



S.K. Shukla and R.I. Bahar (eds.), Nano, Quantum and Molecular Computing, 213-241.

implant millions of dopant atoms, and store state with millions of electrons. We were able to provide a reliable device abstraction by employing the "Law of Large Numbers" (LLN) at the device scale. With billions of atoms, in a structure, we could count on the statistical behavior of billions of atoms to assure that the correct device was built almost every time. With millions of electrons, we could be assured that the likelihood of a spontaneous change in charge state was sufficiently low as to be ignorable.

Consequently, we have successfully built and employed an abstraction hierarchy for component and system design that contains this uncertainty at the device level. We ask our manufacturing to produce devices reliably and consistently, and we ask that those devices operate reliably and deterministically. This has been a reasonable place to ask for reliability in our design hierarchy because we could afford to spend large numbers of atoms and charges to maintain it.

Nonetheless, this state of affairs breaks down if we hope to approach the atomic scale. We could choose to continue to demand the phenomenal reliability in manufacture and device operation we have achieved in the past. If we were to do so, this would put a large limit on our ability to scale down device size. We would be forced to continue to build devices with a sufficiently large number of atoms that we could statistically guarantee manufacture and low variation in operating parameters (*e.g.* resistance, delay).

The alternative is to recognize that we must adapt our design hierarchy if we are to approach these ultimate scales. Devices built from 100's of atoms won't always be fabricated perfectly or will have high variance in electrical proprieties. Devices storing or switching on 10's to 100's of electrons won't always hold their value or switch in the desired manner. Nonetheless, when we use these smaller and denser devices, we can have orders of magnitudes more devices to work with. We will have *large numbers* of devices and *large numbers* of switching events. We can continue to employ the LLN to ensure reliable and predictable behavior of our systems, but we must do so where we have a large numbers of things—that point will now be at higher levels in the system than the device level.

This chapter illustrates many opportunities to employ the "Law of Large Numbers" at the component and system design level. We start by briefly reviewing the "Law of Large Numbers," a few of the statistical phenomena ever present in our electronic systems, and the ways in which the "Law of Large Numbers" has traditionally supported reliable device manufacture and operation (Section 7.2). We then highlight opportunities for employing the LLN above the device level (Section 7.3) and briefly review cases where this is already an accepted part of conventional system design (Section 7.4). We detail specific opportunities for LLN design in device manufacture (Section 7.5), functionality mapping (Section 7.6), nanoscale addressing (Section 7.7), and nanoscale differentiation (Section 7.8). We touch briefly on tolerating faults (Section 7.9), and suggest how these atomic-scale effects and LLN design will impact our abstraction stack for atomic-scale system engineering (Section 7.10).

## 7.2 Background

### Law of Large Numbers

The "Law of Large Numbers" (LLN) is said to hold for a sequence of random variables with finite expected values when the mean value for the sequence converges to the expected value [1] (for a modern treatment see, for example, [13]). That is, for a sequence of independent, identically distributed random variables, $y_i$:

$$\lim_{n\to\infty} \text{Prob}\left[\left|\frac{1}{n}\sum_{i=1}^{n} y_i - E(y)\right| \geq \epsilon\right] = 0 \quad (7.1)$$

This says that, even though each individual $y_i$ is probabilistic in nature, aggregate properties of a large number of the $y_i$'s are quite predictable. As we increase the number of events over which we aggregate, the likelihood of deviating more than a tiny percentage from this mean is smaller and smaller.

### Statistical Phenomena

Most of the physical phenomena we rely upon in electronic systems are statistical at the atomic scale.

In the construction of silicon devices, doping is a prime example. We build transistors in MOS devices by mixing in a percentage of impurities (donors, acceptors) to change the band structure of the devices. We do not place 999 Silicon atoms and then one Boron atom. Instead, we arrange to have a 1000:1 mix of Silicon and Boron atoms in a growth environment, or we arrange to impact the silicon with a given intensity of Boron atoms to replace the Silicon atoms. We don't guarantee exactly where each Boron atom ends up. Nor do we guarantee that every bond in the crystalline lattice is perfectly made.

During operation, we typically think about charges on capacitors and gates and current flows across devices. However, current flow is simply an aggregate view of the behavior of individual electrons. Individual electrons travel across a device or region of silicon probabilistically depending on the fields and the thermal energy. We only know statistically what each electron will do.

Similarly, we can isolate charge on a node such as a memory element or gate input. We can construct electrostatic barriers to hold the charge in place. Nonetheless, thermal energy and quantum tunneling give the individual electrons some probability of surmounting the barrier and leaving, or entering, the node.

### Reliable Manufacturing and Behavior from LLN

The LLN is how we can effectively get very reliable device manufacture and device properties even though each individual atom or electron behaves probabilistically.

Returning to our doping example, while we cannot guarantee that we have exactly 999 Silicon atoms and then one Boron atom in our Silicon crystal, the LLN assures us that we can have high confidence that there are close to $10^6$ Boron atoms in a doped region with $10^9$ crystal cites. The variation from this mean will be small, such that the variation in parameters (*e.g.* resistance, threshold level) is small from device to device. A region with 1000 atoms will have exactly one dopant atom only 36% of the time, while a region with $10^9$ atoms is within 1% of the mean of $10^6$ dopant atoms over 99.9999% of the time. Similarly, while we may have spot defects in our lattice, if the cross-sectional area of a device or wire is sufficiently large, we know that the probability that there are insufficient bonds across any boundary to disconnect the device or significantly change its resistance is very low.

When there are millions of electrons sitting on a device, we know it is highly unlikely that they will all choose not to travel in the direction of the applied field. The behavior of the ensemble, the current flow, will approach very closely to the expected behavior of an individual electron. Consequently, with sufficient electrons, our abstraction of a current flow is very good and we can, with very high certainty, depend on the aggregate electron behavior to be deterministic.

Similarly, when we store charge, we can erect energy barriers that are sufficiently high to hold the charge on the node. While there is a probability that each individual electron may gain enough energy to cross the barrier or may tunnel across the barrier, the probability that a significant fraction of the electrons can cross the barrier can be made sufficiently small as to be negligible.

This manufacturing and device behavior consistency works very well as long as we have a large number of atoms or electrons to begin to approach the large number limit. However, by the time wires have cross-sections on the order of 100's of atoms and 10's to 100's of electrons store important charge state, we no longer have large number guarantees. We can no longer guarantee every device built will have the right number of particles to lie within some reasonable parameter range. We can no longer guarantee that the probability of a transition occurring in a reasonable window of time is sufficiently high to depend upon it for correct device operation.

## 7.3 "Law of Large Numbers" Above the Device Level

These observations lead us to consider how we can exploit the LLN above the device level. We can:

1. Tolerate variations in manufacture by selecting which devices to use (Sections 7.5)
2. Tolerate variation in manufacture by selecting which device to use for what role (Section 7.6)
3. Exploit variations to get differentiation at the nanoscale (Sections 7.7 and 7.8)
4. Tolerate variations in behavior by performing redundant and self-checking computations (Section 7.9)

Before we examine these techniques, we observe that the use of LLN above the device level is already a well established practice in some of the systems we employ every day (Section 7.4).

## 7.4 Component and System Level LLN in Conventional Systems

We do, in fact, already employ LLN as a design principal for some of today's large-scale systems.

DRAM row and column sparing is perhaps the most familiar case of using LLN design to deal with manufacturing defects [28] [19]. These defects may arise from the probabilistic assembly of atomic-scale structures as described above, or, more likely, from the lack of perfect purity and calibration in the design of our manufacturing systems. In either case, if we are unhappy with the probability that our manufacturing process produces a perfect memory bank where we require every row, column, and memory bit to perform perfectly, we add spare rows and columns (See Figure 7.3A). When the expected number of defects per bank is less than one, it is highly unlikely we will see many defects per bank. Providing one or a few spares per bank guarantees a very high probability that each bank can be made perfect. Section 7.5 discusses sparing as a general, system level, LLN design technique and quantifies the benefits.

Error-correcting codes which protect DRAM bits are a familiar example of how we use LLN to tolerate operational faults [24]. To push the density of memory storage, DRAM storage cells are one case where the number of electrons is already becoming small. A typical DRAM cell with a capacitance around 30fF and a small storage voltage around 1.65 volts [19], stores its state with only 300,000 electrons. It is well known that this is too few electrons to preserve the value when the part is impacted with alpha particles or cosmic rays. While these events are improbable, very large memory arrays running for long periods of time will see storage cells disrupted—another LLN effect. Here the LLN tells us that the probability of a group of bits seeing multiple errors is low. We then use single-error-correcting codes (*e.g.* [16]) to identify and correct the erroneous bits. While the use of error-correcting codes has been justified by cosmic ray and particle bombardment, it also serves to protect against statistical noise fluctuations in the silicon and the possibility that a sufficient number of electrons might choose to migrate off of, or on to, a storage cell.

With external electromagnetic noise on wires a consistent reality, our inter-system communications have almost always required that we employ redundancy and LLN effects to achieve reliable communications [29]. We regularly employ one or more forms of error-detecting or error-correcting codes on the wires. Here, again, we use redundancy and exploit the fact that the probability we will see multiple errors is much lower than the probability of seeing one or a few errors. Given sufficiently uncorrelated noise sources, the LLN assures us that, over a sufficiently large block, the number of errors we will see is closely bound near the expected value.

While our primitive devices have been very reliable, individual computer systems have not been. This is in part due to the aggregate effect of a large number of devices. However, the biggest impact for computer system reliability has traditionally been issues outside of a single, hardware component including software faults, operation faults, power faults, and wiring faults [30]. From this composite effect, we found it necessary to design large-scale communication protocols (*e.g.* TCP [26]) and networks to tolerate individual message or component failures. Here again, we rely on the fact that the LLN guarantees that we can get a certain fraction of good messages through a statistically faulty link and that the likelihood we will see sufficient faults to partition the network is very low.

## 7.5 Architectures with Sparing

Row sparing in DRAMs is a special case of M-of-N sparing. That is, we fabricate or assemble $N$ equivalent items in our system but only require that $M$ of them function in order to have a correct system. This way, rather than requiring that $M$ things yield perfectly, we simply require that at least $M$ items out of $N$ yield. Here we use the term *yield* broadly to mean that the device or component has been manufactured adequately to perform its intended role (*e.g.* all the bits on the row or column can be read and written within a defined timing window and will hold their values for a suitable length of time).

### LLN in Sparing

Statistically, if the probability that each item yields is $P$, then the probability that every one of $M$ items will yield is:

$$P_{allyield}(M) = P^M \tag{7.2}$$

We can calculate the probability that exactly $i$ items will yield using a binomial distribution:

$$P_{yield}(N,i) = \left(\binom{N}{i} P^i (1-P)^{N-i}\right) \quad (7.3)$$

That is, there are $\binom{N}{i}$ ways to select $i$ good items from $N$ total items, and the yield probability of each case is: $P^i (1-P)^{N-i}$. We yield an ensemble with $M$ items whenever $M$ or more items yield, so our system yield is actually the cumulative distribution function:

$$P_{MofN} = \sum_{M \le i \le N} \left(\binom{N}{i} P^i (1-P)^{N-i}\right) \quad (7.4)$$

When we want to yield $M$ things with some probability $P_{sysyield}$, we select a sufficiently large $N$ such that $P_{MofN} \ge P_{sysyield}$. Alternately, for a given $N$, we can determine $M_{pmofn}$, the largest $M$ such that $P_{MofN} \ge P_{sysyield}$. As $N$ increases, $M_{pmofn}$ becomes very close to the expected aggregate yield:

$$E(M) = N \times P \quad (7.5)$$

This arises exactly from the LLN effect where the expected value converges to the mean (Equation 7.1). Figure 7.1 plots $\frac{E(M)-M_{pmofn}}{E(M)}$ versus $N$ demonstrating the convergence as $N$ increases for various level of $P_{sysyield}$. This mean difference ratio can be viewed as the overhead cost required to guarantee a minimum level of yield given statistical variations. As we aggregate over larger ensemble, the LLN drives down this overhead. Illustrating this effect, Figure 7.2 plots $P_{MofN}$ versus $M$ for $N = 1000$ and $P = 0.9$, showing the sharp transition around the mean of $M = N \times P = 1000 \times 0.9 = 900$.

## Architectures which Support Sparing

To exploit this M-of-N sparing, we need designs structured with a large number of identical items which are cheaply interchangeable. The crossbar organization used in memory arrays, interconnects, and programmable logic arrays is a prime example of a structure which has this property.

- In a memory all the rows (or columns) are identical. As long as we can program up the addressing for each row and column and configure non-used lines so they do not interfere with operation, we can use any $M$ of the $N$ row lines to serve as our desired row lines (See Figure 7.3A). Schuster [28] and Tammaru and Angell [31] describe early VLSI memory designs with spare rows and columns.
- In a Programmable Logic Array (PLA), all of the programmable terms (*e.g.* product terms) are logically equivalent. We simply need to be able to allocate enough product terms to cover our logic function (See Figure 7.3B).

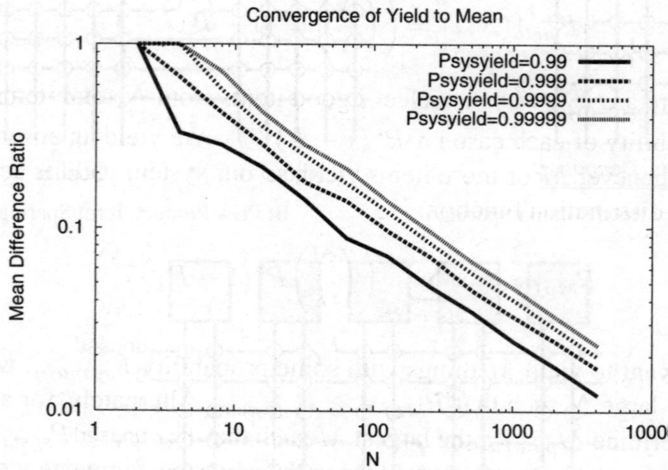


*Figure 7.1.* Convergence to Mean of Yielded Resources. $\frac{E(M)-M_{pmofn}}{E(M)}$ as a function of $N$ for $P = 0.9$ showing the convergence of our M-choose-N calculation to the mean for large numbers $N$.

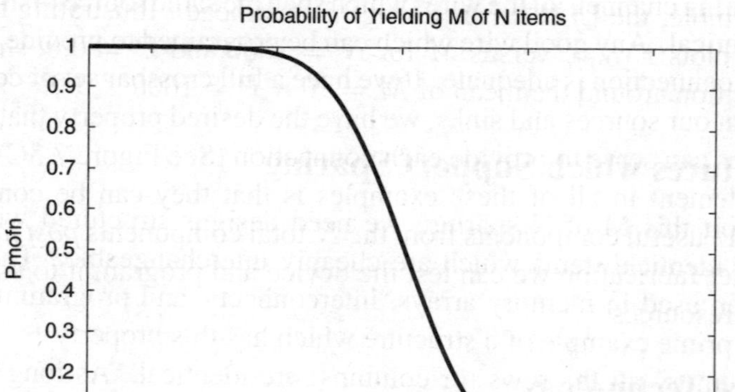


*Figure 7.2.* $P_{MofN}$ versus $M$ for $N = 1000$, $P = 0.9$

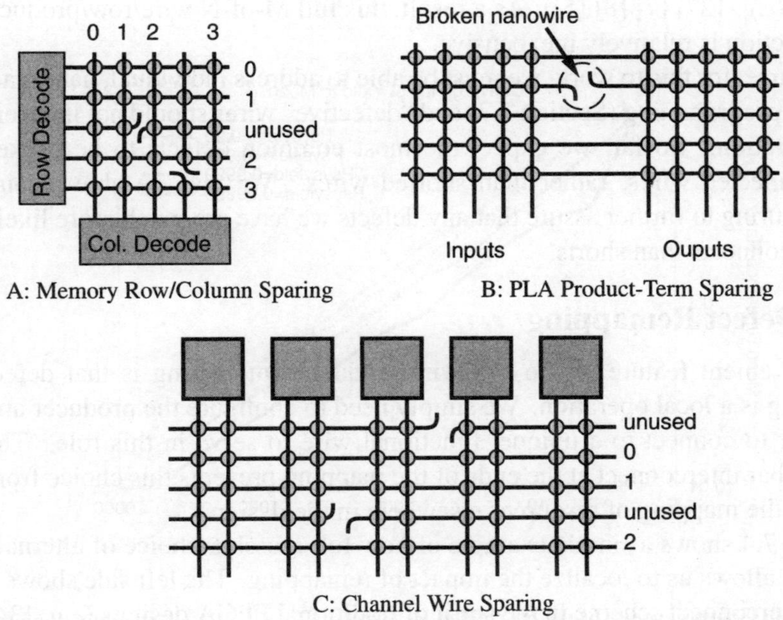


*Figure 7.3.* Crossbar-Based Resource Sparing

Wey *et al.* [35] describe a VLSI PLA design with spare input lines, output lines, and product terms.

- In a routing channel, all the wires which span the same source→sink distance are identical. Any good wire which can be programed to provide the source to sink connection is adequate. If we have a full crossbar set of connections between our sources and sinks, we have the desired property that any $M$ of the lines can serve to provide each connection (See Figure 7.3C).

The key element in all of these examples is that they can be configured to select the $M$ useful components from the $N$ total components *post fabrication*. That is, after fabrication we can test the device and program it to use only the functional resources.

## Nanowire Technology

Fortuitously, large sets of parallel, nanowires assembled into crossbars is one of the things we do know how to build at sublithographic scales (*e.g.* [18] [36] [37] [6] [22]). Further, while the **full** connectivity of the crossbar is quite expensive for interconnect in conventional CMOS where programmable switchpoints are large compared to wire crossings, full connectivity is relatively cheap in many of the emerging nanotechnologies. In particular, several technologies offer the prospect of non-volatile crosspoints that fit within the space of a wire crossing (*e.g.* [27] [7] [8] [5]). As a result, this full M-of-N wire/row/product-term selection is relatively inexpensive.

Of course, for this to work, we must be able to address individual, nanoscale wires for programming (Section 7.7) and "defective" wires should not interfere with operation. So far, we expect the most common defects to be broken or disconnected wires, rather than shorted wires. We may be able to bias manufacturing to further assure that any defects we have are much more likely to be disconnects than shorts.

### Local Defect Remapping

A convenient feature of the row/wire/product-term sparing is that defect remapping is a local operation. We simply need to configure the producer and consumer to connect to a unique, functional wire to serve in this role. The full crossbar interconnect at the ends of the mapping prevents this choice from effecting the mapping of resources elsewhere in the design.

Figure 7.4 shows a simple example of how full crossbar choice of alternate resources allows us to localize the impact of remapping. The left side shows a sparse interconnect scheme in the spirit of traditional FPGA designs (*e.g.* [34] [33]). The right side shows the design with full connectivity for sparing. The middle row shows a broken wire. In the crossbar case, this can be accommodated simply by shifting the net segment which previously used the broken wire to a free track in the same channel. This change is contained locally to this one channel; the segments of the net in different channels do not need to change, nor does the routing of any of the other nets. In the sparse case, however, we are forced to change the track assignment of this A→B net in all the channels it traverses due to the limited switch-box population; we are further forced to reroute the C→D net in order to accommodate this change.

## 7.6 Architectures with Choice

The simple model presented in the previous section is that a resource is simply all good or bad. However, we may not need to use all the potential functionality of a resource. This gives us the opportunity to select the resource which is simply "good enough" to serve for the purpose at hand.

### Technology Example

To be concrete, consider the molecular-switch crosspoints (*e.g.* [7] [4] [8] [5]). The assembly techniques allow us to, statistically, place a number of switchable molecules between a pair of crossed conductors (See Figure 7.5). Any particular junction may get fewer or no molecules in the junction. For example, Chen *et al.* reports that only 85% of the junctions could be programmed [5]. While we certainly expect this percentage to improve as the technology

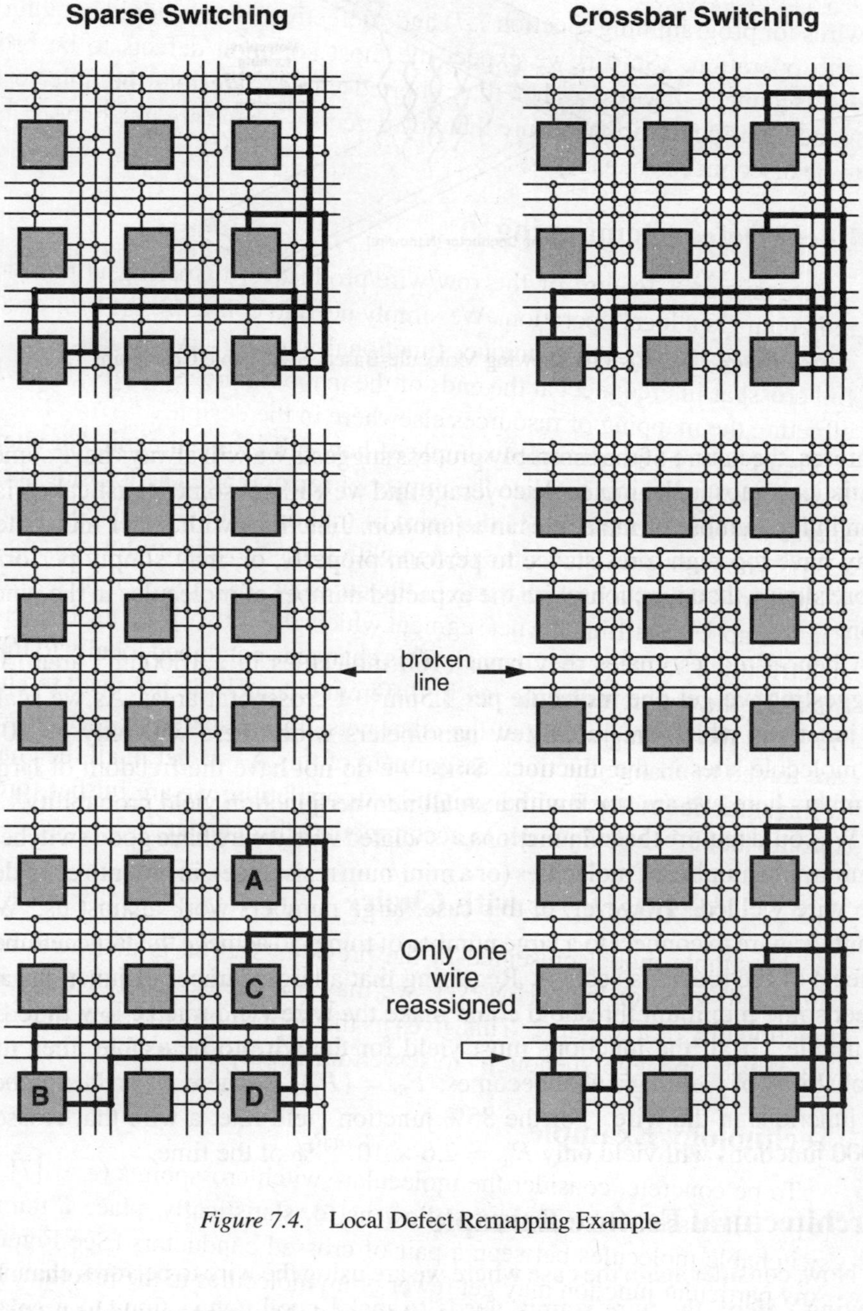


*Figure 7.4.* Local Defect Remapping Example

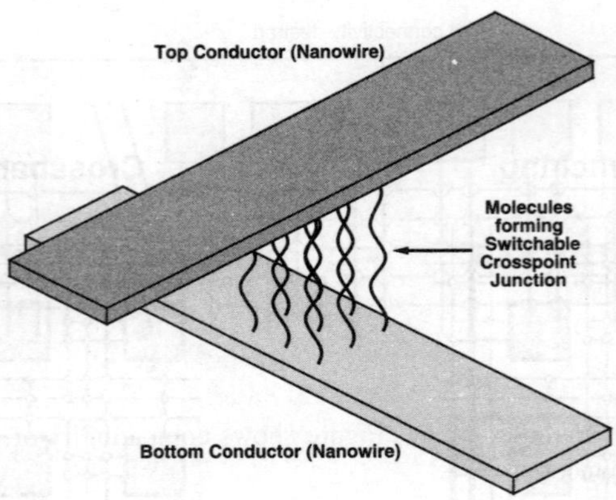


*Figure 7.5.* Cartoon Showing Molecule-Based, Switchable Crosspoint

matures, the nature of the assembly process suggests we will always have some, statistical, gaps in the molecule coverage and we will get some statistical variation in the number of molecules in a junction. Junctions with fewer molecules may have too high a resistance to perform properly, or, may simply perform more slowly than junctions with the expected number of molecules in the junction.

Chen *et al.* estimates they have 1100 molecules in a 1600nm$^2$ area [5], suggesting we get one molecule per 1.5nm$^2$ of crosspoint area. As we scale to junctions which are just a few nanometers wide, there will only be 10's of molecule sites in the junction. Since we do not have the freedom of large numbers here, we are stuck with a small number junction yield probability.

We could require that all junctions associated with a wire have good switches, a minimum number of molecules (or a minimum resistance), in order to consider the wire yielded. However, in this case, large numbers work against us. We want the wire to connect to a large number of things to achieve the large number effects of the previous section. Requiring that a large number of junctions all meet some minimum threshold could make the wire yield rate so low as to be unusable. If all the junctions must yield for the wire to be usable, then the probability of yielding a wire becomes: $P_w = (P_j)^{N_j}$, where $N_j$ is the number of junctions in the wire. For the 85% junction yield rate, a wire that crosses 1000 junctions will yield only $P_w = 2.6 \times 10^{-69}\%$ of the time.

## Architectural Feature Example

Now, consider again the case where we are using the wire to perform channel routing.. Here, the wire simply needs to make good connections to a small number of crossed connections. As shown in Figure 7.6, we might only need

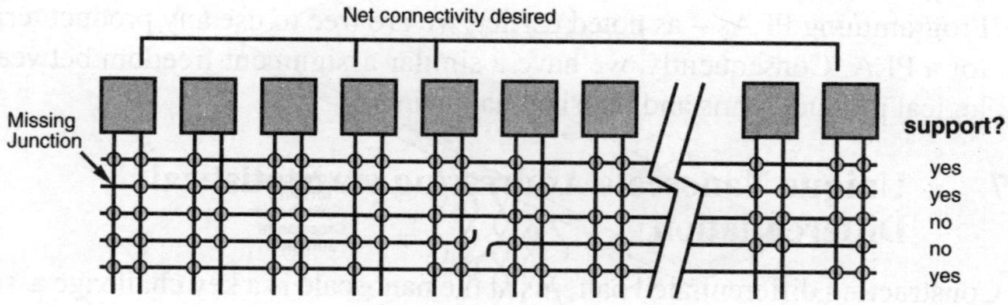


*Figure 7.6.* Channel Route Example. Figure shows compatibility of net with tracks in the presence of defective junctions.

to make four junction connections for a particular net route. If our junction faults are missing molecules, as suggested above, then we only need a wire which has four good junctions; the rest of the junctions can have fewer or no molecules. Now, the probability that a given wire can provide this particular connection is simply $P_{goodenough} = (P_j)^4$. For the aforementioned example, we calculate $P_{goodenough} = 52\%$. In general, a net that requires $N_c$ connections, has $P_{goodenough} = (P_j)^{N_c}$. Since $N_c << N_j$, there can be a significant difference between $P_w$ and $P_{goodenough}$ as the example illustrates. Further, we have the freedom of using any of the $N$ wires in the channel for this connection, so the probability that some wire in the channel can provide this connection is simply a $P_{MofN}$ calculation with $P = P_{goodenough}$ and $M = 1$. In a channel of 10 wires, the probability that at least one wire is good enough is over 99.9%. The whole problem of routing the channel then becomes an assignment problem of nets to wires. For a channel with 100 4-point nets, 13 spare wires are sufficient to guarantee over a 99.99% probability that all 4-point nets find a compatible track in the channel.

This idea can be extended further to deal with timing effects that might arise from fewer molecules in some junctions. In the simplest cases, we demand sufficient molecules to provide a minimum speed on all junctions. However, we could, further, use slow junctions, ones with fewer molecules, fewer dopants, or poor ohmic contacts, off the critical path. This is a familiar problem in resource binding for timing optimization (*e.g.* [2]).

## Other Cases with Choice

This choice of resources also shows up in our other structures:

1. Programming address decoders – when we program the address into a programmable address decoder, we may only need to have good connections between the particular "on"-bits in the address. This provides a scenario similar to the track routing case described above.

2. Programming PLAs – as noted earlier, we are free to use any product term for a PLA. Consequently, we have a similar assignment freedom between logical product terms and physical nanowires.

## 7.7 Unique Nanoscale Addressing via Statistical Differentiation

Constructing differentiated patterns at the nanoscale is a key challenge associated with nanoscale engineering. Our conventional lithographic approaches may not extend economically to the nanometer scale [17]. There are a number of bottom-up assembly techniques that can produce interesting nanoscale features (*e.g.* aforementioned parallel lines and crossbars, periodic lattices, self-assembled monolayers). These techniques, however, can generally only give us one of two things:

1. regular structures such as the crossbars, memory cores, and PLAs noted above
2. statistical structures

The regular, crossbar structures cannot be used without some way of programming their crosspoints—which will require some nanoscale differentiation so that individual nanoscale wires can be addressed. This leads us to explore the large number properties we can extract from the statistical assembly and ask if these can be used profitably in building our desired, nanoscale systems.

### Nanoscale Interfacing

A key challenge is addressing our nanoscale wires from conventional microscale wires. As noted, (1) we do not expect all of our nanoscale components to be perfect, and (2) our nanoscale components may start out as regular, undifferentiated arrays. If we can address our nanoscale wires from reliable, microscale wires, we can test resources, configure the system to use the functional resources, and programmably differentiate the regular structure.

Because of the scale difference between our microscale wires (*e.g.* 100–200nm pitch) and our nanoscale wires (*e.g.* <20nm pitch), it is not desirable to directly connect each nanoscale wire to a single microscale wire. A direct connection would prevent us from packing the nanoscale wires at their tight pitch. A natural solution to bridging between the micro- and nano-scale is to use a demultiplexer. The demultiplexer decodes a densely coded input and uses that to address one of its outputs. Using a demultiplexer here allows a small number of reliable, microscale wires to address a large number of nanoscale wires. From an information theoretic standpoint, the demultiplexer only needs $\log_2(N)$ input wires in order to specify which of the $N$ nanowires it should address. Since $\log_2(N)$ will be much smaller than $N$, for sufficiently large

$N$, this allows us to minimize the cost of the interface microscale wires and maintain the density benefit of the nanoscale wires.

### Statistical Assemble of Nanoscale Interfaces

Williams and Kuekes argue that a statistical scheme can be sufficient to construct such a demultiplexer [38]. They observe that we can use a physical process to produce a random distribution of gold particles in a layer between the microscale and nanoscale wires. These particles effectively provide sparse, random addresses for the nanowires. Using a sufficient number of microscale address wires, $5\log(N)$, they can uniquely address each of $N$ nanowires. This is an example of using the LLN along with our statistical effects to engineer a desired property.

DeHon, Lincoln, and Savage introduced a different technique for statistically achieving unique nanowire addressability using LLN effects [11]. They note that chemists have begun to demonstrate ways to differentiate individual nanowires [15] [39] [3]. In particular, it is now possible to vary the doping profile or material composition along the length of a nanowire, making some regions field-effect gateable while other regions act as wires and are not gateable. This allows us to give each nanowire an address. They further note that we can assemble a small number of these coded wires randomly selected from a large code space to achieve a unique set of wires with very high probability. A small example of this address decoder is shown in Figure 7.7. The result, reviewed in the following section, shows that we need a little over $2\log_2(N)$ address wires in order to uniquely address $N$ nanowires.

## 7.8 Generalizing Statistical Assembly

Since the previous section showed that large-number, statistical assembly can be a valuable tool for microscale to nanoscale interfacing, it is interesting to ask where else we can use this tool and what other properties we can achieve using statistical assembly of a large numbers of resources. We start by reviewing the details of the unique nanowire guarantee introduced in the previous section.

### Unique Wire Review

Once we had a physical mechanism to provide differentiated wires (*e.g.* nanowires with different doping profiles), and a suitable coding scheme (*e.g.* N/2-hot), the question became: How large of a code space, $C$, do we need so that a random selection of $N$ codes is unique? This turns out to be a generalized instance of the well known "Birthday Problem" (*e.g.* see p. 126 in [9]). In the birthday problem, the question is usually framed as: How many people must be in a room before there is a certain probability that at least two of them have the same birthday. Here, our code space, $C$, takes the place of the days in the

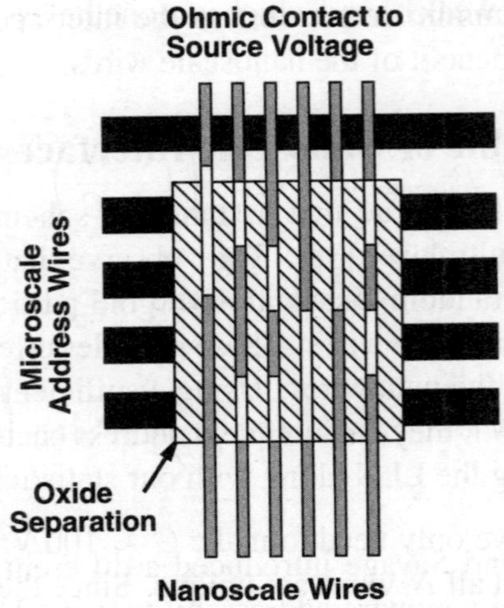


*Figure 7.7.* Stochastic Address Decoder using Coded Nanowires. Each nanowire is given an address by varying the doping profile along its length. Only the lightly doped (white) regions are gateable. The small example shown here uses a 2-hot code—every codeword has two high address bits and two low address bits. A nanowire only conducts when all of its lightly doped regions are gated by microscale control wires driven to low voltages. This example is shown small for clarity; more typical examples will use 10–30 microscale address wires and control 100-1000 nanowires.

year, the space we are sampling from, and our $N$ takes the place of the number of people.

Assume that we have a sufficiently large supply of each nanowire type so that selecting a nanowire of a given type does not change our sampling probability; that is, we always select each particular code with probability $1/C$ on each selection. Consider selecting nanowires one at a time. The first nanowire is definitely unique. The probability of selecting a unique nanowire on the second selection is $\frac{C-1}{C}$. Assuming we continue to select unique nanowires, the probability we select a unique nanowire on the $i$th selection is $\frac{C-i+1}{C}$. Thus, the probability of selecting $N$ unique nanowires from a code space of size $C$ is:

$$
\begin{aligned}
P_{unique}(C,N) &= 1 \cdot \left(\frac{C-1}{C}\right) \cdot \left(\frac{C-2}{C}\right) \cdots \left(\frac{C-N+1}{C}\right) \\
&> \left(\frac{(C-N)}{C}\right)^N
\end{aligned}
\tag{7.6}
$$

The last inequality is a weak bound which is, nonetheless, simple for estimation. Rearranging terms:

$$P_{unique}(C,N) > \left(1 - \frac{N}{C}\right)^{N} \tag{7.7}$$

Since $(1-x)^N \geq (1-Nx)$ when $x > 0$ and $N \geq 1$:

$$P_{unique}(C,N) > \left(1 - \frac{N}{C}\right)^{N} > 1 - \left(\frac{N^2}{C}\right) \tag{7.8}$$

Equation 7.8 tells us we only need to make $C \approx 100N^2$ in order to guarantee over a 99% chance that all $N$ wires are unique. Since the lower bound is weak, the actual guarantee is much tighter.

Using a dense code (*e.g.* $N/2$-hot code in [11]), the number of address wires we need is roughly the logarithm of the size of the code space we want to support. From this, we see that making $C = 100N^2$ means we need a little over twice as many address wires as a dense code (*i.e.* $\log(100N^2) = 2\log(N) + \log(100)$).

## Allowing Duplication

In the previous section we actually demanded a very strong property when we asked that *every* nanowire be unique. In practice, there are many applications where we can tolerate or even benefit from duplications. Duplication may result in multiple nanowires logically acting in tandem; this will lower the effective resistance of the replicated code or resources. The lower resistance of a parallel group of nanowires may reduce the delay on the logical group and provides some greater tolerance to contact faults. These lower resistance cases may be employed to accelerate the performance during resource assignment as described at the end of Section 7.6. However, variable ganging of multiple nanowires in parallel does have the effect of increasing the variance in device resistance which may be undesirable in some applications. A notable example where large variations in decoder resistance could be problematic are cases where we store memory bits in differential pairs to tolerate variations in junction off/on resistance. Duplicated nanowires can also raise the capacitance which must be driven when selecting a nanowire, which can, in turn, increase the energy requirements and increase the variance in power consumption. Consequently, the application and system context will be necessary to determine when duplication is allowable.

When duplications are allowed, we might want to know: Given that we select $N$ wires from $C$ different wire types, how many different wires types, $u$, should we expect to see?

Allowing duplication, we can derive a recurrence relationship for calculating the probabilities for all cases of $C$, $N$, and $u$:

$$\begin{aligned} P_{different}(C,N,u) = & \\ & \left(\frac{C-(u-1)}{C}\right) \times P_{different}(C,N-1,u-1) \\ & + \left(\frac{u}{C}\right) \times P_{different}(C,N-1,u) \end{aligned} \tag{7.9}$$

The recurrence in Equation 7.9 says we can extend the probability distribution from $N-1$ to $N$ in one of two ways. That is, either:

1. we have $u-1$ things in the $N-1$ case, and we select a new item not in the $u-1$ things already selected;
2. or we already have $u$ different things in the $N-1$ case, and we extend that by selecting among the $u$ different things we already have.

The base cases are:

- $P_{different}(C,1,1) = 1$ (if we pick one thing, we get one different thing)
- $P_{different}(C,1,u \neq 1) = 0$ (if we pick one, we will get exactly one different thing)
- $P_{different}(C,0,0) = 1$ (if we pick nothing, we get nothing)
- $P_{different}(C,0,u > 0) = 0$ (if we pick nothing, we get nothing)
- $P_{different}(C,N,u < 0) = 0$ (we cannot have less than nothing)

This recurrence counts each code once even if it appears multiple times, which is what we want if we allow duplications.

Since we are generally interested in achieving at least a certain number of different wires, we are interested in the cumulative distribution function (CDF) for the probability that we have at least a certain number of codes. We calculate this:

$$P_{atleast}\left(C,M,u\right) = \sum_{i=u}^{M} P_{different}\left(C,M,i\right) \tag{7.10}$$

We can now identify four basic regions of operation for stochastic population:

1. $N > C$
2. $N \approx C$
3. $N < C$
4. $N << C$

The $N << C$ case is what we saw with the address decoder and is useful when we simply want uniqueness of all resources. When duplications are allowed the other cases can be quite useful as well, as described below.

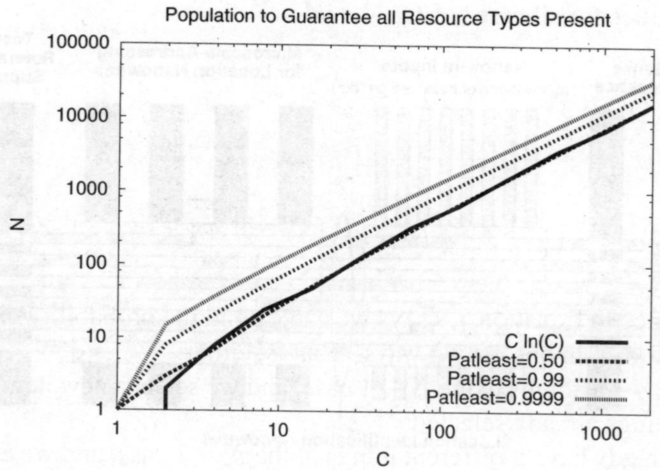


*Figure 7.8.* Number of Random Resources, $N$, needed to Guarantee "At Least One" of Each Resource Type, $C$

## At Least One of Each ($N > C$)

In the aforementioned address decoder, we were able to uniquely address each nanowire, but this required that we populate addresses sparsely. A natural question is: What would it take to guarantee all, or almost all, of the addresses were present? Naturally, in this case, we will need to assemble $N > C$ wires to have any chance of achieving this goal.

This is an instance of the "Coupon Collectors Problem" [21]. In the classic coupon collectors problem, the customer gets one coupon (or trading card) from a set of $C$ with each purchase of some item (*e.g.* bubble gum, cereal, cigarettes). The problem asks how many items the collector will have to buy before he can expect to have a full set? Maunsell and others show that the collector needs $N = C\ln(C)$ in order to have a 50% likelihood of gathering a complete set [21]. Gojman *et al.* revisit this directly for this nanowire decoding case [14]. Figure 7.8 plots the $N$ necessary to achieve various guarantees for the presence of all $C$ wire types based on our recurrence relation (Equation 7.9).

A logarithmic factor overhead to guarantee the presence of all wire types could be too expensive in many scenarios. However, in cases where the particular $C\ln(C)$ for the decoder is dominated by other area factors in the device, this may be reasonable. Notably, this could be useful for test access or bootstrap addressing.

As an illustration, consider the case where we want to find the location of the addressed nanowires in an array of $M$ nanowires. This might arise using our unique nanowire coding above. After we've found that a given address

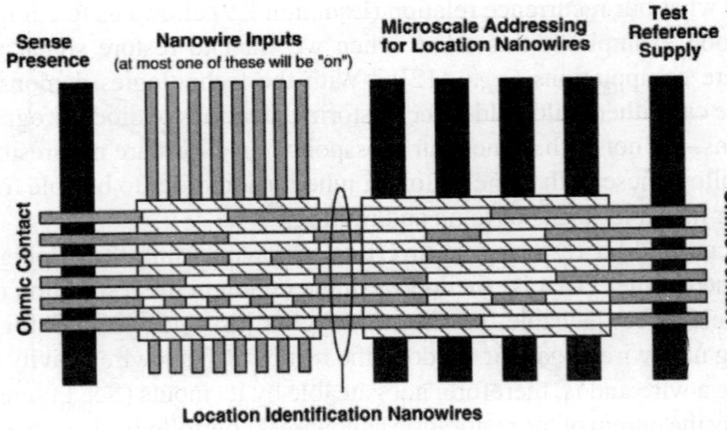


*Figure 7.9.* Ideal Construction of Test Structure to Locate the Position of a Single, Activated Nanowire. The location nanowires come in pairs to resolve one bit of the selected nanowire's position. Using the microscale address wires, we can enable a single location nanowire. Each nanowire will allow conduction from the reference voltage to the sensing line only when it is both enabled by the microscale addresses and the selected line does not overlap with its set of controllable inputs. By sequencing through all the location address wires, we can read off the location of the the single, selected input nanowire. To assemble this location array, we populate it with enough location nanowires to guarantee a high probability that each distinct location nanowire appears at least once as described in the text.

is in the array, we might want to know where it is in the array. If we could code up a set of nanowires which had different halves of their inputs active (field-effect controllable), it would only take $C_l = 2\log_2(M)$ such nanowires to uniquely determine the location of the single "on" nanowire in the array (See Figure 7.9). As before, we have the challenge that we cannot select exactly the right set of nanowires to place in the array. From the coupon collector result, we know that if we substitute $C = C_l = 2\log_2(M)$, into the above relation we need roughly $(2\log_2(M) \times \ln(2\log_2(M)))$ nanowires. Using the direct recurrence relation (Equation 7.9), we see that testing a $1024 \times 1024$ nanowire array ($C_l = 2\log_2(1024) = 20$)) in this manner will require that we populate the ends of the array with 149 of these randomly selected test wires to guarantees a 99% probability that we have one copy of each of the 20 nanowire types necessary for complete localization.

## Restoration Array ($N \approx C$)

The preceding suggests that demanding a full set of $C$ items could be prohibitively expense. We might instead want to know what fraction of the different resource types we would get with a given population, $N$. This, of course, is exactly what our recurrence relation (Equation 7.9) allows us to compute.

A good example of this arises when we want to restore signal values in nanowire computations (*e.g.* [12]). With the technologies demonstrated to date, we can either build field-effect restoring junctions or diode-programmable junctions, but not both. Since our crosspoints junctions are non-restoring, we must follow these with some restoring junctions in order to be able to cascade together multiple stages of logic and routing.

We can build a restoring or inverting nanowire using the doping profiles mentioned earlier. That is, we dope a single region in the restoration nanowire to be field-effect gateable, making this region just wide enough for a single crossing nanowire to gate it; we dope the rest of the nanowire heavily so that it acts like a wire and is, therefore, not gateable by its inputs (See Figure 7.10A). This way the output of the restoration nanowire is controlled only by the intended input. Given an array of nanowires with $N$ input nanowires, we would, ideally, like to have $N$ restored output nanowires, each coded to restore one of the inputs (See Figure 7.10B). As discussed, we cannot simply pick out one of each of the $N$ types of restoring wires and put them in the restoring plane. Instead, we stochastically populate the plane (See Figure 7.10C) and ask how many of the inputs we restore.

Figure 7.11 plots the fraction of unique nanowires, $u/N$ we get when $C = N$ for various $N$, showing that we get roughly 60% unique nanowires. This says a restoration scheme like this requires that we only populate 1.7 times as many raw nanowires as we want to have restored. We can, of course, tradeoff the number of restored inputs with the number of populated restoration wires. Figure 7.12 plots $u/C$ (fraction of restored outputs) versus $N/u$ (population factor) for $C = 1000$.

Figure 7.13 shows a complete nanoPLA from [12]. This design combines stochastic addressing (Section 7.7) for programming, M-of-N sparing (Section 7.5) to tolerate defects in the OR terms (horizontal nanowires), and restoration arrays based on this stochastic population scheme. The nanowires are all organized as two orthogonal sets of parallel nanowires making the design compatible with the known, regular assembly techniques (*e.g.* [37]).

## Mostly Unique Addressing ($N < C$)

In the preceding, we saw that we needed a moderately large code space ($C > 100N^2$) to have a high probability of obtaining completely unique addresses. However, if we are willing to accept a few duplicates, we can shrink the code space considerably. This may be useful in reducing the number of distinct nanowire types we need to manufacture and in reducing the number of microscale address wires. Figure 7.14 plots the fraction of resource duplicates

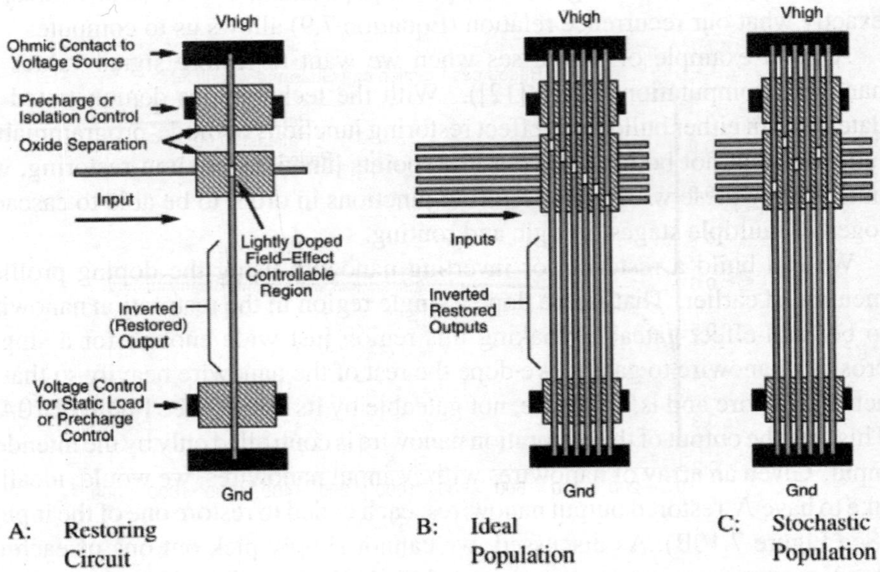


*Figure 7.10.* Nanowire Restoration Array. Using field-effect gating in the nanowires, we can isolate an output from an input and restore, possibly inverting, the input. Part A shows the basic arrangement for inverting a single input. The input and output nanowire are separated by a thin oxide so that the electrical field from the input (horizontal) nanowire can potentially effect the output (vertical) nanowire. The output is heavily doped along its length so that it is not effected by most input fields. However it is lightly doped in the region of the designated input so that the voltage on the input nanowire can gate the output nanowire. If the output nanowire is a depletion-mode P-type nanowire, then a low input will allow the output nanowire to conduct current from the high supply ($V_{high}$) and pull the output to a high voltage level. A high input will deplete the carriers in the lightly doped control region and cutoff conduction from the high supply; the weak pulldown or precharge load at the bottom of the structure pulls the output to low in this case. The structure in A thus behaves as an inverter. Note that the output draws current from the supplies, not from the input nanowire. DeHon shows that these field-effect nanowire gates have sufficient non-linearity to provide gain [10]. B and C show how this structure is assembled in an array to invert a set of inputs.

we can expect as we increase the $C/N$ ratio. For $N = 1000$, this suggests we only need $C = 8N = 8000$ to achieve over 90% (900) unique resources over 99% of the time. To achieve 99% (990) unique resources 99% of the time, we need $C = 128N = 128,000$. Both numbers are, of course, much lower than the $C = 100N^2 = 10^8$ unique resources or code words which we require to achieve guaranteed non-duplication according to our weak uniqueness bound (Equation 7.8).

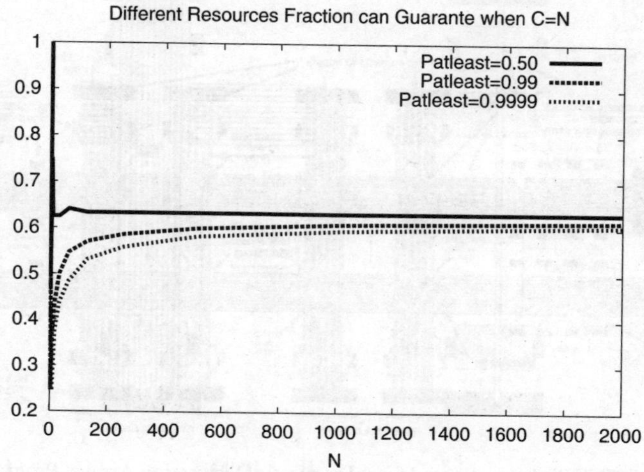


*Figure 7.11.* Fraction of Distinct Resource Guaranteed when $C = N$

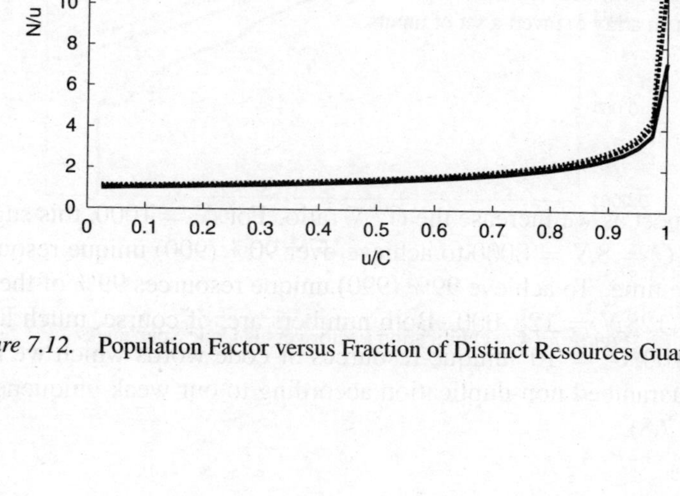


*Figure 7.12.* Population Factor versus Fraction of Distinct Resources Guaranteed

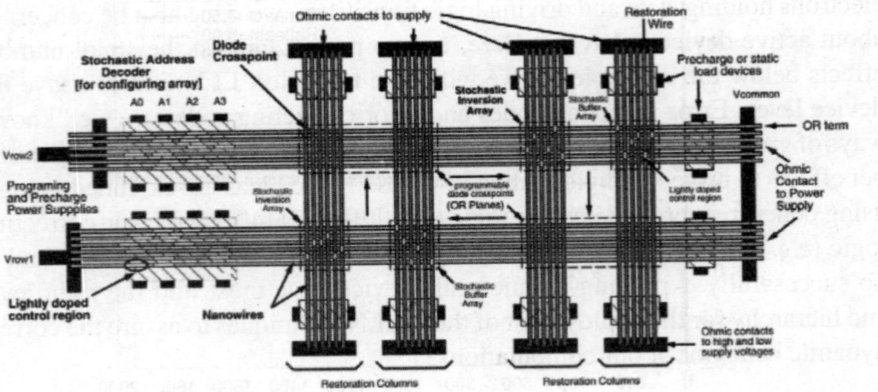


*Figure 7.13.* NanoPLA constructed from Stochastic Restoration Arrays, Programmable Crosspoints, and Stochastic Addressing

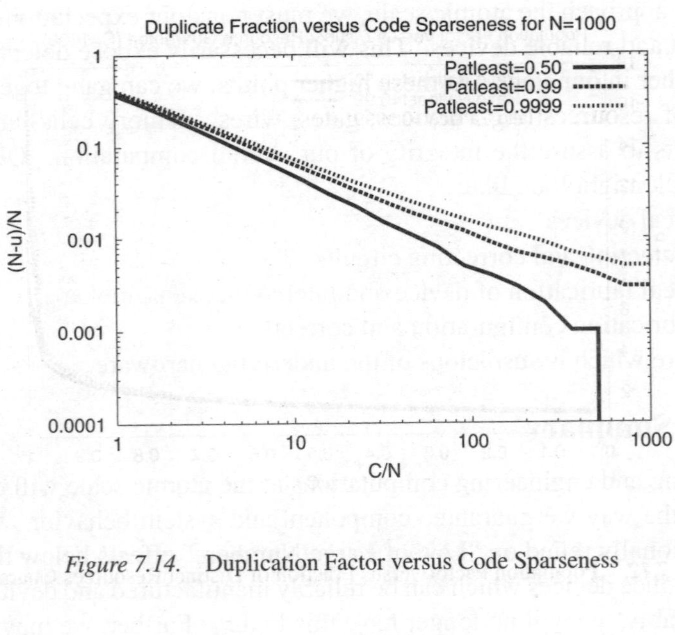


*Figure 7.14.* Duplication Factor versus Code Sparseness

## 7.9 Fault Tolerance

In the preceding sections, we have focused on the statistical nature of fabrication and assembly at the nanoscale. Since we will also have a small number of electrons holding state and driving logic transitions, we must also be concerned about active device behavior. Here, too, to protect against the small number effects below the device level, we will have to exploit LLN effects above the device level. Error-detecting codes and error-correcting codes are well known ways of grouping many data elements together so that we can exploit large number effects to protect individual bits or wires. These ideas are extended to logic using concurrent error-detecting logic (*e.g.* [20] [32] [40]) and error-correcting logic (*e.g.* [23] [25]). Error-correcting logic is reviewed in Chapters 2 and 6. To successfully exploiting atomic-scale devices, we must find the right level and hierarchy for the deployment of these LLN techniques to assure the correct dynamic behavior of our computations.

## 7.10 Atomic-Scale System Stack

Our traditional system stack has consisted of:

1. Reliable devices
2. Gates
3. Perfect and deterministic fabrication of interconnected gates and memories
4. Reliable architecture abstraction
5. Software which assumes perfect hardware

In order to approach the atomic scale, we must relax our expectation of perfect fabrication and reliable devices. This will necessarily expose defect and fault effects higher in our stack. At these higher points, we can gang together large numbers of resources (*e.g.* devices, gates, wires, memory cells) and exploit LLN effects to assure the integrity of our overall computation. Our revised system stack might look like:

1. Statistical devices
2. Fault detecting and correcting circuits
3. Statistical fabrication of device and interconnect ensembles
4. Post fabrication configuration and correction
5. Software which is suspicious of the underlying hardware

## 7.11 Summary

Designing and engineering computations at the atomic scale will demand a change in the way we guarantee component and system behavior. While we have traditionally relied on "Law of Large Numbers" effects below the device level to produce devices which can be reliably manufactured and devices which behave reliably, we will no longer have this luxury. Further, we may not have the ability to deterministically specify how a structure is built at the atomic scale. We may be forced to build regular structures and accept that the only differentiation we can get at the atomic scale will be statistical in nature.

Nonetheless, we can still exploit "Law of Large Numbers" statistical techniques to build reliable components and systems. We simply need to apply these LLN techniques above the device level. Sections 7.5–7.8 suggest that combining LLN statistical yield and differentiation with programmability is a sufficiently powerful tool for the reliable construction of nanoscale devices despite the fact that we cannot guarantee the yield of individual devices or the exact placement or selection of resources. Large number statistical differentiation gives us strong enough properties to construct the gross structure of a system and achieve sufficiently unique nanoscale addressing. With this access and post-fabrication crosspoint configurability, we can test and program the device to avoid the defective portions and to behave in a desired, deterministic manner despite the statistical nature of the initial assembly and device yield rate.

Atomic-scale design will necessitate a shift in our system engineering approach, where the statistical effects are exposed to higher levels of the system stack. This allows us to solve problems at the appropriate level, applying LLN where large numbers of components can participate.

## 7.12 Acknowledgments

This research was funded in part by the DARPA Moletronics program under grant ONR N00014-01-0651. This chapter synthesizes ideas from a growing community of researchers, including Seth C. Goldstein, Philip J. Kuekes, John Kubiatowicz, Charles M. Lieber, Patrick D. Lincoln, Helia Naeimi, John E. Savage, Howard E. Shrobe, and Michael J. Wilson. Michael Wrighton and Sandeep Shukla provided valuable feedback on early manuscripts which allowed us to improved the presentation in this chapter.

## References

[1] Jakob Bernoulli. *Ars Conjectandi*. Impensis Thurnisiorum, fratrum, Basel, Switzerland, 1713.

[2] Vaughn Betz, Jonathan Rose, and Alexander Marquardt. *Architecture and CAD for Deep-Submicron FPGAs*. Kluwer Academic Publishers, 101 Philip Drive, Assinippi Park, Norwell, Massachusetts, 02061 USA, 1999.

[3] M. T. Björk, B. J. Ohlsson, T. Sass, A. I. Persson, C. Thelander, M. H. Magnusson, K. Depper, L. R. Wallenberg, and L. Samuelson. One-dimensional Steeplechase for Electrons Realized. *Nano Letters*, 2(2):87–89, February 2002.

[4] Christopher L. Brown, Ulrich Jonas, Jon A. Preece, Helmut Ringsdorf, Markus Seitz, and J. Fraser Stoddart. Introduction of [2]Catenanes into Langmuir Films and Langmuir-Blodgett Multilayers. A Possible Strategy for Molecular Information Storage Materials. *Langmuir*, 16(4):1924–1930, 2000.

[5] Yong Chen, Gun-Young Jung, Douglas A. A. Ohlberg, Xuema Li, Duncan R. Stewart, Jan O. Jeppesen, Kent A. Nielsen, J. Fraser Stoddart, and R. Stanley Williams. Nanoscale Molecular-Switch Crossbar Circuits. *Nanotechnology*, 14:462–468, 2003.

[6] Yong Chen, Douglas A. A. Ohlberg, Xuema Li, Duncan R. Stewart, R. Stanley Williams, Jan O. Jeppesen, Kent A. Nielsen, J. Fraser Stoddart, Deirdre L. Olynick, and Erik Anderson. Nanoscale Molecular-Switch Devices Fabricated by Imprint Lithography. *Applied Physics Letters*, 82(10):1610–1612, 2003.

[7] C. Collier, G. Mattersteig, E. Wong, Y. Luo, K. Beverly, J. Sampaio, F. Raymo, J. Stoddart, and J. Heath. A [2]Catenane-Based Solid State Reconfigurable Switch. *Science*, 289:1172–1175, 2000.

[8] C. P. Collier, E. W. Wong, M. Belohradsky, F. M. Raymo, J. F. Stoddard, P. J. Kuekes, R. S. Williams, and J. R. Heath. Electronically Configurable Molecular-Based Logic Gates. *Science*, 285:391–394, 1999.

[9] Thomas Cormen, Charles Leiserson, and Ronald Rivest. *Introduction to Algorithms*. MIT Press, 1990.

[10] André DeHon. Array-Based Architecture for FET-based, Nanoscale Electronics. *IEEE Transactions on Nanotechnology*, 2(1):23–32, March 2003.

[11] André DeHon, Patrick Lincoln, and John Savage. Stochastic Assembly of Sublithographic Nanoscale Interfaces. *IEEE Transactions on Nanotechnology*, 2(3):165–174, 2003.

[12] André DeHon and Michael J. Wilson. Nanowire-Based Sublithographic Programmable Logic Arrays. In *Proceedings of the International Symposium on Field-Programmable Gate Arrays*, pages 123–132, February 2004.

[13] Alvin W. Drake. *Fundamentals of Applied Probability Theory*. McGraw-Hill, Inc., 1988.

[14] Benjamin Gojman, Eric Rachlin, and John E. Savage. Decoding of Stochastically Assembled Nanoarrays. In *Proceedings of the 2004 International Symposium on VLSI*, February 2004.

[15] Mark S. Gudiksen, Lincoln J. Lauhon, Jianfang Wang, David C. Smith, and Charles M. Lieber. Growth of Nanowire Superlattice Structures for Nanoscale Photonics and Electronics. *Nature*, 415:617–620, February 7 2002.

[16] Richard W. Hamming. Error Detecting and Correcting Codes. *Bell Systems Technical Journal*, 29(2):147–160, April 1950.

[17] Lloyd R. Harriott. Limits of Lithography. *Proceedings of the IEEE*, 89(3):366–374, March 2001.

[18] Yu Huang, Xiangfeng Duan, Qingqiao Wei, and Charles M. Lieber. Directed Assembley of One-Dimensional Nanostructures into Functional Networks. *Science*, 291:630–633, January 26 2001.

[19] Bent Keeth and R. Jacob Baker. *DRAM Circuit Design: A Tutorial*. Microelectronic Systems. IEEE Press, 2001.

[20] Javad Khakbaz and Edward J. McCluskey. Concurrent Error Detection and Testing for Large PLA's. *IEEE Journal of Solid-State Circuits*, 17(2):386–394, April 1982.

[21] F. G. Maunsell. A Problem in Cartophily. *The Mathematical Gazette*, 22:328–331, 1937.

[22] Nicholas A. Melosh, Akram Boukai, Frederic Diana, Brian Gerardot, Antonio Badolato andPierre M. Petroff, and James R. Heath. Ultrahigh-Density Nanowire Lattices and Circuits. *Science*, 300:112–115, April 4 2003.

[23] John Von Neumann. Proabilistic Logic and the Synthesis of Reliable Organisms from Unreliable Components. In Claude Shannon and John McCarthy, editors, *Automata Studies*. Princeton University Press, 1956.

[24] Fazil I. Osman. Error-Correction Technique for Random-Access Memories. *IEEE Journal of Solid State Circuits*, 17(5):877–881, October 1982.

[25] Nicholas Pippenger. Developments in 'The Synthesis of Reliable Organisms from Unreliable Components'. In *Proceedings of the Symposia of Pure Mathematics*, volume 50, pages 311–324, 1990.

[26] (Ed.) Jon Postel. Transmission Control Protocol – DARPA Internet Program Protocol Specification. RFC 793, USC/ISI, Information Sciences Institute, University of Southern California, 4676 Admiralty Way, Marina del Rey, California, 90291, September 1981.

[27] Thomas Rueckes, Kyoungha Kim, Ernesto Joselevich, Greg Y. Tseng, Chin-Li Cheung, and Charles M. Lieber. Carbon Nanotube Based Nonvolatile Random Access Memory for Molecular Computing. *Science*, 289:94–97, 2000.

[28] Stanley E. Schuster. Multiple Word/Bit Line Redundancy for Semiconductor Memories. *IEEE Journal of Solid State Circuits*, 13(5):698–703, October 1978.

[29] Claude E. Shannon. A Mathematical Theory of Communication. *Bell Systems Technical Journal*, 27:379–423, 623–656, 1948.

[30] Daniel P. Siewiorek and Robert S. Swarz. *Reliable Computer Systems: Design and Evaluation*. Digital Press, Burlington, MA, 2nd edition, 1992.

[31] Enn Tammaru and James B. Angell. Redundancy for LSI Yield Enhancement. *IEEE Journal of Solid State Circuits*, 2(4):172–182, December 1967.

[32] Nur A. Touba and Edward J. McCluskey. Logic Synthesis of Multilevel Circuits with Concurrent Error Detection. *IEEE Transactions on Computed-Aided Design for Integrated Circuits and Sytems*, 16(7):783–789, July 1997.

[33] Stephen Trimberger. *Field Programmable Gate Arrays*. Kluwer Academic Press, 1992.

[34] Stephen Trimberger. A Reprogrammable Gate Array and Applications. *Proceedings of the IEEE*, 81(7):1030–1041, July 1993.

[35] Chin-Long Wey, Man-Kuan Vai, and Fabrizio Lombardi. On the Design of Redundant Programmable Logic Array (RPLA). *IEEE Journal of Solid State Circuits*, 22(1):114–117, February 1987.

[36] Dongmok Whang, Song Jin, and Charles M. Lieber. Nanolithography Using Hierarchically Assembled Nanowire Masks. *Nanoletters*, 3(7):951–954, July 9 2003.

[37] Dongmok Whang, Song Jin, Yue Wu, and Charles M. Lieber. Large-Scale Hierarchical Organization of Nanowire Arrays for Integrated Nanosystems. *Nanoletters*, 3(9):1255–1259, September 2003.

[38] Stanley Williams and Philip Kuekes. Demultiplexer for a Molecular Wire Crossbar Network. United States Patent Number: 6,256,767, July 3 2001.

[39] Yiying Wu, Rong Fan, and Peidong Yang. Block-by-Block Growth of Single-Crystalline Si/SiGe Superlattice Nanowires. *Nano Letters*, 2(2):83–86, February 2002.

[40] Chaohuang Zeng, Nirmal R. Saxena, and Edward J. McCluskey. Finite State Machine Synthesis with Concurrent Error Detection. In *Proceedings of the International Test Conference*, pages 672–679, 1999.

# NANO-SCALE QUANTUM COMPUTING

III

# NANO-SCALE QUANTUM COMPUTING

# Preface

*How can we simulate the quantum mechanics? ...Can you do it with a new kind of computer — a quantum computer? It is not a Turing machine, but a machine of a different kind.*

The above quote from Richard Feynman made in 1981, summarizes the motivation that gave rise to the idea of a quantum computer. The scale of quantum physical phenomena is so vast, that even a super computer built on von Neumann style computing cannot realistically model quantum physics at the atomic and sub-atomic level. On the other hand, quantum computers, which mimic the quantum physics themselves, are capable of vast parallelism and could theoretically simulate such phenomenon. In 1985 seminal work by Deutsch showed that quantum computers can create a quantum superposition of states, allowing each state to follow a distinct computational path until a final output is obtained. Such free access to parallelism is unprecedented using a classical model of computation.

With the advent of new nanoscale technologies, we see quantum computing as more than a source of large scale parallelism. It is likely that in the near future we can exploit quantum entanglement, and quantum mechanical reversible transformations to build new kinds of computing systems.

In the last decade, theoretical results have led to practical concerns as to the vulnerability of systems dependent on cryptographic secrecy. Currently, cryptography often relies on the difficulty of factoring large number using traditional models of computation. However, in 1994 Peter Shor described a polynomial-time algorithm for factoring large numbers on a hypothetical quantum computer.

Another motivating factor in quantum computing comes from the possibility of power reduction inherent in reversible computing. In 1973, Charles Bennett's article entitled "The thermodynamics of computation" discussed reversibility in quantum computation steps. This work motivates us to ask if we can we actually realize computing logic that is reversible and hence would free us from the growing power concerns in CMOS-based computing.

All this said, we present in this part three very interesting chapters that are all concerned with the physical realization of quantum models of computation, and reliability issues inherent therein. Overall, we believe this part of the



*S.K. Shukla and R.I. Bahar (eds.), Nano, Quantum and Molecular Computing*, 245-246.

book will familiarize the readers with the basics of Quantum Computing models, Quantum-Dot Cellular Automata, and some of the system level statistical phenomena and reliability issues relevant to quantum computing.

# Chapter 8

# CHALLENGES IN RELIABLE QUANTUM COMPUTING

Diana Franklin
*Department of Computer Science*
*California Polytechnic State University, San Luis Obispo*
dkeen@csc.calpoly.edu

Frederic T. Chong
*Department of Computer Science*
*University of California, Davis*
chong@cs.ucdavis.edu

**Abstract** Quantum computing is a new and promising technology with the potential of exponentially powerful computation - if only a large-scale one can be built. There are several challenges in building a large-scale quantum computer - fabrication, verification, and architecture. The power of quantum computing comes from the ability to store a complex state in a single bit. This also what makes quantum systems difficult to build, verify, and design. Quantum states are fragile, so fabrication must be precise, and bits must often operate at very low temperatures. Unfortunately, the complete state may not be measured precisely, so verification is difficult. Imagine verifying an operation that is expected to not always get the same answer, but only an answer with a particular probability! Finally, errors occur much more often than with classical computing, making error correction the dominant task that quantum architectures need to perform well. We provide a basic tutorial of quantum computation for the system designer and examine the fundamental design and verification issues in constructing scalable quantum computers. We find the primary issues to be the verification of precise fabrication constraints, the design of quantum communication mechanisms, and the design of classical control circuitry for quantum operation.

**Keywords:** quantum, fabrication, verification, error correction, nanotechnology



*S.K. Shukla and R.I. Bahar (eds.), Nano, Quantum and Molecular Computing*, 247-266.

## Introduction

While the ideas for quantum computing have been gestating for years, it has gained momentum in the past decade due to key discoveries in potential, feasibility, and implementation. In 1994, Shor presented an algorithm that, in polynomial time, can factor very large numbers. The fastest known classical algorithm requires exponential time. With Shor's algorithm, our current cryptographic techniques would be rendered useless, if a quantum computer could be built. Building a large-scale computer became more feasible with the Threshold Theorem [2], which essentially showed (in theory) an error correction technique that can correct errors as quickly as they can be introduced into the system. Finally, several groups have successfully built 5-10 quantum bit computers [30, 19, 23].

As promising as these developments are, we are still a long way from building a reliable, large-scale quantum computer. The physical properties of quantum matter make it very difficult to build a reliable quantum computer. There are several aspects to this - error correction, communication and fabrication.

Quantum error correction is both more challenging and more cumbersome than in classical computing. The power of quantum computation lies in the fact that each quantum bit stores much more information than a classical bit. Unfortunately, this means that a simple bit-flip is just one of many errors that can occur. This, alone, would not be a large problem except that the quantum data can not be copied exactly, nor measured precisely. In fact, a direct measurement alone changes the state of a quantum bit! This leads to a solution that, in the best case, requires seven physical bits for every logical bit. This seven to one encoding is applied hierarchically - each time it is applied, the error rate is squared (reducing it). In essence, quantum computers are trading off space for time - they take less time to compute, but error-correction requires polynomial times more space for the extra bits []. In addition, an unreliable computation must be run more times to increase the chances of getting the desired result. The original error rate is higher as the length of computation increases and the underlying technology is less reliable. The hidden advantage of error correction taking so many more resources than the computation itself is that, regardless of how future quantum algorithms behave, an architecture optimized for error-correction will be the most efficient design. We need not wait for a benchmark suite to be developed to know how to design the architecture - error-correction is the only benchmark we need.

Optimizing for error-correction means dealing with the fact that quantum circuits will be very large (since so many bits are required). In order to have these bits interact, they need to travel long distances. Thus, quantum computers need to be optimized for communication. Traditional computation uses a wire to transport a copy of a value to a new location. In quantum computing, we have neither wires nor the ability to make copies. The bit must be moved, not copied.

A highly reliable fabrication technology reduces the amount of error correction necessary, so verification is vital. Because the quantum elements are so small and fragile, incredible precision is required, much more so than with classical computing. While this could be solved through rigorous testing, the test itself is difficult because, given the same operations, many quantum operations are not expected to always produce the same measured result because there is no way to precisely measure the complete quantum state. As much as possible, classical methods must be used to test the fabricated chip.

We begin with an introduction to quantum computation, including some proposed implementations and basic algorithms. Error correction is presented in Section 8.2. Section 8.3 focuses on implementing a quantum computer in two technologies - silicon and ion traps. Architectural difficulties in communication are presented in Section 8.5, and we give conclusions in Section 8.6.

## 8.1 Quantum Computation

Just as classical systems can be represented in boolean algebra, with no specification of the implementation, quantum systems have a mathematical representation to describe their states and operations on those states. We begin by presenting quantum bits and operators in this mathematical abstraction, followed by several possible implementations. Finally, to give an idea of the power of quantum computing, we present two famous quantum algorithms. This is just a short introduction, and more information can be found in [22].

### Quantum Bits and Operators

The basic building block in quantum computing is a *quantum bit*, or *qubit* for short. The main difference between a bit and a qubit is that a bit is restricted to the state 0 or 1. A qubit, on the other hand, is not restricted to its analogous states $|0\rangle$ and $|1\rangle$; It is also possible to form *linear combinations* of states

$|\psi\rangle = \alpha|0\rangle + \beta|1\rangle$, which can also be expressed $\begin{bmatrix} \alpha \\ \beta \end{bmatrix}$

$\alpha$ and $\beta$ are complex numbers, but thinking of them as real numbers is usually sufficient for our purposes.

Initially, this seems like a huge advantage over classical computing - the ability to store complex numbers within a single bit of data. There are two basic operations that we take for granted in classical computing that are difficult in quantum computing - measurement and replication.

There is no way to precisely measure the state of the quantum bit (i.e. determine $\alpha$ and $\beta$). Instead, when we measure the qubit, the outcome is either $|0\rangle$ or $|1\rangle$. The significance of the $\alpha$ and $\beta$ values is that the probability of measuring a $|0\rangle$ is $\alpha^2$, and the probability of $|1\rangle$ is $\beta^2$. Because there are only two possible outcomes, it must hold that $\alpha^2 + \beta^2 = 1$. Furthermore, the direct measurement of a qubit alters its state. Even if the qubit has the actual state $\begin{bmatrix} \frac{1}{2} \\ \frac{\sqrt{3}}{2} \end{bmatrix}$ and happens to be measured as a $|0\rangle$, from then on, it will always measure $|0\rangle$. The act of measurement changes $\alpha$ and $\beta$ (in this case, $\alpha$ becomes 1, and $\beta$ becomes 0).

The problem of measurement affecting the state would not be nearly as bad, surely, if we could copy the original qubit before we measured it. Then, we could measure the copy and retain the original for future computations. Unfortunately, qubits can not be copied without the destruction (through measurement) of the original bit. In classical computing, voltage can be applied and split off into an arbitrary number of wires, allowing one input bit to be outputted to several destinations. Quantum bits do not work this way. Instead, we would need to find a gate that takes as input the qubit to be copied and some known qubit, and output two identical qubits that match the first qubit. This would be analogous to having an or gate and knowing that if you input x and 0, you get x as the output. Unfortunately, for reasons we will see later, no such quantum gate can exist. These two issues - lack of measurement accuracy and the inability to copy bits - make it difficult to test fabrication and provide error correction in the classical way.

Now that we have seen that a bit is no ordinary binary bit, but rather a vector that has the probability of expressing different results, what sort of operations occur on such a vector, and how does this turn into meaningful computation?

Because the qubit is expressed as a vector, single qubit operators, or *quantum gates*, are expressed as 2x2 matrices. In order to be a valid operator, the result must still conform to $\alpha^2 + \beta^2 = 1$. It turns out that any matrix which transforms a source vector with that property to a result vector with that same property is said to be *unitary*, that is $U^tU = \mathrm{I}$, where:

$$U = \begin{bmatrix} a+bi & c+di \\ e+fi & g+hi \end{bmatrix} \text{ and } U^t = \begin{bmatrix} a-bi & e-fi \\ c-di & g-hi \end{bmatrix}$$

The requirement that the operator be unitary greatly restricts what can be done in quantum computing. Most classical operations are not unitary, because you can not get back the original value once the operation is performed. In fact, any operation that takes two bits and produces one is not unitary.

The ability to reverse computation has practical implications. This creates an explosion in the number of bits required to perform computation, because the information necessary to reverse all operations performed must be part of the computation. This is often in the form of requiring extra bits from the very

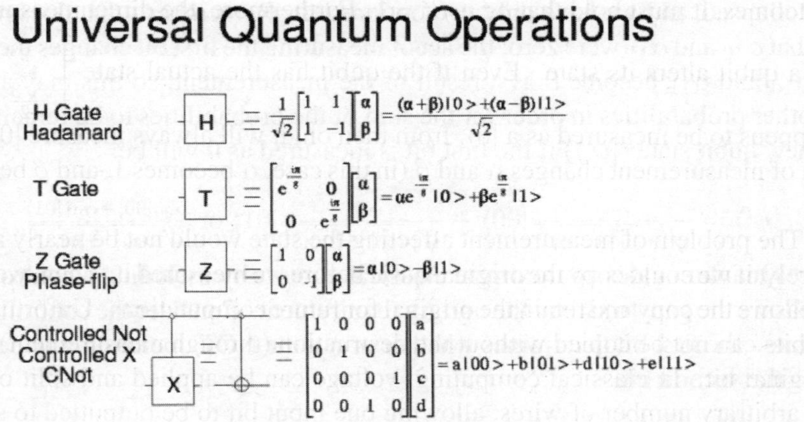


*Figure 8.1.* **Important 1-bit quantum gates**

beginning. While each bit encodes much more information than in classical computing, this unitary requirement requires extra space, greatly increasing the number of bits required to execute an algroithm.

Let us play around with a few quantum gates. The first gate we present is the quantum NOT gate, which is traditionally called X. A classical not gate flips a 0 to a 1 and a 1 to a 0. The quantum equivalent flips the coefficients of $|1\rangle$ and $|0\rangle$.

$$\alpha|0\rangle + \beta|1\rangle \rightarrow \beta|0\rangle + \alpha|1\rangle.$$

$$X = \begin{bmatrix} 0 & 1 \\ 1 & 0 \end{bmatrix} \text{ which has the effect: } X \begin{bmatrix} \alpha \\ \beta \end{bmatrix} = \begin{bmatrix} \beta \\ \alpha \end{bmatrix}$$

Some other important 1-bit quantum gates are shown in Figure 8.1

Things get a little more complicated when we add more qubits to the system. Each additional gate doubles the number of possible outcomes. Now, as before, the probabilities must sum to 1. The qubit state is:

$$|\psi\rangle = \alpha_{00}|00\rangle + \alpha_{01}|01\rangle + \alpha_{10}|10\rangle + \alpha_{11}|11\rangle$$

where

$$\alpha_{00}^2 + \alpha_{01}^2 + \alpha_{10}^2 + \alpha_{11}^2 = 1$$

If we measure, say, the first bit, and it happens to be a 0, then our system has collapsed, and we have only two options left. Remember - that does not mean that $\alpha_{10}$ and $\alpha_{11}$ were zero; the act of measuring the first bit changes the system. $\alpha_{10}$ and $\alpha_{11}$ become 0 as a result of the measurement, so this may affect the other probabilities in order for the sum of the probabilities to be remain 1. The new qubit state ($|\psi'\rangle$) if the first bit is measured as 0 will be:

$$|\psi'\rangle = \frac{\alpha_{00}}{\sqrt{|\alpha_{00}|^2+|\alpha_{01}|^2}}|00\rangle + \frac{\alpha_{01}}{\sqrt{|\alpha_{00}|^2+|\alpha_{01}|^2}}|01\rangle \text{ or } \frac{\alpha_{00}|00\rangle+\alpha_{01}|01\rangle}{\sqrt{|\alpha_{00}|^2+|\alpha_{01}|^2}}$$

Quantum gates approximate classical circuits by including a control bit that allows the gate to satisfy the unitary requirement. Consider the controlled-NOT gate - the first bit determines whether or not the NOT operation will be applied to the second bit.

$$|00\rangle \rightarrow |00\rangle; |01\rangle \rightarrow |01\rangle; |10\rangle \rightarrow |11\rangle; |11\rangle \rightarrow |10\rangle;$$

which corresponds to:

$$\alpha_{00}|00\rangle + \alpha_{01}|01\rangle + \alpha_{10}|10\rangle + \alpha_{11}|11\rangle \rightarrow \alpha_{00}|00\rangle + \alpha_{01}|01\rangle + \alpha_{11}|10\rangle + \alpha_{10}|11\rangle$$

Although there is a mapping of classical algorithms to quantum computer, this is not the true power of quantum computing. The key is in restricting the set of possible outputs and making the different bits have a relationship to each other. For example, one could built a circuit that puts 8 bits through a series of transformations such that, for each of the possibilities for the first 4 bits (x), the second 4 bits was the solution to some function f(x). Upon measurement, you would find out one (x,f(x)) pair, though you would not know which one. If you were to somehow restrict the values of the first 4 bits, you could further control which (x,f(x)) pair you measured.

## An Example Prototype

Now that you have some idea of the mathematical properties of quantum computing, we describe a concrete implementation to give us some physical way to think about the computations.

The largest, and perhaps most complete, prototype is a *bulk-spin NMR* machine which used the nuclei of a synthetic molecule to represent 7 quantum bits [30]. Each nucleus has a different resonant frequency, which allows the NMR machine (nuclear magnetic resonance) to both manipulate and measure the qubits. Current NMR machines, however, do not have the resolution to manipulate or measure a single molecule. Consequently, the experiment is done in bulk, where the same computation is simultaneously performed on a solution of many molecules and a histogram of the measured resonances is used to infer the result of a computation on a single molecule. This approach was used to demonstrate Shor's algorithm (described in the next section) factoring the number 15 into 3 and 7. Although this sounds trivial, it is actually a very significant step towards viable quantum computers.

## Quantum Algorithms

Designers of quantum algorithms must be very clever about how to get useful answers out of their computations. One method is to iteratively skew probability amplitudes in a qubit vector until the desired value is near 1 and the other values are close to 0. This is used in Grover's algorithm for searching an unordered list of $n$ elements [12]. The algorithm goes through $\sqrt{n}$ iterations, at which point a qubit vector representing the keys can be measured. The desired key is found with high probability.

Another option in a quantum algorithm is to arrange the computation such that it does not matter which of many random results is measured from a qubit vector. This method is used in Shor's algorithm for prime factorization of large numbers [25], which is built upon the quantum Fourier transform, an exponentially fast version of the classical discrete Fourier transform. The quantum bits are manipulated so that they contain values with a certain period, r, that holds the key to our factorization. A Fourier transform is used to transform the series into one with period k/r. This period is a fraction, so many of the values in the series have r as their denominator. We now measure the result and use continued fraction expansion to determine r. If we happen to measure an integer rather than a fraction, we can repeat the calculation. Since prime factorization of large numbers is the basis of nearly all modern cryptographic security systems, Shor's algorithm has received much attention.

Additional algorithms include adiabatic solution of optimization problems [6]; precise clock synchronization [15, 7]; quantum key distribution [5]; and very recently, Gauss sums [29] and Pell's equation [13].

# 8.2 Error correction

Perhaps the single most important concept in devising a quantum architecture is quantum error correction. Unlike classical systems, where error correction can be performed by brute-force, signal-level restoration in every transistor, correction of quantum states requires a much more subtle and involved strategy. In fact, localized errors on a few qubits can have a global impact on the exponentially large state space of many qubits. As you will see later in quantum teleportation, some quantum operations require a certain relationship between two bits. Consider, for example, two qubits that have the state $|\Psi\rangle = \frac{1}{\sqrt{2}}(|00\rangle + |11\rangle)$. It is vital that the states $|01>$ and $|10>$ have probability 0. If one bit's state is corrupted, the relationship is changed, and its pair is also useless. In this section, we describe the basic idea of quantum error correction, and how it can be applied recursively to provide fault-tolerant quantum computation. We summarize problems with this traditional approach, and identify opportunities for optimization and key points for architectural design improvements.

The need for error correction is made manifest by a simple calculation. If a single qubit gate fails with probability $p = 1 - e^{-\lambda}$, then in the absence of error correction, the failure probability after $n$ gates is at worst $1 - e^{-n\lambda}$. Thus, to have a circuit fail with at most probability $\epsilon$, the computation would have to be less than $-[\ln(1-\epsilon)]/\lambda$ gates in length, in general. Given the technological considerations of Section 8.1, this would be a very short computation.

The difficulty of error-correcting quantum states arises from two obstacles. First, errors in quantum computations are distinctly different from errors in classical computing. Quantum bits hold much more complex state than classical binary bits. Binary bits have only one possible error - a bit flip. The probability amplitudes of qubit states are parameterized by continuous degrees of freedom, so errors can be continuous in nature, and minor shifts in the superposition of a quantum bit cannot be discriminated from the desired computation. Also, quantum bits may suffer phase flip errors, since the signs of their amplitudes can be negative as well as positive. Second, quantum states must be corrected without measuring the state, because that would collapse the very superpositions we desire to preserve. Instead of measuring the state, extra bits are used that are measured to determine properties of the state rather than the exact value.

Quantum error correction codes successfully address these problems by simultaneously utilizing two classical codes to redundantly protect against both bit and phase errors. An $[n, k]$ code encodes $k$ qubits of data using $n$ qubits. The encoding circuit takes the $k$ data qubits as input, together with $n-k$ *ancilla* qubits each initialized in the state $|0\rangle$. The decoder does the reverse: it takes in an encoded $n$ qubit state, and outputs $k$ (possibly erroneous) qubits, together with $n-k$ qubits which specify which error occurred, with high probability. A recovery circuit then corrects the error on the data, by performing one of $2^{n-k}$ operations. This model assumes that qubit errors are independent and uniformly distributed, just as does classical error correction. Deviations from this model can be treated similarly to classical strategies.

The effect of quantum error correction is powerful and subtle. Without this step the "correctness" (technically, the *fidelity*) of a physical qubit decays exponentially and continuously with time. Quantum error correction transforms this exponential error model into a linear one, to leading order: a logical qubit encoded in a quantum error correcting code and undergoing periodic error measurement suffers only linear discrete amounts of error.

Not all codes are suitable for fault-tolerant computation — many different quantum codes have now been discovered, and among these the largest class, *stabilizer codes*, play a special role: computation on these encoded states can be performed *without* decoding the data, and without overly propagating errors. For this reason, we focus on the [7, 1] Steane code, which encodes one qubit using seven physical qubits and is nearly optimal (the smallest perfect quantum code is [5, 1] [27]). This code has the marvelous property that an important set of single qubit operations, and the controlled-NOT operator (used in the quantum ALU discussed in the next section) can all be performed on the encoded qubit simply by applying the operations to each individual physical qubit.

The cost of error correction is the overhead needed to compute on encoded states and to perform periodic error correction steps (note that the correction can also be done without decoding the data[21]). Each such step will be called a *fault-tolerant* operation. For the Steane code, about 153 physical gates are required to construct a fault-tolerant single qubit operation.

Despite this substantial cost, the 7 qubit error correcting code dramatically improves the situation for quantum computing. The probability of a logical qubit error occurring during a single operation changes from $p$ to $c \cdot p^2$[27] where $c$ is a constant determined by the number of unique points in the error measurement, recovery and the transform being applied to the logical qubit state, where two or more failures can occur and propagate to the output of the logical qubit. For a single logical gate application, $c$ is about 17,446. For a physical qubit transform failure rate of $p = 10^{-6}$ this means the 7 qubit Steane code improves the probable logical qubit transform failure rate to $1.7 \cdot 10^{-8}$. However, more significantly, error correction transforms the exponentially decaying success probability $e^{-\lambda \cdot t}$ to a linear one of $1 - t \cdot p$.

## Recursive Error Correction

The most impressive and important application of quantum codes to computation is a recursive construction[2] which exponentially decreases error probabilities with only polynomial effort. This is crucial because even with the 7 qubit error correction that gives us an error probability of $c \cdot p^2$ is too high for most interesting quantum applications. The construction can be understood with the following example: The Steane code transforms the physical qubit error rate $p$ to a logical qubit error rate $c \cdot p^2$ but requires some number of physical qubit gates per logical qubit gate operation. However, suppose instead that each of those physical gates were again implemented as a logical gate on a 7 qubit code. Each of those gates would have a logical gate accuracy of $c \cdot p^2$, and hence the overall logical gate error rate would become $c \cdot (c \cdot p^2)^2$. For a technology with $p = 10^{-6}$ each upper level gate would have an error rate of roughly $4.3 \cdot 10^{-10}$. The key observation is that as long as $cp^2 < p$, then error probabilities decrease *exponentially* with only a *polynomial* increase in overhead. Asymptotically, this gives rise to the Threshold Theorem, according to which quantum computation can be sustained for any finite length of time

| Recursion level (k) | Storage overhead $7^k$ | Operation overhead $153^k$ | Min. time overhead $5^k$ | Error probability |
|---|---|---|---|---|
| 0 | 1 | 1 | 1 | $10^{-6}$ |
| 1 | 7 | 153 | 5 | $1.7 \cdot 10^{-8}$ |
| 2 | 49 | 23,409 | 25 | $5.3 \cdot 10^{-12}$ |
| 3 | 343 | 3,581,577 | 125 | $4.9 \cdot 10^{-19}$ |
| 4 | 2,401 | 547,981,281 | 625 | $4.2 \cdot 10^{-33}$ |
| 5 | 16,807 | 83,841,135,993 | 3125 | $3.1 \cdot 10^{-61}$ |

*Table 8.1.* Overhead of recursive error correction for a single qubit operation, c=17,446, p=$10^{-6}$

so long as the underlying technology has a reliability greater than $1/c$. The relevant inequality may be expressed as:

$$\frac{(c \cdot p)^{2^k}}{c} \leq \frac{\epsilon}{p(n)} \tag{8.1}$$

This relates the overall failure rate $\epsilon$ and space-time complexity $p(n)$ of an algorithm to the individual physical qubit error rate $p$, logical transform complexity $c$, and recursive error correction level $k$ required to satisfy the inequality. Table 8.1 summarizes the costs of recursive error correction.

Clearly, due to the high cost of such error correction, a quantum computer architecture should not choose the error correction method and recursion level indiscriminately. For a given algorithm and data size, the minimum recursion level should be chosen in order to reduce the overhead.

## 8.3 Quantum Computing Technologies

A variety of technologies have been used to successfully construct quantum computing prototypes [30, 19, 23]. For quantum machines to scale to thousands or even hundreds of thousands of quantum bits, howevever, more scalable technologies are necessary. We focus on the long-term vision of ions implanted in silicon. Nearer term, we expect much to be learned from micromachined ion traps. In fact, both technologies share fundamental design and verification challenges. We shall see that we can think of both technologies in terms of the same system-level abstractions.

### Ions Implanted in Silicon

The Kane [17, 26] schemes of phosphorus in silicon builds upon modern semiconductor fabrication and transistor design, drawing upon understood physical properties.

Kane proposes that the nuclear spin of a phosphorus atom coupled with an electron embedded in silicon under a high magnetic field and low temperature

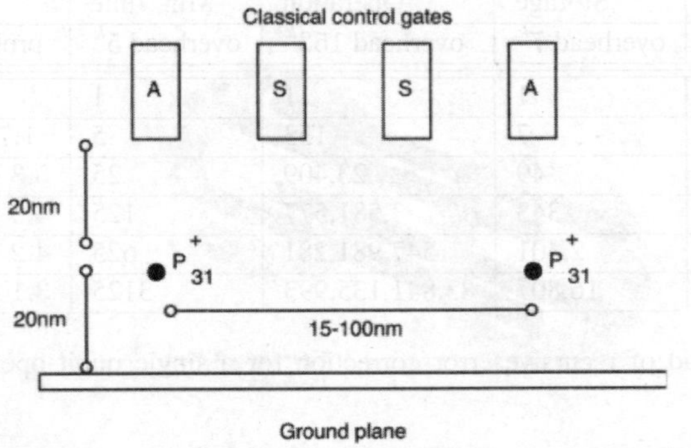


*Figure 8.2.* The basic quantum bit technology proposed by Kane [26]. Qubits are embodied by the nuclear spin of a phosphorus atom coupled with an electron embedded in silicon under high magnetic field at low temperature.

can be used as a quantum bit, much as nuclear spins in molecules have been shown to be good quantum bits for quantum computation with nuclear magnetic resonance [10]. This quantum bit is classically controlled by a local electric field. The process is illustrated in Figure 8.2. Shown are two phosphorus atoms spaced 15-100 nm apart. This inter-qubit spacing is currently a topic of debate within the physics community, with conservative estimates of 15 nm, and more aggressive estimations of 100 nm. What is being traded off is noise immunity versus difficulty of manufacturing.

Twenty nanometers above the phosphorus atoms lie three classical wires that are spaced 20 nm apart. By applying precisely timed pulses to these electrodes Kane describes how arbitrary one- and two-qubit quantum gates can be realized. Four different sets of pulse signals must be routed to each electrode to implement a universal set of quantum operations.

## Micromachined Ion Traps

Nearer term, ion traps technologies provide a means to scale to perhaps thousands of quantum bits. Ion traps are one of the best understood technologies, with extensive experimental data describing their characteristics.

A typical ion trap, shown in Figure 8.3 contains up to half a dozen ions arranged in a linear array and trapped by a magnetic field. The ions are individually manipulated by hitting them with lasers of the appropriate frequency. This allows both quantum operations and measurement. Measurement occurs when a quantum bit is excited into either a phosphorescent state and a photodetector detects the photons emited.

The key to scaling ion traps is the ability to move ions between traps via a series of electrodes and magnetic fields. As we shall discuss later, this motion

*Figure 8.3.* An ion trap containing ions arranged in a linear array and trapped by a magnetic field.

is both slow and error prone, but allows us to build chips of micromachined ion traps with up to perhaps 100s or 1000s of bits [18].

### Common Abstractions

At a basic level, both silicon-embedded ions and ion traps involve the spatial layout of quantum bits and control lines to implement quantum algorithms and circuits. Both will face challenges in fabrication and verification of the precisely placed bits and control signals, as well as in the communication of quantum data across each system. These are the open issues we will explore in the next two sections.

## 8.4 Fabrication and Test Challenges

Perhaps the most obvious difficulty in fabricating quantum computers is the small scale of the components and precision with which they must be placed in the system. Since reliable quantum operations are already challenging (discussed further in the next section) given a fabricated system with perfect spacing and alignment, variations should be minimized and probably need to be detected. Furthermore, the use of quantum operations to test components should also be minimized.

### Manufacturing

For the Kane technology, the first hurdle is the placement of the phosphorus atoms themselves. The leading work in this area has involved precise ion implantation through masks and manipulation of single atoms on the surface of silicon [16]. For applications where substantial monetary investment is not an issue, slowly placing a few hundred thousand phosphorus atoms with a probe device [11] may be possible. For bulk manufacturing, the advancement of DNA or other chemical self-assembly techniques [1] may need to be developed. Note, while new technologies may be developed to enable precise placement, the key for this discussion is only the spacing (60 nm) of the phosphorus atoms themselves and the number of control lines (3) per qubit. The relative scale of quantum interaction and the classical control of these interactions is what will lead to the fundamental constraints on quantum computing architectures.

A second challenge is the scale of classical control. Each control line into the quantum datapath is roughly 10 nm in width. While such wires are difficult to fabricate, we expect that either electron beam lithography [3], or phase-shifted masks [24] will make such scales possible.

A remaining challenge is the temperature of the device. In order for the quantum bits to remain stable for a reasonable period of time, the device must be cooled to less than one degree Kelvin. The cooling itself is straightforward, but the effect of the cooling on the classical logic is a problem. Two issues arise. First, conventional transistors stop working as the electrons become trapped near their dopant atoms, which fail to ionize. Second, the 10 nm classical control lines begin to exhibit quantum-mechanical behavior such as conductance quantization and interference from ballistic transport [9].

Fortunately, many researchers are already working on low-temperature transistors. For instance, single-electron transistors (SET's) [20] are the focus of intense research due to their high density and low power properties. SET's, however, have been problematic for conventional computing because they are sensitive to noise and operate best at low temperatures. For quantum computing, this predilection for low temperatures is exactly what is needed! Tucker and Shen describe this complementary relationship and propose several fabrication methods in [28].

## Testing

Once fabricated, qubits and control will be difficult to test. Tolerances are tight, and it may be necessary to avoid using qubits in the system that are spaced incorrectly or have control signals that are misaligned.

It is likely that the most effective test for the spacing of control signals is to inspect, using an SEM or other device, the pattern of small 10 nm vias above each ion before they are covered by subsequent layers of metal. Connectivity from wide control wires to the vias will have to be verified via a quantum test program.

The spacing and alignment of the ions that implement the qubits is also problematic. Defects could be caught via quantum test programs, but the test would have difficulty distinguishing between ion spacing errors, misalignment between control vias and ions, and control via spacing errors. Efficient test patterns will be needed to test individual qubits and the two-qubit operations between neighboring qubits.

## 8.5 Architectural Challenges

We now look at how to take these individual quantum components and structure them into a working, large-scale quantum computer. We have two fundamental difficulties. First, because quantum computing requires very low temperatures, and classical circuits (required for control of quantum gates) are designed for higher temperatures, the design must be adjusted to allow classical circuits to work at very low temperatures. Second, because quantum operations are so error-prone, and error correction circuits themselves so large, a reliable communication mechanism is required. This material is derived from our previous work [14, 8].

### Basic Geometric Constraints

The quantum-mechanical behavior of the control lines presents a subtle challenge that is often overlooked. At low temperatures, and in narrow wires, the quantum nature of electrons begins to dominate over normal classical behavior. For example, in 100 nm wide polysilicon wires at 100 millikelvin, electrons propagate ballistically like waves, through only one conductance channel, which has an impedance given by the quantum of resistance, $h/e^2 \approx 25$ k$\Omega$. Impedance mismatches to these and similar metallic wires make it impossible to properly drive the AC current necessary to perform qubit operations.

Avoiding such limitations mandates a geometric design constraint: narrow wires must be short and locally driven by nearby wide wires. Using 100 nm as a rule of thumb[1] for a minimum metallic wire width sufficient to avoid undesired quantum behavior at these low temperatures, we obtain a control gate structure such as that depicted in Figure 8.4. Here, wide wires terminate in 10 nm vias that act as local gates above individual phosphorus atoms. Note how access lines quickly taper into upper layers of metal and into control areas of a classical scale. These control areas can then be routed to access transistors that can gate on and off the frequencies (in the 10's to 100's of MHz) required to apply specific quantum gates.

### Quantum Communication

We examine the problem of scaling a quantum system by focusing on perhaps the primary task of quantum computer – error correction. Recall that error

---

[1]This value is based on typical electron mean free path distances, given known scattering rates and the electron Fermi wavelength in metals.

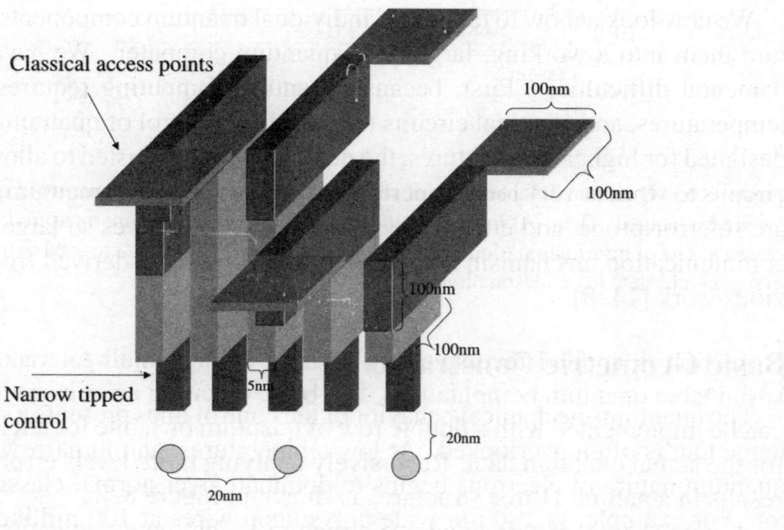


*Figure 8.4.* Instead of narrow wires throughout the system, wide wires are required for long distances, and narrow wires may only be used for short distances in order to minimize the quantum effects in the classical control.

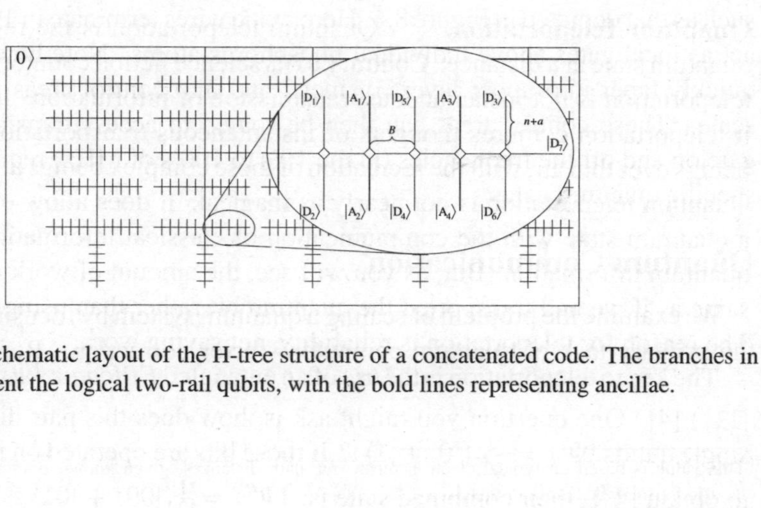


*Figure 8.5.* Schematic layout of the H-tree structure of a concatenated code. The branches in the inset represent the logical two-rail qubits, with the bold lines representing ancillae.

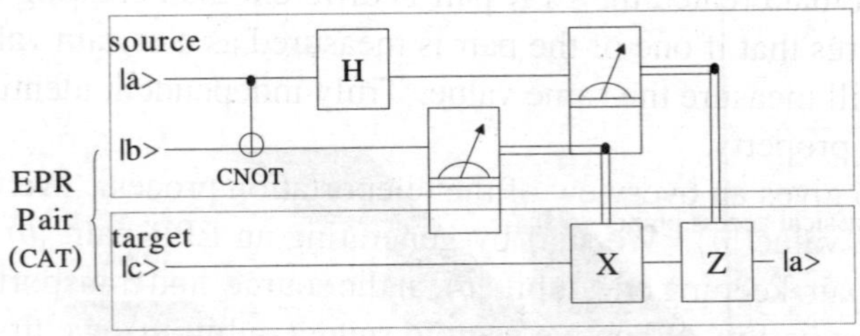


*Figure 8.6.* Quantum Teleportation of state $|a\rangle$ over distance. First, *entangled* qubits $|b\rangle$ and $|c\rangle$ are exchanged. Then, $|a\rangle$ interacts with $|b\rangle$, and both $|a\rangle$ and $|b\rangle$ are measured to obtain two *classical* bits of information (double lines). After transport, these bits are used to manipulate $|c\rangle$ to regenerate state $|a\rangle$ at destination.

correction is applied recursively to achieve enough fault tolerance to allow sustainable quantum computation. The basic quantum error correction circuit can be implemented with a double row of quantum bits, one for ancilla and one for the actual quantum data. Recursively applying more levels error correction results in a natural H-tree structure, as shown in Figure 8.5.

A crucial trait of this recursive structure is that communication distances increase as we approach the root of the tree for any substantial level of recursion. In fact, the naive approach of swapping quantum data from bit to bit becomes untenable. There are too many swaps to accomplish without error correction, yet we are trying to construct the basic error correction circuit! In fact, it is possible to apply intermediate error correction, but this would substantially increase the overhead of an already costly process. Instead, we examine another method of achieving quantum communication over long distance – quantum teleportation.

**Quantum Teleportation.** Quantum teleportation is the re-creation of a quantum state at a distance. Contrary to its science fiction counterpart, quantum teleportation is not instantaneous transmission of information. In fact, calling it teleportation conjures thoughts of instantaneous transportation of complex beings over thin air, with the recreation of these complex beings at the other end. Quantum teleportation is not nearly as magical. It does allow one to recreate a quantum state with the communication of classical information rather than quantum information. But, as you will see, the amount of work is actually the same as if we had transported the quantum bit rather than using teleportation. The reason for teleportation is reliability, not saving work.

The key to teleportation is the use of an entangled *EPR pair*, $|\Psi\rangle = \frac{1}{\sqrt{2}}(|00\rangle + |11\rangle)$ [4]. One question you might ask is, how does this pair differ from two single qubits $|\Psi'\rangle = \frac{1}{\sqrt{2}}(|0\rangle + |1\rangle)$? If those bits are operated on independently to obtain $|\Psi'\rangle$, their combined state is: $|\Psi''\rangle = \frac{1}{2}(|00\rangle + |01\rangle + |10\rangle + |11\rangle)$, which is clearly a different equation. What is the core difference? Somehow, the operation that created the EPR pair is different than creating two identical bits - it requires that if one of the pair is measured as a certain value, the other of the pair will measure the same value. Truly independent identical qubits do not have this property.

Figure 8.6 gives an overview of the teleportation process. We want to communicate the value $|a\rangle$. We start by generating an EPR pair, $|b\rangle$ and $|c\rangle$. We separate the pair, keeping one qubit, $|b\rangle$, at the source, and transporting the other, $|c\rangle$, to the destination. When we want to send a qubit, $|a\rangle$, we first interact $|a\rangle$ with $|b\rangle$ using a CNOT gate. We then measure $|a\rangle$ and $|b\rangle$ in the computational basis, and send the two one-bit classical results to the destination, and use those results to re-create the correct phase and amplitude in $|c\rangle$ such that it takes on the original state of $|a\rangle$. The re-creation of phase and amplitude is done with $X$ and $Z$ gates, whose application is contingent on the outcome of the measurements of $|a\rangle$ and $|b\rangle$. Intuitively, since $|c\rangle$ has a special relationship with $|b\rangle$, interacting $|a\rangle$ with $|b\rangle$ makes $|c\rangle$ resemble $|a\rangle$, modulo a phase and/or amplitude error. The two measurements allow us to correct these errors and re-create $|a\rangle$ at the destination. Note that the original state of $|a\rangle$ is destroyed when we take our two measurements. This is consistent with the "no-cloning" theorem, which states that a quantum state cannot be copied.

Why bother with teleportation when we end up transporting $|c\rangle$ anyway? Why not just transport $|a\rangle$ directly? As you might notice, teleportation is not a huge savings in effort, because to accomplish this, two special quantum bits must be created in the same location and then travel to the source and destination locations, which is the same amount of actual work as transporting the original bit. First, we can pre-communicate EPR pairs with extensive pipelining without stalling computations. Second, transportation has the possibility of errors, and we are constructing communication for use in error correction. We need this communication to be error-free. But what about if $|b\rangle$ and $|c\rangle$ have a failure? That is okay, because as long as we can detect the error, we throw them out and use another pair. Since $|b\rangle$ and $|c\rangle$ have known properties, we can employ a specialized procedure known as *purification* to turn a collection of pairs partially damaged from transport into a smaller collection of asymptotically perfect pairs. If we accidentally destroy $|a\rangle$, we need to restart the whole quantum computation. Thus, teleportation does not actually save work, it just makes sure that important qubits use reliable, classical communication for long-distance travel rather than the error-prone swapping method.

## 8.6 Conclusions

Quantum computing has unique obstacles to overcome before a large-scale machine can be built. Our goal has been to provide some basic understanding of the mechanisms and issues involved. Since quantum technologies are still under development, the key is to develop design and verification techniques that are somewhat technology independent, much like the communication and placement mechanisms discussed. Our hope is that the right abstractions and techniques will help pave the way to realizing a scalable quantum computer in the near future.

## 8.7 Acknowledgements

We would like to thank our co-authors in research that is described in this chapter. A good summary of our work can be found in [14] and [8].

This work was supported in part by the DARPA Quantum Information Science and Technology Program, and by an NSF CAREER and a UC Davis Chancellor's Fellowship to Fred Chong. Diana Franklin's faculty position is supported by a Forbes Endowment.

## References

[1] Leonard Adleman, *Toward a mathematical theory of self-assembly*, USC Tech Report, 2000.

[2] D. Aharonov and M. Ben-Or, *Fault tolerant computation with constant error*, Proceedings of the Twenty-Ninth Annual ACM Symposium on the Theory of Computing, 1997, pp. 176–188.

[3] E.H. Anderson, V.Boegli, M.L. Schattenburg, D.P. Kern, and H.I. Smith, *Metrology of electron beam lithography systems using holographically produced reference samples*, J. Vac. Sci. Technol. **B-9** (1991).

[4] John S. Bell, *On the Einstein-Podolsy-Rosen paradox*, Physics **1** (1964), 195–200, Reprinted in J. S. Bell, *Speakable and Unspeakable in Quantum Mechanics*, Cambridge University Press, Cambridge, 1987.

[5] Charles H. Bennett and Gilles Brassard, *Quantum cryptography: Public key distribution and coin tossing*, Proceedings of the IEEE International Conference on Computers, Systems, and Signal Processing, 1984, pp. 175–179.

[6] A. M. Childs, E. Farhi, and J. Preskill, *Robustness of adiabatic quantum computation*, Phys. Rev. A (2002), no. 65.

[7] Isaac L. Chuang, *Quantum algorithm for clock synchronization*, Phys. Rev. Lett. **85** (2000), 2006.

[8] D. Copsey, M. Oskin, F. Impens, T. Metodiev, A. Cross, F. Chong, I. Chuang, and J. Kubiatowicz, *Toward a scalable, silicon-based quantum computing architecture*, Journal of Selected Topics in Quantum Electronics, To appear.

[9] David K. Ferry and Stephen M. Goodnick, *Transport in nanostructures*, Cambridge Studies in Semiconductor Physics & Microelectronic Engineering, 6, Cambridge University Press, Cambridge, 1997.

[10] N. Gershenfeld and I.L. Chuang, *Quantum computing with molecules*, Scientific American (1998).

[11] Al Globus, David Bailey, Jie Han, Richard Jaffe, Creon Levit, Ralph Merkle, and Deepak Srivastava, *Nasa applications of molecular nanotechnology*, Journal of the British Interplanetary Society **51** (1998).

[12] L. Grover, Proc. $28^{th}$ Annual ACM Symposium on the Theory of Computation (New York), ACM Press, 1996, pp. 212–219.

[13] Sean Hallgren, Quantum Information Processing '02 Workshop (2002).

[14] N. Isailovic, M. Whitney, D. Copsey, Y. Patel, F. Chong, I. Chuang, J. Kubiatowicz, and M. Oskin, *Datapath and control for quantum wires*, ACM Transactions on Architecture and Compiler Optimization, To appear.

[15] R Jozsa, DS Abrams, JP Dowling, and CP Williams, *Quantum atomic clock synchronization based on shared prior entanglement*, Phys. Rev. Lett. (2000), 2010–2013.

[16] B. E. Kane, N. S. McAlpine, A. S. Dzurak, R. G. Clark, G. J. Milburn, He Bi Sun, and Howard Wiseman, *Single spin measurement using single electron transistors to probe two electron systems*, arXive e-print cond-mat/9903371 (1999), Submitted to Phys. Rev. B.

[17] Bruce Kane, *A silicon-based nuclear spin quantum computer*, Nature **393** (1998), 133–137.

[18] D. Kielpinsky, C. Monroe, and D.J. Wineland, *Architecture for a large-scale ion trap quantum computer*, Nature **417** (2002), 709.

[19] E. Knill, R. Laflamme, R. Martinez, and C.-H. Tseng, *A cat-state benchmark on a seven bit quantum computer*, arXive e-print quant-ph/9908051 (1999).

[20] Konstantin K. Likhareve, *Single-eletron devices and their applications*, Proceedings of the IEEE **87** (1999).

[21] M. A. Nielsen and I. L. Chuang, *Quantum computation and quantum information*, Cambridge University Press, Cambridge, UK, 2000.

[22] M.A. Nielsen and I.L. Chuang, *Quantum computation and quantum information*, Cambridge University Press, Cambridge, England, 2000.

[23] C.A. Sackett, D. Kielpinsky, B.E. King, C. Langer, V. Meyer, C.J. Myatt, M. Rowe, Q.A. Turchette, W.M. Itano, D.J. Wineland, and C. Monroe, *Experimental entanglement of four particles*, Nature **404** (2000), 256–258.

[24] Michael Sanie, Michel Cote, Philppe Hurat, and Vinod Malhotra, *Practical application of full-feature alternating phase-shifting technology for a phase-aware standard-cell design flow*, (2001).

[25] P. Shor, *Algorithms for quantum computation: Discrete logarithms and factoring*, Proc. $35^{th}$ Annual Symposium on Foundations of Computer Science (Los Alamitos, CA), IEEE Press, 1994, p. 124.

[26] A. Skinner et al., *Hydrogenic spin quantum computing in silicon: a digital approach*, quant-ph/0206159 (2002).

[27] A. Steane, *Error correcting codes in quantum theory*, Phys. Rev. Lett. **77** (1996).

[28] J. R. Tucker and T.-C. Shen, *Can single-electron integrated circuits and quantum computers be fabricated in silicon?*, International Journal of Circuit Theory and Applications **28** (2000), 553–562.

[29] W. van Dam and G. Seroussi, *Efficient quantum algorithms for estimating gauss sums*, quant-ph (2002), 0207131.

[30] Lieven M.K. Vandersypen, Matthias Steffen, Gregory Breyta, Costantino S. Yannoni, Richard Cleve, and Isaac L. Chuang, *Experimental realization of order-finding with a quantum computer*, Phys. Rev. Lett. **December 15** (2000).

Chapter 9

# ORIGINS AND MOTIVATIONS FOR DESIGN RULES IN QCA

Michael T. Niemier
*College of Computing*
*Georgia Institute of Technology*
mniemier@cc.gatech.edu

Peter M. Kogge
*Department of Computer Science and Engineering*
*University of Notre Dame*
kogge@cse.nd.edu

**Abstract** This chapter introduces the quantum-dot cellular automata (QCA), what constructs can be made from it, and what constructs could be implementable in the near-to-midterm. It will also explain a systems-level research component that complements work in physical science. One objective of the systems-level track is to compile a set of design rules to not only help system designers become more involved with the evolution of emergent, nano-scale devices geared for computation, but it should also help us to reach computationally interesting, nano-scale systems in an accelerated time frame. The motivations for, and the origins of design rules for QCA (and other emergent technologies) will be explained here.

**Keywords:** Quantum-dot Cellular Automata, QCA, Design Rules

## Introduction

The real purpose of this chapter is to present a set of parameters and a methodology that will assist in closing the gap of understanding between individuals researching physical devices, and those interested in building systems from those devices. With many classes of nano-scale devices emerging, often the



*S.K. Shukla and R.I. Bahar (eds.), Nano, Quantum and Molecular Computing*, 267-293.

two groups speak two different languages. Electrical engineers and chemists do not have the background required to design larger-scale and computationally useful systems from the devices that they are building. Similarly, while system architects have this knowledge, they usually lack an understanding of what can physically be built.

The goal of this chapter is to introduce a set of fundamental parameters for the computer engineer which, if followed, should help eliminate the problem just mentioned for one emergent nano-scale device - the Quantum-dot Cellular Automata (QCA). Carver Mead and Lynn Conway's work will be used as a basis and a framework for a set of QCA design rules. Overall, this work should help to further a goal of using systems-level research to help drive device development. (In other words, systems designers can provide physical scientists with computationally interesting schematics to physically build, and identify what device characteristics are most important to implement in order to perform computationally interesting tasks.) If the device and design communities are better able to communicate, more proof-of-concept, realistic, and computationally useful circuits and systems should become more physically realizable at an accelerated time scale.

Specifically, this chapter (from a computer engineering perspective) will begin by providing a detailed background of QCA devices. It will discuss the basic logical properties associated with QCA - how a 1 or a 0 is represented, how wires are formed, how logical gates are formed, etc. Next, it will focus on what the current state of the art is with regard to QCA demonstrations. We will discuss possible "real" QCA cell implementations. Additionally, the role that a "clock" will play in QCA circuits will be introduced at a conceptual and implementable level. We will then review the historical precedence for design rules. It will be followed by a short section that considers the initial questions that laid the ground work for design rules in QCA. Next we will present a compilation of all of the information a computer engineer will want to know, and a list of all of the questions that a computer engineer will need to answer when attempting to design larger-scale systems of QCA cells. Finally, a brief example of what a design rule might look like for a specific device technology will be discussed.

## 9.1 The Basic Device and Circuit Elements

We will first present a very conceptual view of QCA - essentially showcasing the building blocks available for the design of logical circuits and systems. Possible implementations will be discussed later.

## A "Generic" 4-dot QCA Cell

A high-level diagram of a "candidate" four-dot metal QCA cell appears in Figure 9.1. It depicts four quantum dots that are positioned to form a square. Quantum dots are small semi-conductor or metal islands with a diameter that is small enough to make their charging energy greater than $k_bT$ where $k_b$ is Boltzmann's constant and $T$ is the operating temperature. The charging energy is the potential energy needed to overcome the electrostatic repulsion from the other electrons in the dot – or in other words, the energy required to add an electron to a dot. If this energy is greater than the thermal energy of the environment ($k_BT$), dots can trap individual charges.

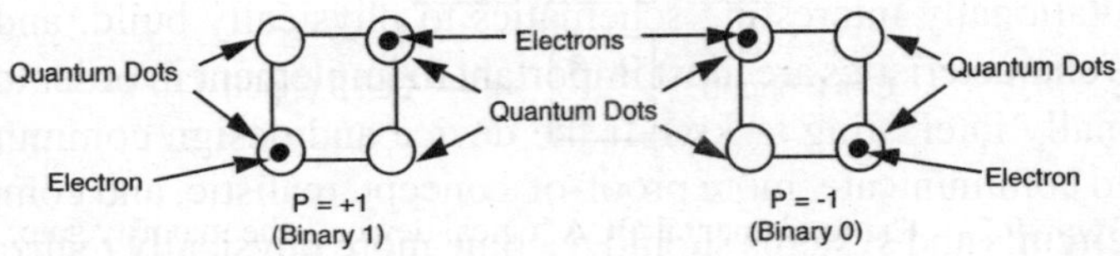


*Figure 9.1.* QCA cell polarizations and representations of binary 1 and binary 0.

Exactly two mobile electrons are loaded into this cell and can move to different quantum dots by means of electron tunneling. Tunneling paths are represented by the lines connecting the quantum dots in Figure 9.1. Coulombic repulsion will cause "classical" models of the electrons to occupy only the corners of the QCA cell, resulting in two specific polarizations (again, see Figure 9.1). These polarizations are configurations where electrons are as far apart from one another as possible, in an energetically minimal position, without escaping the confines of the cell. Here, electron tunneling is assumed to be completely controllable by potential barriers that can be raised and lowered between adjacent QCA cells by means of capacitive plates parallel to the plane of the dots [15].

It is also worth noting that in addition to these two "polarized" states, there also exists a decidedly non-classical unpolarized state. Briefly, in an unpolarized state, inter-dot potential barriers are lowered to a point which removes the confinement of the electrons on the individual quantum dots, and the cells exhibit little or no polarization as the wave functions of two electrons smear themselves across the cell [7].

## The Majority Gate

The fundamental QCA logical gate is the three-input majority gate which appears in Figure 9.2 [15]. Computation is performed with a majority gate by driving the device cell (cell 4 in the figure) to its lowest energy state, which will occur when it assumes the polarization of the majority of the three input cells (1, 2, and 3). We define an input cell simply as one that is changed by a logical signal propagating toward the device cell. The device cell will always assume the majority of the polarizations of the input cells because in that polarization, the electrostatic repulsion between the electrons in the three input cells and the electrons in the device cell will be at a minimum.

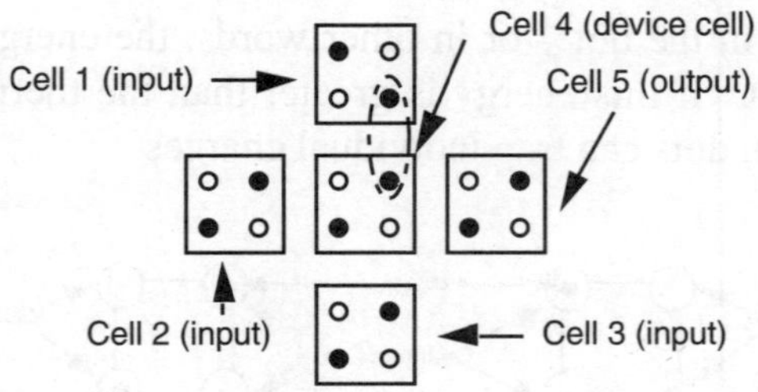


*Figure 9.2.* The fundamental QCA logical device – the majority gate.

## A Wire

Figure 9.3 illustrates what is called a "90-degree" wire. (The wire is called "90-degrees" as the cells from which it is made up are oriented at a right angle). The wire is a horizontal row of QCA cells and a binary signal propagates from left-to-right because of electrostatic interactions. Initially, cell 1 has polarization P = -1 and cell 2 has polarization P = +1. It is assumed that charges in cell 1 are trapped in polarization P = -1 but those in cells 2-9 are not. Because the driving cell is "trapped", there is no danger that this wire could "reverse directions" and have a polarization propagate in a direction from which it came. Initially, electron repulsion between cell 1 and 2 will cause cell 2 to change polarizations. Then, electron repulsion between cell 2 and 3 will cause cell 3 to change polarizations. This process will continue down the length of the QCA "wire". When electrons in all cells settle in an energetically minimal position, the system is said to be in a ground state. ("Energetically minimal positions" simply means that electrons are in positions such that the Coulombic repulsions between them are as low as possible).

## A 45-degree Wire

It is also possible to form what is called a "45-degree wire" [15]. Illustrated in Figure 9.4, a binary value propagates down the length of such a wire, alternating between polarization P = +1 and polarization P = -1. It is this orientation of

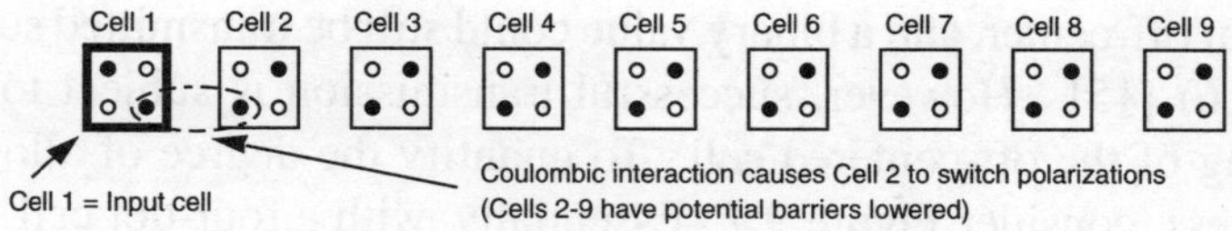


*Figure 9.3.* A QCA "wire".

electrons within QCA cells that represents the minimum energy configuration for each cell. Interestingly, with this orientation of wire, both a complemented or uncomplemented signal value can be ripped off of the wire by placing a 90-degree "ripper" cell at the proper location between 45-degree cells (see Figure 9.5).

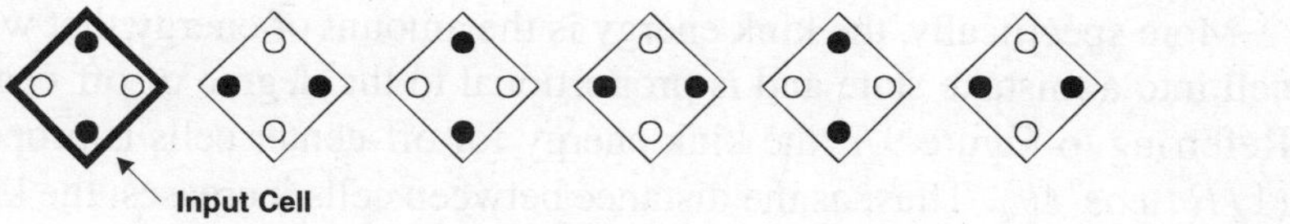


*Figure 9.4.* A 45-degree wire.

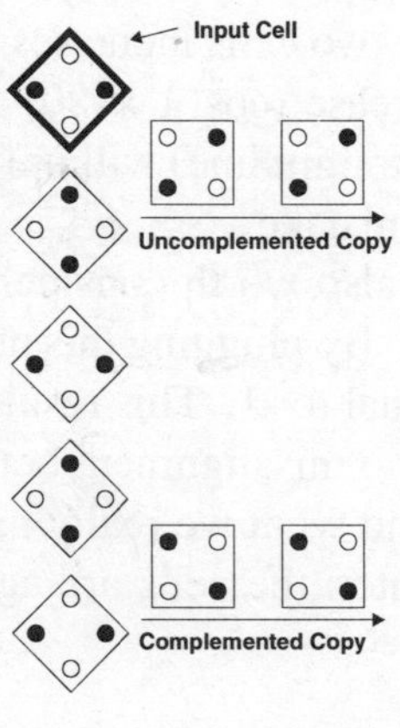


*Figure 9.5.* Ripping off a binary 0 and 1 from a 45-degree wire.

## Off-center Wires

In theory, QCA cells do not have to be exactly aligned to transmit binary signals correctly. Cells with a 90-degree orientation could be placed next to one another but off-center, and a binary value could still be transmitted successfully (Figure 9.6) [15]. However, successful transmission is subject to the exact positioning of the off-centered cell. To quantify the degree of allowable off-centeredness, consider Figure 9.7. Essentially, with a four-dot cell, if the two quantum dots of the middle cell ("polariztion okay" in the figure) are below the center lines of its neighboring cells, then the polarization of the next cell of the wire will be weak or indefinite which is undesirable. If the cell is above the imaginary center line, the value should be transmitted successfully. It should also be noted that different implementations of QCA cells (i.e. metal versus molecular) will be subjected to different allowable degrees, ranges, and types of "off-centeredness" with regard to propagating a signal correctly. The last portion of Figure 9.7 illustrates one possible defining rule. External energy can cause a cell in a wire or a system to switch into a mistake state (defined as $E_{kink}$).

More specifically, the kink energy is the amount of energy that will excite a cell into a mistake state and is proportional to the degree of off-centeredness. Referring to Figure 9.7, the kink energy for off-center cells is proportional to $(1/R^5)cos(4\theta)$. Thus, as the distance between cells increases, the kink energy will decrease indicating that a smaller amount of external energy could excite a cell into a mistake state. Intuitively, this makes sense as one cell that is supposed to drive another is now farther away from the cell that it is supposed to drive. Additionally, if two cells are placed exactly in line, the angle of their off-centeredness would simply be 0 ($cos(0)$ is 1). However, if the angle of offcenteredness between the two cells increases for example to 20-degrees, the kink energy will again decrease (cos($4 \times 20$) is approximately 0.17). Thus disorder (i.e. cells not in a straight line) will only lower the amount of external energy required to create a mistake.

(Before continuing, it is also worth considering what happens if two cells are off-center by 45 degrees. By plugging this number into the above equation, $cos(180)$ falls out and is equal to -1. This results in a kink energy identical to that for two cells that have no misalignment between them. The negative sign indicates signal inversion and what we really have is a 45-degree wire. If the cells are 90 degrees off-center, the cells are again in-line, but in the vertical direction.)

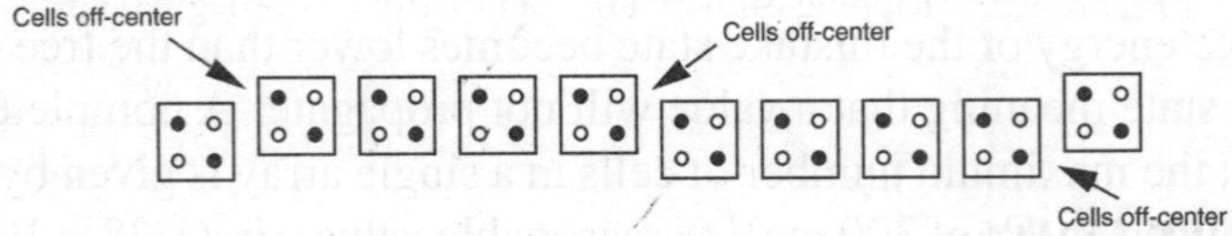


*Figure 9.6.* An off-center binary wire.

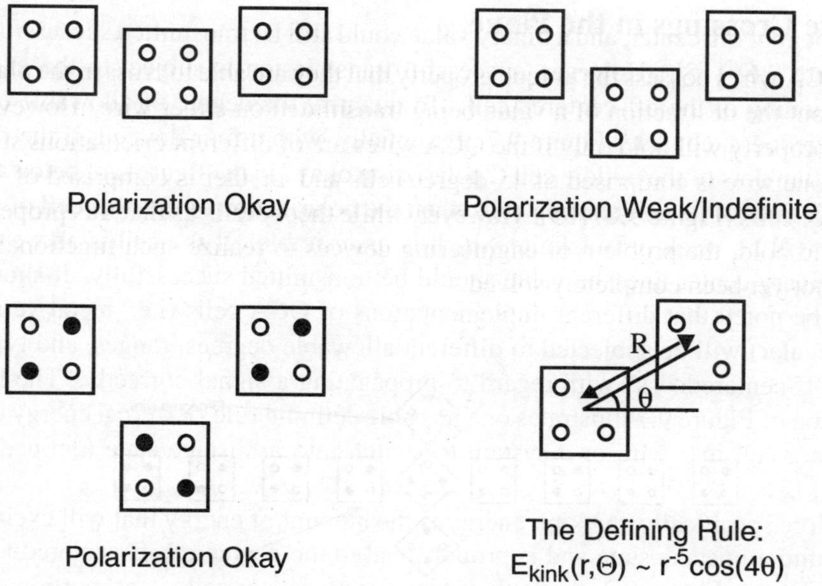


*Figure 9.7.* The restrictions on an off-center binary wire.

Finally, considering wire length in general, while it is to some extent a function of implementation technology, wire length is also largely a function of kink energy. As an example, consider a linear array of *N* cells that form a wire that we want to transmit a logical 1. The ground state for this configuration would be all of the cells switching to the same polarization as that of the driving cell – namely a line of cells in the logical '1' polarization. The first excited (mistake) state of this array will consist of the first *m* cells polarized in a representative binary 1 state and *N-m* cells in the binary 0 state. The excitation energy of this state ($E_k$) is the energy required to introduce a "kink" into the polarization of the wire. This energy is independent of where the kink occurs (i.e. the exact value of *m*). As the array *N* becomes larger, the kink energy $E_k$ remains the same. However, the entropy of this excited state increases as there are more ways to make a "mistake" in a larger array. When the array size reaches a certain size, the free energy of the mistake state becomes lower than the free energy of the correct state meaning that a value will not propagate. A complete analysis reveals that the maximum number of cells in a single array is given by $e^{E_k/k_BT}$ [7]. Thus, given an $E_k$ of 300 meV (a reasonable value), $k_b$ ($1.38 \times 10^{-23}$ J/K), close to room temperature operation (300K), and that 1 J = $1.6 \times 10^{-19}$ eV, arrays of cells on the order of $10^5$ are not unreasonable.

## Wire Crossings in the Plane

QCA wires possess the unique property that they are able to cross in the plane without the destruction of a value being transmitted on either wire. However, this property will hold only if the QCA wires are of different orientations such that one wire is comprised of 45-degree cells and another is comprised of 90-degree cells (Figure 9.8) [15]. However, while theory tells us that this property should hold, the problem of engineering devices to realize such functionality has not yet been completely solved.

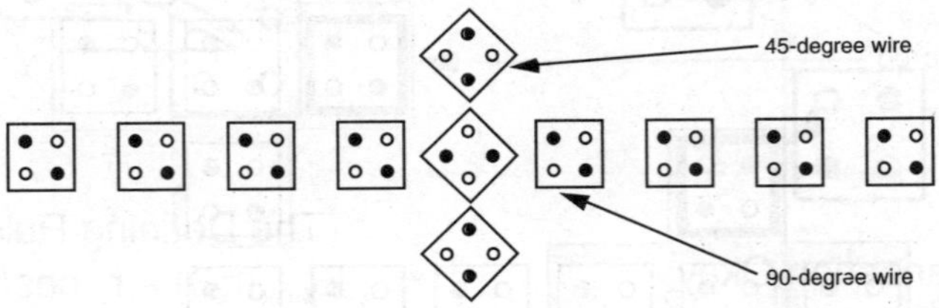


*Figure 9.8.* Two wires crossing in the plane.

## An Example

To implement more complicated logical functions, a subset of simple logical gates will be required. For example, it would be impossible to implement certain circuits in QCA with just majority gates (at least without inverters). Earlier, we have shown that a value's complement can be obtained simply by ripping a signal value off of a 45-degree wire at the proper location (it is also possible to make an inverter with only 90-degree cells [15]). Implementing the logical AND and OR functions is also quite simple. The logical function the majority gate performs is: $Y = AB + BC + AC$.

The AND function can be implemented by setting one value (A, B, or C) in the majority gate equation to a logical 0. Similarly, the OR function can be implemented by setting one value (A, B, or C) in the majority gate equation to a logical 1. This results in the logical AND/OR equations. It is worth noting that because this property exists, and given the fact that it is possible to obtain the inverse of a signal value, the QCA logic set is functionally complete, and any logical circuit can theoretically be generated with only QCA devices.

More complex logical circuits (such as the multiplexor in Figure 9.9) can then be constructed from majority-gate converted AND gates, OR gates, and inverters, if not more clever combinations of simply majority gates. (Note: QCA cells labeled "anchored" in Figure 9.9 are considered to have their electron polarization permanently frozen to successfully implement the AND and OR functions).

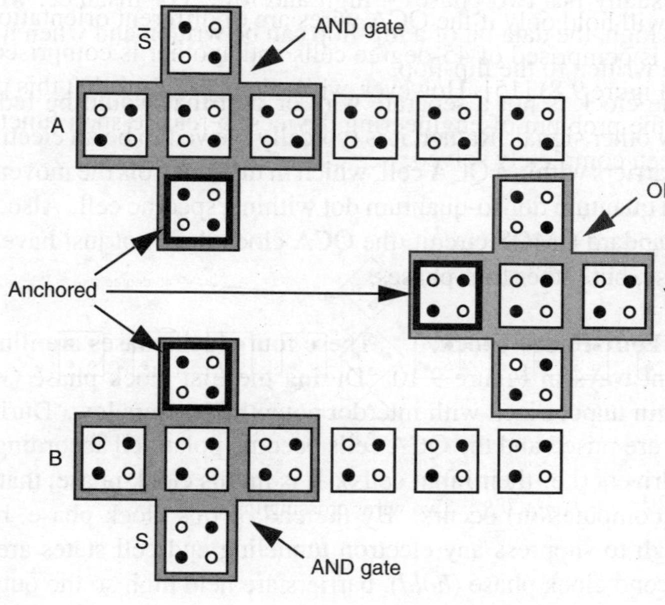


*Figure 9.9.* A 2x1 QCA multiplexor with logical equation: Y = AS' + BS.

# The Clock in QCA

This section will discuss the role that a clock will play in circuits and systems of QCA cells. This will be followed by a generic explanation of what effect a clock will have on data movement which is of specific importance to those worried about the design of computational systems. A brief example will follow and the section will conclude with a discussion of clocking mechanisms geared toward implementation.

**The Purpose of the Clock.** Without providing specific or implementation related detail (later sections will do that) it is nevertheless important to note that the specific role that the "clock" plays in circuits and systems of QCA cells is to provide power gain. Inevitably, in a long wire of cells, QCA signal energy could be lost to irreversible processes in a circuit's environment and somehow must be replaced. Gain must come from some source external to the QCA cells themselves, and is necessary to ensure that the binary 0s and 1s that the cells encode propagate through a circuit as the specific data signals that they were intended to represent [14].

**A CMOS Clock versus the "Clock" in QCA.** In standard CMOS, the clock is generally considered to be a signal that precisely controls the time at which data bits are transferred to or from memory elements (i.e. flip-flops). A typical clock signal usually has two phases – high and low. For instance, when the clock signal is high, the data bit of a flip-flop can be written and when it is low, no data can be written to the flip-flop.

In QCA, the clock is not a separate wire or port that would be fed into a circuit like any other signal. Rather, it is typically viewed to be an electric field that controls barriers within a QCA cell, which in turn controls the movement of electrons from quantum dot-to-quantum dot within a specific cell. Also, unlike a clock in a standard CMOS circuit, the QCA clock does not just have a high and a low phase, but rather four phases.

**A "Generic" Four-Phase Clock.** These four clock phases are illustrated in two different ways in Figure 9.10. During the first clock phase (*switch*), QCA cells begin unpolarized with interdot potential barriers low. During this phase barriers are raised and the QCA cells become polarized according to the state of their drivers (i.e. their input cells). It is in this clock phase, that actual switching (or computation) occurs. By the end of this clock phase, barriers are high enough to suppress any electron tunneling and cell states are fixed. During the second clock phase (*hold*), barriers are held high so the outputs of the subarray that has just switched can be used as inputs to the next stage. In the third clock phase, (*release*), barriers are lowered and cells are allowed to relax to an unpolarized state. Finally, during the fourth clock phase (*relax*), cell barriers remain lowered and cells remain in an unpolarized state [7].

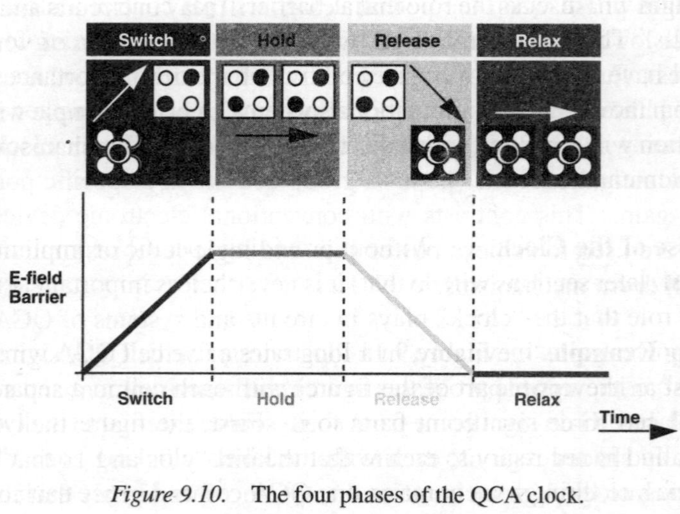


*Figure 9.10.* The four phases of the QCA clock.

Individual QCA cells need not be clocked or timed separately. The wiring required to clock each cell individually could easily overwhelm the simplification won by the inherent local interconnectivity of a QCA architecture [7]. However, a physical array of QCA cells can be divided into *zones* that offer the advantage of multi-phase clocking and group pipelining. For each zone, a single potential would modulate the inter-dot barriers in all of the cells in the given zone [7].

When a circuit is divided into different zones, each zone may be clocked differently from others. In particular, this difference is important when discussing neighboring, or physically adjacent, zones. Such a clocking scheme allows one zone of QCA cells to perform a certain calculation, have its state frozen by the raising of interdot barriers, and then have the output of that zone act as the input to a successor zone. It is this mechanism that provides the inherent self-latching associated with QCA. During the calculation phase, the successor zone is kept in an unpolarized state so it does not influence the calculation or result in a signal propagating back upon itself.

In an example circuit, clocking zones are partitioned using three rules. First, there are four "colors" of clocking zones, with all cells in each zone marked as having exactly one color. Each of the four clocking zones corresponds to one of four different clock phases. All zones with the same color receive the same phase clocks at the same time. Second, no two zones that touch can have the same color. Third, physically, neighboring zones concurrently receive temporally neighboring clock phases [7].

Finally, it is important to stress exactly what is meant when referring to the QCA clock. As mentioned above, the QCA clock has more than a high and a low phase but it is not a "signal" with four different phases either. Rather, the clock changes phase when the potential barriers that control a clocking zone are raised or lowered or remain raised or lowered (thus accounting for the four clock phases). Furthermore, all of the cells within a clocking zone are said to be in the same phase. One clock cycle occurs when a given clocking zone (electric field generating mechanism) cycles through the four different clock phases. Most importantly, the clock "traps" a group of cells in a specific polarization to provide gain. This contrasts with conventional electronic devices where transistors are used to achieve power gain and logic-level restoration through pull-up/pull down mechanisms.

**A Clocking Example.** Figure 9.11 illustrates a five cell QCA wire (labeled "schematic" in the upper part of the figure) with each cell in a separate zone. Figure 9.11 has three significant parts to it. First, the figure is divided into five vertically shaded regions, each with the label "clocking zone *x*". In this example, each clocking zone contains one QCA cell and hence each cell exists in a different clock phase. Second, the state of the wire is shown during five different time steps. Third, the state transitions for cells that make up the wire are illustrated for each time step. The number of cells that will have a meaningful change of state (during a given time step) with regard to the ongoing computation (or in other words, data movement) is equal to the time step number. Thus, in Time Step 1, the first cell (in switch) will acquire a binary state. In Time Step 2, the first cell is latched (in hold) and the second cell (in switch) changes state accordingly. Nevertheless, other cells in the wire must start in specific clock states to ensure that they are in the switch state when computed data arrives. As can be seen from this example, clocking zones clearly "latch" data, as it is transferred from cell-to-cell.

Two items are worthy of further note. First, for detail on zone partitioning, please see Chapter 10. Second, it may be worthwhile to investigate globally ascynchronous, locally synchronous systmes (GALS). More specifically, different parts of a QCA system could be clocked differently/have different clock layouts, and a protocol could be developed so that these portions of a system could "talk" or share data when needed.

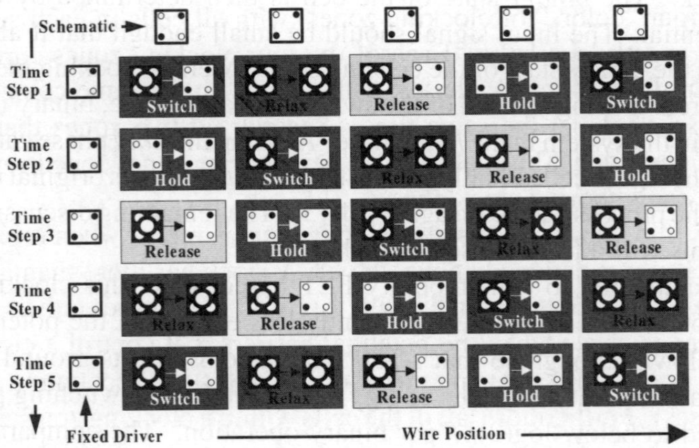


*Figure 9.11.* An example of QCA clock transitions.

**A Clocking Mechanism.** Again, the role that the clocking mechanism plays in a QCA circuit or system is to provide a means for power gain and to ensure that a QCA cell does not settle into a metastable state. In a metastable state a cell's polarization might encode a binary '1' as opposed to the binary '0' that it should represent (or vice versa). The restoring energy provided by the clock works to ensure power gain and prevent metastable states.

Up until now we have described the clock in a very generic way to underscore what affects it will have on computer architectures. However, it is now important to consider more implementation-related detail to ensure that as the areas of QCA design and QCA fabrication mature, resulting circuits and systems generated are not only are geared toward being buildable, but also so that design work can continue to be used to help drive device development [14], [4].

The theory behind an implementable QCA clock will actually apply to either metal or molecular QCA cells, and is based inherently in the cyclical manipulation of quantum wells to perform binary operations and move binary data. Specifically, research focuses on systems that can be cyclically transformed between monostable and bistable states (one stable state and two stable states respectively), with the QCA clock controlling whether or not the system is monostable or bistable. In other words, the clock will control whether or not a device is active or null. This allows some external input or driver (ideally another QCA cell) to control whether or not the cell is a 1 or a 0 [5].

In detail, because of the clock, at the start of a computation QCA cells (a.k.a. the "system") would begin in a monostable, *restore* state. During the *switch* phase, an input potential is applied and the system is then converted from a monostable state to a bistable state by means of some external energy source – the clock. The *binary* state of the cell is then determined by the applied input potential. The input signal should be small enough that it alone cannot determine the binary state of the cell and some external clocking mechanism is also required. During the end of the *switch* clocking phase, binary information is latched in the system in the *hold* phase. A cell in the *hold* phase can be used to drive another QCA cell. Finally, the system is *restored* to its original monostable state and the process can begin again [5], [3]. The remaining discussion of QCA experiments will refer to these clock phases.

When compared to contemporary FET based logic, this clocking scheme has the potential to dissipate significantly less energy. If the potential profile changes slowly enough so that a cell remains close to its ground state at all times (quasi-adiabatically), the energy dissipated in the switching process can be lowered to below $\log(2)kT$ per binary operation. The comparable energy for FET-based logic is $10^6 kT$. Overall, the induced inherent pipelining in QCA helps to alleviate problems related to the size of the non-clocked, edge driven QCA architecture imposed by thermodynamic considerations which can lead to metastability and also provides gain [6].

## 9.2 Implementable QCA

To better explain how CAD can have a positive effect on a given buildability point for QCA, we will now explain how physical scientists envision the constructs discussed in the first section being built. Specifically, the discussion will center on molecular QCA.

## QCA Devices

In contrast to metal-dot QCA cells, the small size of molecules (1-5 nm) means that Coulomb energies are much larger, so room temperature operation is possible [6]. Also, power dissipation from QCA switching would be low enough that high-density molecular logic circuits and memory are feasible. In molecular QCA, the role of a "dot" will be played by reduction-oxidation (redox) sites within a molecule. A redox center is capable of accepting an electron and donating an electron without breaking any of the chemical bonds that are used to bind the atoms of the molecular device together. Molecules with at least two redox centers are desired for QCA. It is possible to build molecules with 2, 3, and 4 "dots" [8].

Molecular QCA is discussed further in the context of 3-dot cells. The molecule forms a "v"-shape and charge can be localized on any one of the three dots at the "points" of the "v". If charge is localized on one of the top two dots, the cell will encode a binary 1 or 0. Whether or not charge is in the top two dots (the *active* state of the molecule) or the lower dot (the *null* state of the molecule) can be determined by an electric field that will raise or lower the potential of the central dot relative to the top two dots [5]. This idea will be fundamental to a clock that can be used to control a circuit made from molecular QCA cells and will be discussed further below. (Note that in Figure 9.12 the hole is represented by a dot in the cell "schematic". Open circles in Figure 9.12 are indicative of electrons – or the absence of the hole). 2 and 4-dot cells can also have null states. When considering basic cell-to-cell interactions, binary 1s and 0s are physically represented by the dipole moments of QCA molecules. Dipole moments are formed by the way that charge is localized within certain sites of a QCA molecule, and how that charge can tunnel between these sites [10]. In the presence of a strong dipole, a larger $E_{kink}$ is required to excite a cell into a mistake state [5].

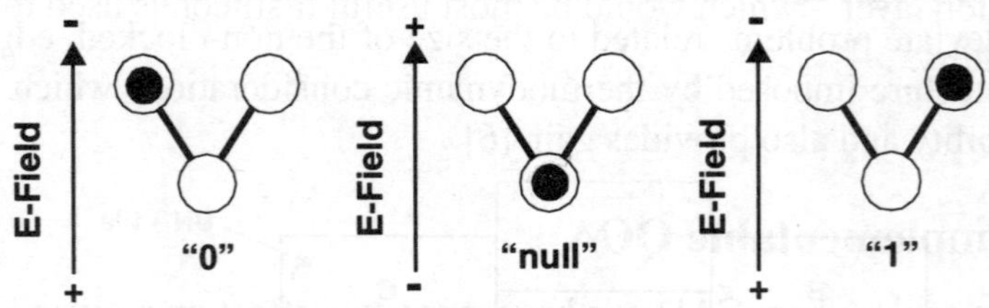


*Figure 9.12.* A "schematic" view of a three-dot molecular QCA cell. An electric field determines if the cell is active or null.

## Wires

Earlier, circuit elements were shown and described in terms of 4-dot QCA cells. Assuming molecular QCA, a 4-dot cell could simply be formed by two adjacent 3-dot or 2-dot cells, but are also being engineered as explicit molecules. 4-dot cells are ideal because of symmetry. Binary information is stored and moved with quadropole moments and all the constructs/circuit elements shown in the first section are theoretically possible.

## Substrates

A pitch matching problem exists between the substrates to which molecular QCA devices could attach, and the devices themselves [12], [2]. Molecular devices are at most a few nanometers in length or width. However, lithography that might be used to etch attachment paths is limited to 200 *nm* and 10 *nm* in the cases of optical or x-ray/e-beam lithography. Given these current resolutions, it would be either most difficult or impossible to create detailed patterns desired for computationally interesting circuits.

One mechanism that might allow for selective cell placement is DNA tiles. First proposed by Seeman, et al, DNA tiles can form rigid, stable junctions with well defined shapes, and can further self-assemble into more complex patterns. Additionally each DNA tile could also contain several points to which something (i.e. a QCA cell) could attach. Lieberman et. al. have developed a DNA raft (37 *nm* long and 8 *nm* tall) built from $4nm \times 12.5nm \times 2nm$ individual DNA tiles. Each individual tile could hold 8 QCA cells [9].

Each portion of a raft has a different DNA sequence. Consequently, molecular recognition could be used to differentiate locations on the raft to which individual molecules could attach – forming a "circuit board" for molecular components. Individual tiles self-assemble according to the affinities shown in Figure 9.13 and parts of the circuit board could self-assemble in a similar manner. Finally, DNA rafts could be attached to silicon wafers using a thick polyadhesion layer – which would be most useful if silicon is used to form the clock circuitry.

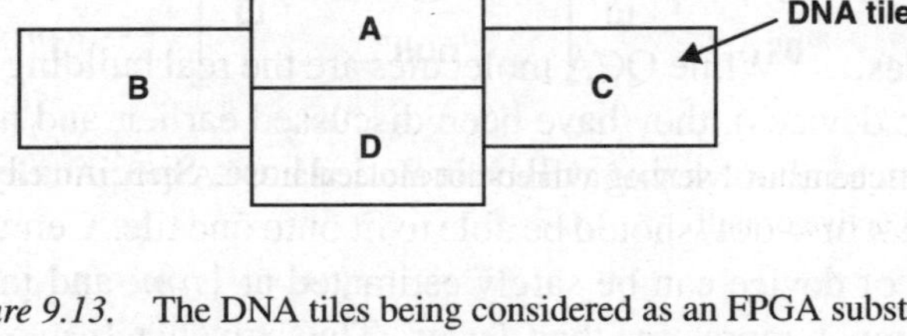


*Figure 9.13.* The DNA tiles being considered as an FPGA substrate.

## The Clock

For molecular QCA, the four phases of a clock signal could take the form of time-varying, repetitious voltages applied to silicon wires embedded underneath a substrate to which QCA cells were attached. Every fourth wire would receive the same voltage at the same time [3]. Neighboring wires see delayed forms of the same signal. The charge and discharge of the clocking wires will move the area of activity across the molecular layer of QCA cells. Computation occurs at the "leading edge" of the applied electric field in a continuous "wave" (see Figure 9.14).

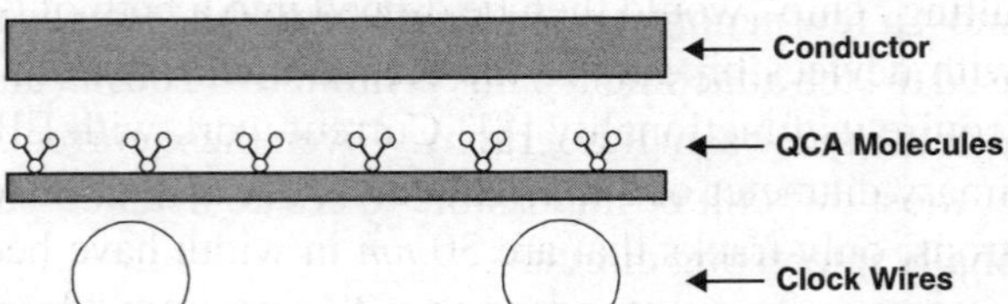


*Figure 9.14.* A schematic showing buried silicon wires that could provide the needed electric field to implement the monostable-bistable-monostable clocking scheme.

## Error

Error will be discussed further below in the section QCA Fabrication Sources of Error

## Building Blocks

We briefly summarize the technological building blocks that will constrain our designs. These could be used to build small systmes in the near-to-midterm. Substrates (DNA tiles), devices (QCA molecules), and support mechanisms (a silicon clock) will all be considered.

**QCA Molecules.** While QCA molecules are the real building blocks of any circuit (i.e. the devices), they have been discussed earlier, and no more that a few brief sentences about sizing will be included here. Specifically, 8 molecular QCA cells (2-dot or 4-dot) should be able to fit onto one tile. Center cell spacing for either type of device can be safely estimated at $1nm$, and individual cells do not require much more area than $1nm^2$. Thus, given a $4nm \times 12.5nm$ tile, 8 cells per tile is not an unreasonable number.

**Substrates.** DNA tiles are currently the most realistic. However, one other possible attachment mechanism currently under investigation involves the use of electron beam lithography (EBL) to create tracks along the surface of some substrate. Ideally, these tracks would dictate attachment points for QCA cells which would then inherently form logical and computational QCA circuits with functionality determined by the pattern of assembly [8]. The overall process envisioned is as follows: first, a molecular QCA cell would be engineered that will pack and assemble properly on a self-assembled monolayer (SAM) on top of a silicon surface. (A SAM is a uni-directional layer formed on a solid surface by spontaneous organization of molecules.) Second, I/0 structures would be constructed lithographically. Third, tracks would be etched into the self-assembled monolayer (SAM) on top of a silicon surface with EBL. Finally, the resulting "chip" would then be dipped into a bath of QCA cells for self-assembly with devices binding to the etched tracks. In this manner, the "chip" would acquire its functionality [8]. Current work with EBL involves a beam with a primary diameter of 5 *nm*. However, because of the dispersion of secondary electrons, only tracks that are 30 *nm* in width have been produced. Also, the tracks are seven Angstroms deep on a 20 Angstrom "deep" monolayer [8].

**Silicon Underneath.** The final major component required for a functional design is some kind of clocking mechanism. A current vision for the implementation of a clock for a QCA circuit calls for silicon wires embedded underneath a DNA substrate to which actual devices would have attached. Charge would then move from wire-to-wire to actually generate propagating electric fields and when compared to the overall power budget, the power dissipated by this silicon circuitry should be quite low. Adiabatic switching was previously discussed.

## 9.3 Design Rules

### Historical Precedence

Before considering design rules in the context of QCA, a brief discussion of what impact they have had on the computer engineering community is worthwhile. This will take place largely in the context of the original rules Mead and Conway proposed for MOS.

### Motivation for Mead and Conway's Design Rules

In the pre-Mead/Conway era (1960's-1970's), chip development flow usually had system architects express a design at a high level (such as Boolean equations), and then turn it over to logic designers, who converted the designs into "netlists" of basic circuits. Fabrication experts would then lay out implementations of the individual logic blocks, and just "wire them together."

Interaction between the architects and the fabrication experts was limited. In terms of technology, MOSFETs were considered "slow and sloppy" and the real design was in sophisticated bipolar devices. However, with the advent of VLSI electronics, the way that digital systems were structured, the procedures used to design them, trade-offs between hardware and software, and the design of computational algorithms for integrated electronics all changed [11]. In a sense, integrated electronics required integrated design, and a common language for system architects and fabrication experts was required.

## What Mead and Conway Did

When considering chip lithography, it is important to note that there is a clear separation between the fabrication of wafers and the design work that generates the patterns to be placed on them. However, for this separation to be possible, the designer will (at least indirectly) require specific knowledge about the resolution and performance of a given processing line. In other words, the designer needs information about how physical devices are actually made. With MOS circuits, this information was (and still is) usually provided in the form of geometries. If a circuit's layout conforms to certain patterns, a designer can be assured that a particular layout will conform to the resolution of a particular fabrication process. Consequently, a fabricated chip should work as intended [11], [13]. With MOS, geometries are usually specified in the form of allowable widths, separations, extensions, and overlaps between various components of a circuit. The values used to specify these parameters are usually a function of a given process, take into account lithography limitations, and usually add a margin for error.

Otherwise stated, design rule geometries exist to minimize the risk of fabrication errors, and are usually specified as multiples of a unit length $\lambda$. For example, when designing a circuit, two polygons representing two metal wires might have to be '$n$' $\lambda$ apart to ensure that a fabricated version of this circuit will actually work. $\lambda$s for various physical integrated circuit components presented by Mead and Conway were originally abstracted from a conglomeration of processes. Values given to them continued to scale downward over time as lithography improved and define *micron rules*. A resulting benefit was that, if specified in $\lambda$s and lithography scaling held, a design targeted for one process could be transferable to another.

The above paragraph provides a general description of $\lambda$ design rules. It is also worth a brief discussion of micron rules. When scaling in the sub-micron range, relationships between layers can scale non-linearly (they are usually valid in the 1-3 micron range). Additionally, scaling rules are often (necessarily) conservative. Thus, with maximum densities desired by industry, $\lambda$ rules are usually avoided and are replaced by *micron rules* which specify explicit distances associated with parts of a CMOS circuit. However, both provide the computer engineer with knowledge that we seek to duplicate for QCA, and would be beneficial to develop for other emergent devices [13].

## Technical Merits of Mead and Conway's Work

At a high-level, by developing a set of design rules and abstractions that a computer architect could use in the circuit design process, Mead and Conway changed the focus of design from considering a chip "in cross section", to "an overhead view." They reduced the physics-dependent device descriptions to a scale-independent set of parameters based largely on area and shape, with some simple rules for first order modeling of how such devices would interact in combination with each other. They also introduced some simple but useful circuit primitives that changed the discussion from isolated logic gate performance to interconnect. This allowed architects (who became experts in hierarchical designs) to extend their hierarchies one level down to potentially new basic structures. This in turn allowed architects to take advantage of these structures in implementing larger and larger systems. The introduction and use of clever circuits using pass transistors is just one example of such an insight.

When considering design rules, one VLSI-design text author writes, "Turning a conceived circuit into an actual implementation also requires a knowledge of the manufacturing process and its constraints. The interface between the design and processing world is captured as a set of design rules that act as prescriptions for preparing the masks used in the fabrication process of integrated circuits. The goal of defining a set of design rules is to allow for a ready translation of a circuit concept into an actual geometry in silicon. The design rule acts as the interface between the circuit designer and the process engineer. As processes have become more complex, requiring the designer to understand the intricacies of the fabrication process is a sure road to trouble [13]."

The above paragraph was originally included in a discussion of Mead and Conway-esq design rules in Rabaey's VLSI text. It is included here as it summarizes the purpose of design rules for MOS, and because words such as "masks" and "silicon" could easily be swapped with words related to a specific nano-scale device – and the ideas would still apply equally well. System designers are generally interested in the densest, most efficient designs possible (in whatever technology), while a process engineer or device designer is mainly concerned with a methodology that offers a high percentage of chips that work, and can be fabricated in an efficient, reproducible manner [13]. Design rules are viewed as the compromise to satisfy both groups. Given this, working to develop design rules for emergent devices will hopefully allow computer architects to become involved with nano-scale devices. This should help us toward our goal of working nano-systems sooner.

## Components of a CMOS Design

Before discussing relevant background for QCA designs, as well as the design rules themselves, we will first review the components for CMOS circuits and the design rules for them. How design rules capture sources of error will also briefly be discussed.

## Basic Structures

Generating a design in CMOS requires at least an implicit knowledge about the set of masks that are used to actually fabricate it. The exact details of the fabrication process are abstracted so that a designer can just think of a CMOS circuit as being comprised of the following circuit elements:

- *Substrates* for *wells* (p-type for NMOS, n-type for PMOS).
- *Diffusion regions* which define areas where transistors can be formed. Diffusions of an inverse type are required to form contacts to the wells on substrates. These are called *select regions*.
- *Polysilicon* forms the gate electrodes of transistors and also serves as interconnect.
- One or more layers of *metal* (also used for interconnect).
- *Contacts* to provide connections between lines and circuit elements in various layers.

Thus, a design layout will consist of a combination of polygon geometries all of which represent one of the items listed above. The shapes are attached to a specific layer and, along with interplay between objects in different layers, specify the functionality of the circuit. For example, a MOS transistor is formed when polysilicon crosses a diffusion layer [13].

## Types of Rules

When considering design rules for CMOS, all of the entities discussed above, as well as the interactions between them, can be specified by assigning ranges to their allowable geometries. The ranges apply within a layer and are: the minimum width of an entity to guarantee current flow (i.e. in poly or metal), the minimum spacing between entities on the same layer (i.e. to prevent a short circuit between two wires), and the required overlap to create devices, contacts, etc. [13]. As an example, when considering metal, $\lambda$s include the minimum metal width (2-3 $\lambda$), the minimum spacing between metal lines (3 $\lambda$), and the minimum metal overlap of a contact [16].

## Sources of Error

In Mead and Conway's original work, unit length $\lambda$ was equal to the fundamental resolution of the process itself. It captures the distance by which a polygon (and hence an actual circuit component) could stray from another polygon on the same (or another) layer and still function correctly. $\lambda$ also encompasses an additional "margin of error" as well as other processing factors. Using the above and the last two sections as context, $\lambda$ design rules should help computer engineers avoid (or at least minimalize) errors inherent to the fabrication process. For CMOS, these include: over etching, misalignment between mask levels, distortion of the silicon wafer ("runout") due to high temperature processing, and overexposure or underexposure of resist. Or, from a circuit designer's perspective: if a structure is not where its intended to be, if a structure is too wide or too narrow, the diffusion rates of values, and the heights of structures in circuits with multiple layers of metal [11].

## The Beginnings of Design Rules for QCA

The previous sections detailed why design rules have been historically useful for computer engineers, and also what they accomplished for CMOS. Using this as a guide, the purpose of this section is to detail the initial questions that led us to the list of parameters (to be presented below) that would be of interest to a computer engineer, in the context of QCA. Specifically:

- What is the QCA equivalent of $\lambda$?
- What factors will govern $\lambda$?
- What sources of error will affect design?
- Will the QCA clock (or another component of fabrication) create the need for a second $\lambda$? Will it affect an original $\lambda$?
- How will the switching times for QCA cells and the QCA clock be related?
- How does the idea of floorplanning factor into design rules?
- Will attachment and substrates govern cell placement?
- What about I/O to the non-nano world?

Using these questions as a base, we will begin to define the parameters that must eventually be compiled and enumerated and will propose some sought after generalizations by attempting to answer some of the questions listed above. We will generate a "process independent" list of values that must be defined when given a specific fabrication methodology – just as Mead and Conway did for MOS. However, like we did here with CMOS, we will first specify the basic structures that a circuit designer would need to use when generating a QCA schematic intended for fabrication. Sources of error related to fabricating systems of molecular QCA cells will also be considered.

## QCA Device Types

A CMOS circuit can be constructed from substrates, wells, diffusion regions, select regions, polysilicon, metal, and contacts. The building blocks for circuits of molecular QCA cells will be detailed here. (Note that most of these entities have been considered in detail previously and unless appropriate, will otherwise just be listed here).

- *90-degree molecular QCA cells*: see the discussion on "Implementable QCA", particularly two-dot, three-dot, and four-dot implementations.

- *45-degree molecular QCA cells*: again, refer to the sections just listed. Also, remember that this will be a function of attachment as well.

- *"permanent cells"*: In some of the QCA circuit constructs (i.e. the small multiplexor), some kind of cell permanently engineered to represent a binary 1 or 0 was used. A similar type of device was mentioned when discussing the simple multiplexor when explaining how a majority gate could implement AND/OR functionality.

(It is worth briefly reconsidering two-dot versus four-dot cells in the context of molecular QCA. Recall that, while researchers are currently working to develop four-dot QCA molecules, much of the early work with specific device implementations has focused on two-dot QCA molecules. However, a designer could simply place two, two-dot cells adjacent to one another and gain the functionality of a four-dot cell.)

## Clock Structures

Currently, silicon is envisioned as a means for implementing a clock for molecular QCA cells. Why the clock is useful, as well as possible implementations for it, was discussed above.

## Bases

Two substrates envisioned for systems of molecular QCA cells were also discussed when considering attachment: etching patterns to which cells would attach with EBL, and using DNA tiles to hold devices.

## QCA Fabrication Sources of Error

As with CMOS, for QCA, we eventually want to help qualify and quantify potential sources of error for the circuit designer in the form of design rules. For molecular QCA, most sources of error arise from issues of placement, offcenteredness, and the self-assembly process. All are shown graphically in Figure 9.15. In particular, Figure 9.15a and Figure 9.15b illustrate that if a cell does not attach where it should (a), the required distance between two QCA cells (b) might be affected. Other sources of error related to off-centeredness include cells that are not exactly aligned (c), cells with the wrong rotation (d), and cells that are off-center with regard to the y-dimension of a circuit(e). While not yet defined, and not an explicit part of a design rule, we will eventually want to quantify rates for each one of these "defects". This will give the system designer an initial notion of error rates and circuit functionality when provided with a finished product – and also could allow the designer to think about required redundancies [9].

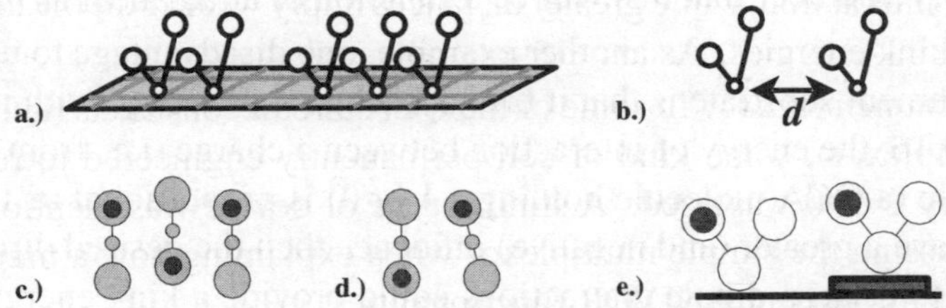


*Figure 9.15.* Possible errors that can occur with molecular QCA cells – cells not attaching (a), improper distance between cells (b), cell misalignment (c), improper cell rotation (d), cells not level (e).

It is also important to begin to quantify what affects the sources of error listed above will have on the storage and movement of binary data. For this reason, we provide a short qualitative discussion of possible electrostatic interactions between molecules. Possible interactions are listed and illustrated in Figure 9.16 and include: interactions between the charges, interactions between a charge and a dipole, charge induced dipole interactions (an anion or cation may induce a dipole in a polarizable molecule and thus be attracted to it), dipole induced dipole interactions (the same as before but now the anion or cation is a permanent dipole), dispersion (when 2 molecules are very close together, their charge fluctuations are synchronized and have an attractive force), and the van der Waals radius (the distance between two molecules such that there outer electron orbitals begin to overlap creating a mutual repulsion between the molecules). To what degree each of the interactions just listed is a function of the distance between entities (*d*) is also shown in Figure 9.16 [10].

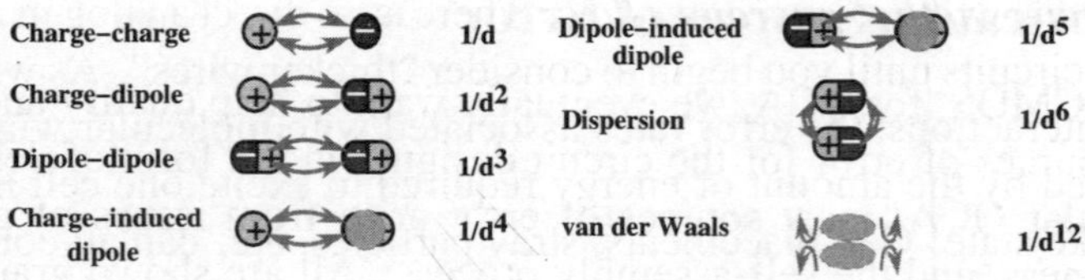


*Figure 9.16.* Possible electrostatic interactions – all could positively or negatively affect interaction/communication between molecular QCA cells.

To provide a flavor of how some of these interactions can affect QCA cells, we can consider kink energy as a function of driver dipole strength. In the presence of a stronger driver dipole, the energy required to induce a mistake state increases; and from our above discussion, strong dipoles are a function of the distance between them (*d*). Thus, intuitively, closer dipoles imply stronger energies of interaction, while greater distances imply weaker dipole interactions and lower kink energies. As another example, one disadvantage to using DNA as an attachment substrate is that it brings background charge with it. As seen in Figure 9.16, the energy of interaction between a charge (i.e. from the DNA) and a dipole (a QCA molecule holding a 1 or 0) is proportional to $1/d^2$ – and thus will have a greater (and negative) influence then the desired dipole-dipole interactions. Charge-dipole interactions could provide a kink energy and as a result cause a mistake.

All of the electrostatic interactions illustrated in Figure 9.16 (whether positive, negative, or applicable at all), must be considered in the context of the specific components of a molecular QCA circuit implementation and will eventually help quantify error tolerances in a schematic.

Note that other nano-scale technologies are also subject to error and defects. (See Chapters 5, 8, 4, 11). This is a common problem.

# QCA Design Rules

This section will describe a preliminary set of design rules for QCA. We will first consider any analogs to CMOS and then specify more specific rules for molecular QCA.

## How they are Analogous to CMOS

The general defining rules for components (i.e. polygons) in a CMOS layer were listed earlier and included specifying minimum widths, minimum wire spacings, and required overlaps. Each will be considered in turn in an effort to provide some initial "guides" for each QCA design rule.

- *Minimum width to guarantee flow:* There is no direct analog in molecular QCA circuits until you begin to consider "thicker wires". As with simple cell interactions, the error rates associated with molecular wires will be affected by the amount of energy required to excite one cell in it into a mistake state. Cell placements, stray particles, etc. can all contribute to kink energy. The designer should be aware that "thicker" wires (i.e. 2 or 3 cells thick) can raise kink energy. For example, for a wire one cell wide, simulations show that with molecular QCA cells the energy required to excite the system into the mistake state is about 600 meV. However, for a wire 3 cells thick, this excitation energy rises to approximately 1.5 eV – a much more robust system [1].

- *Minimum wire spacing for separations:* As with metal wires in CMOS circuits, molecular QCA wires will also have to be a certain distance apart from one another to ensure that there is no cross talk or short circuits between them. Distance between individual cells in a wire will also have to be defined to ensure a value is propagated. Additionally, clock wires must be laid out as well to generate required electric fields. (CMOS design rules would obviously apply here if silicon is used to build the clock circuitry).

- *Overlap rules to create devices and contacts:* When considering overlap, we must ensure that all cells are "clocked" by an electric field and thus space silicon wires accordingly. QCA analogs to overlap also include crossovers between 45-degree and 90-degree cells and the inputs of a majority gate.

Given a possible self-assembly process and the potential for defects, we should also consider what the maximum possible spacing is between cells that will still allow for signal propagation.

## A Specific Rule

We will now consider an example design rules for molecular QCA. This will help to qualify what a circuit designer will need to consider when building a schematic of QCA cells. We are currently finishing the specifications for a complete set of QCA design rules.

Our first design rule(s) (1A and 1B in Figure 9.17), consider(s) spacing between two molecular QCA cells. Specifically, what is the maximum allowed and minimum required distance between two cells such that they will still transmit data? In Figure 9.17, these distances are labeled $x_{max}$ and $x_{min}$ and specific values would be governed by: substrates to which QCA cells can attach, $E_{kink}$ (i.e. background charge with energy of interaction proportional to $1/d^2$ could cause it), and dipole interactions between cells (proportional to $1/d^3$). Also, $x_{min}$ will provide an initial upper bound on maximum device densities.

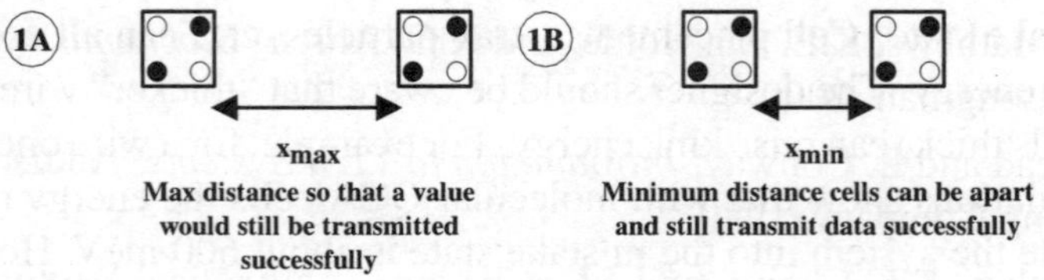


*Figure 9.17.* Design rules 1A and 1B for molecular QCA cells spacing.

## 9.4 Wrap up

Historically, design rules have helped system designers to become more involved in the process of bringing computationally interesting systems to realization. We are applying this idea to QCA, and it has had a most positive effect on interactions between system designers and physical scientists. Both groups now better understand the needs and constraints of the other. Applying these ideas and this "process" to other emergent technologies should be most beneficial.

## References

[1] E. Blair and C. Lent. *Proceedings of the nth IEEE Conference on Nanotechnology*, 2003.

[2] A. Dehon. Array-based Architectures for FET-based Nano-scale Electronics. *IEEE Trans. on Nano.*, 2(1): 23-32, March 2003.

[3] K. Hennessy and C.S. Lent. Clocking of molecular quantum-dot cellular automata. *J. of Vac. Sci. and Tech. B*, 19(5):1752-1755, Sep-Oct 2001.

[4] R.V. Kummamuru, J. Timler, G. Toth, C.S. Lent, and R. Ramasubramaniam. Power gain in a quantum-dot cellular automata latch. *Applied Physics Letters*, 81(7):1332-1334, 2002.

[5] C.S. Lent, B. Isaksen, and M. Lieberman. Molecular Quantum-dot Cellular Automata. *J. of the Am. Chem. Soc.*, 125:1056-1063, 2003.

[6] C.S. Lent, G.L. Snider, G. Berstein, W. Porod, A. Orlov, M. Lieberman, T. Fehlner, M. Niemier, and P. Kogge. Quantum-Dot Cellular Automata. *Electron Transport in Quantum Dots*, 397-433, 2003.

[7] C.S. Lent and P.D. Tougaw. A Device Architecture for Computing with Quantum Dots. *Proceedings of the IEEE*, 85:541, 1997.

[8] M. Lieberman, S. Chellamma, B. Varughese, Y. Wang, and C. Lent. Quantum-dot Cellular Automata at a Molecular Scale. *An. of the New York Ac. of Sci.*, 960:225-239, April 2002.

[9] Personal communciations with Marya Lieberman, 2003.

[10] C.K. Mathews, K.E. van Holde, and K.G. Ahren. *Biochemistry*. Addison Wesley Longman, 2000.

[11] C. Mead and L. Conway. *Introduction to VLSI Systems*. Addison-Wesley Publishing Company, 1980.

[12] M. Oskin, F.T. Chong, and I.L. Chaung. A Practical Architecture for Reliable Quantum Computers. *IEEE Computer*, January:79-87, 2002.

[13] J.M. Rabaey. *Digital Integrated Circuits: A Design Perspective*. Prentice Hall Electronics, 1996.

[14] J. Timler and C.S. Lent. Power gain and dissipation in quantum-dot cellular automata. *Jounral of Applied Physics*, 91(2):823-831, Jan. 15, 2002.

[15] P.D. Tougaw and C.S. Lent. Logical Devices Implemented Using Quantum Cellular Automata. *Journal of Applied Physics*, 75:1818, 1994.

[16] N.H.E. Weste and K. Eshraghian. Principles of CMOS VLSI Design: A Systems Perspective. Addison-Wesley Publishing Company, 1985.

Chapter 10

# PARTITIONING AND PLACEMENT FOR BUILDABLE QCA CIRCUITS

Sung Kyu Lim
*School of Electrical and Computer Engineering*
*Georgia Institute of Technology*
limsk@ece.gatech.edu

Mike Niemier
*College of Computing*
*Georgia Institute of Technology*
mniemier@cc.gatech.edu

**Abstract** Quantum-dot Cellular Automata (QCA) is a novel computing mechanism that can represent binary information based on spatial distribution of electron charge configuration in chemical molecules. It has the potential to allow for circuits and systems with functional densities that are better than end of the roadmap for CMOS, but also imposes new constraints on system designers. In this article, we present the first partitioning and placement algorithm for automatic QCA layout. The purpose of zone partitioning is to initially partition a given circuit such that a single clock potential modulates the inter-dot barriers in all of the QCA cells within each zone. We then place these zones as well as individual QCA cells in these zones during our placement step. We identify several objectives and constraints that will enhance the buildability of QCA circuits and use them in our optimization process. The results are intended to: (1) define what is computationally interesting and could actually be built within a set of predefined constraints, (2) project what designs will be possible as additional constructs become realizable, and (3) provide a vehicle that we can use to compare QCA systems to silicon-based systems.

**Keywords:** Nanotechnology, Quantum-dot Cellular Automata, partitioning, placement



*S.K. Shukla and R.I. Bahar (eds.), Nano, Quantum and Molecular Computing*, 295-317.

## Introduction

Nanotechnology and devices will have revolutionary impact on the Computer-Aided Design (CAD) field. Similarly, CAD research at circuit, logic and architectural levels for nano devices can provide valuable feedbacks to nano research and illuminate ways for developing new nano devices. It is time for CAD researchers to play an active role in nano research. We will discuss CAD related issues for QCA (see Chapter 8.7 for QCA background).

Our goal in this article is to explain how CAD can help research to move from small circuits to small systems of quantum-dot cellular automata (QCA) devices. We leverage our ties to physical scientists who are working to build real QCA devices. Based upon this interaction, a set of near-term *buildability constraints* has evolved - essentially a list of logical constructs that are viewed as implementable by physical scientists in the nearer-term. Until recently, most of the design optimizations have been done by hand. Then these initial attempts to automate the process of removing a single, undesirable, and unimplementable feature from a design were quite successful. We now intend to use CAD, especially physical layout automation, to address all undesirable features of design that could hinder movement toward a "buildability point" in QCA. The net result should be an expanded subset of computationally interesting tasks that can be accomplished within the constraints of a given buildability point. CAD will also be used to project what is possible as the state-of-the-art in physical science expands.

In this article, we present the first partitioning and placement algorithm for automatic QCA layout. The purpose of zone partitioning is to initially partition a given circuit such that a single clock potential modulates the inter-dot barriers in all of the QCA cells within each zone. We then place these zones as well as individual QCA cells in these zones during our placement step. We identify several objectives and constraints that will enhance the buildability of QCA circuits and use them in our optimization process. The results are intended to: (1) define what is computationally interesting and could actually be built within a set of predefined constraints, (2) project what designs will be possible as additional constructs become realizable, and (3) provide a vehicle that we can use to compare QCA systems to silicon-based systems.

## 10.1 Preliminaries

This section provides a detailed comparison between QCA and CMOS technologies. We also discuss the need for QCA CAD research, especially the physical layout automation.

## QCA Wins

As QCA is being considered as an alternative to silicon-based computation, it is appropriate to enumerate what QCA's "wins" over silicon-based systems could be (as well as its potential obstacles). We begin by listing obstacles to CMOS-based Moore's Law design (Table 10.1), their effects on silicon based systems, and how they will affect QCA.

Based on the information in Table 10.1 it is apparent that QCA faces some of the same general problems as silicon-based systems (timing issues, lithography resolutions, and testing), that QCA does not experience some of the same problems as silicon-based systems (quantum effects and tunneling), and that silicon-based systems can address one problem better than QCA currently can (I/O) However, if the I/O problem is resolved, QCA can potentially offer significant "wins" with regard to reduced power dissipation and fabrication. Additionally, QCA can also offer orders of magnitude in potential density gains when compared to silicon-based systems. When examining the existing design of an ALU for a simple processor [11], one version is potentially 1800 times more dense (assuming deterministic cell placement) than an end of the CMOS curve equivalent (0.022 micron process). If based on a more implementable FPGA (whose logic cell is a single NAND gate), the ALU is no less dense than a fully custom, end of the CMOS curve equivalent [12]. Clearly, realizable and potential QCA systems warrant further study.

## Builability Analysis via QCA CAD

One might argue that it would be premature to perform any systems-level study of an emergent device while the physical characteristics of a device continue to evolve. However, it is important to note that many emergent, nano-scale devices are targeted for computational systems - and to date, most system-level studies have been proposed by physical scientists, and usually end with a demonstration of a functionally complete logic set or a simple adder. Useful and efficient computation will involve much more than this, and, in general, it is important to provide scientists with a better idea of how their devices should function. This coupling can only lead to an accelerated development of functional and interesting systems at the nano-scale. More specifically, with QCA, physicists are currently preparing to test the self-assembly process and its building blocks. Thus, our work can help provide the physicists with computationally interesting patterns (that could be fabricated) - the real and eventual desired end result.

Our toolset will focus on the following undesirable design schematic characteristics associated with a near-to-midterm buildability point: large amounts of deterministic device placement, long wires, clock skew, and wire crossings. We will use CAD to: (1) identify logic gates and blocks that can be duplicated

*Table 10.1.* Comparing characteristics of silicon-based systems to QCA based systems

| Obstacle | Effect on CMOS circuits | How it relates to QCA |
|---|---|---|
| Quantum Effects and Tunneling | A gate that controls the flow of electrons in a transistor could allow them to tunnel through small barriers - even if the device is supposed to be off [14]. | No effect; QCA devices are charge containers, not current switches and actually leverage this property. |
| High power dissipation | Chips could melt [15] [9] unless problems are overcome for which SIA roadmap says "there are no known solutions". 2014 projection: chip with $10^{10}$ devices dissipates 186W of power. | $10^{11}$ QCA devices with $10^{-12}$ switching times dissipate 100W of power. QCA's silicon-based clock will also dissipate power. Still, clocking wires should move charge adiabatically [5], greatly reducing power consumption. |
| Slow wires | Wires continue to dominate the overall delay [6]. Also, projections show that for 60 nm feature sizes, less than 10% of the chip is reachable in 1 clock cycle [4]. | The inherent pipelining caused by the clock make global communication and signal broadcast difficult [13]. Problems are similar to silicon-based systems but for different reasons. |
| Lithography resolutions | Shorter wavelengths and larger apertures are needed to provide finer resolutions for decreased feature sizes. | QCA's clock wiring is done lithographically, which is subject to the same constraints as silicon-based systems. However, closely spaced nanowires could also be used [8]. |
| Chip I/O | I/O count continue to increases as the technology advances (Rent's rule), but pin counts do not scale well. With more processing power, we will need more I/O [15]. | I/O remains under investigation with one approach being to include "sticky ends" at the ends of certain DNA tiles in order to bind nano-particles or nano-wires |
| Testing | Even if designs are verified and simulated, defects caused by impurities in the manufacturing process, misalignment, broken interconnections, etc. can all contribute to non-functional chips. Testing does not scale well [15]. | We must find and route around defects caused by self-assembly and/or find new design methodologies to make circuits robust. Defects for self-assembled systems could range from 70% to 95%. Structures such as thicker wires could help. |
| Cost | Fabrication facility cost doubles approximately every 4.5 years [17], and could reach 200 billion dollars in 2015. | Self-assembly could be much more inexpensive. |

to reduce wire crossings, (2) rearrange logic gates and nodes to reduce wire crossings, (3) create shorter routing paths to logical gates (to reduce the risk of clock skew and susceptibility to defects and errors), and (4) reduce the area of a circuit (making it easier to physically build). Some of these problems have been individually considered in existing work for silicon-based VLSI design, but in combination, form a set of constraints unique to QCA requiring a unique toolset to solve them.

## CMOS vs QCA Placement

Although QCA and CMOS have considerable technological differences, CMOS VLSI placement algorithms [2, 7, 19] have been modified to satisfy the design constraints imposed by QCA physical science. There are many reasons for using this approach. Notably, VLSI design automation algorithms work on graph-based circuits, and it has been found to be advantageous to represent QCA circuits as graphs – especially because at present, only two dimensional circuits have been proposed and are seen as technically feasible. Existing algorithms can be fine-tuned to meet QCA's constraints and objectives. Additionally, physical design issues for CMOS have been widely studied, optimized, and proven to be NP-complete [3]. Thus, it makes sense to leverage this existing body of knowledge and apply it to a new problem. Finally, because so few design automation tools and methodologies exist for QCA, using VLSI algorithms as a base will allow us to compare and set standards for our place and route methodologies.

More specifically, we note the following similarities and differences between CMOS and QCA placement.

- Similarities: In CMOS placement, partitioning, floorplanning, and placement are performed in this order (a hierarchical approach) to efficiently handle the design complexity. We use a similar approach in QCA placement: zone partitioning, zone placement, and cell placement. The following objectives are common in both CMOS and QCA partitioning and work to solve the same goal: cutsize and performance. The area, performance, congestion, and wire length objectives are common to both CMOS and QCA placement.

- Differences: two major differences are QCA clocking and QCA single-layer routing resource. Minimizing the total number of QCA wire crossings is critical in QCA placement as QCA layouts must be done in a single layer (unlike the multi-layer CMOS layout). Thus, node duplication in CMOS targets area and performance optimizations while QCA duplication targets minimizing wire crossings to conform to QCA's clocking requirements, we use *k-layered bipartite graphs* to represent an original and partitioned netlist. This in turn requires QCA partitioning to minimize area increases (after the bipartite graph construction). In addition, the length of all reconvergent paths from the same partition should be balanced (to be discussed in detail later) and cyclic dependencies are not allowed.

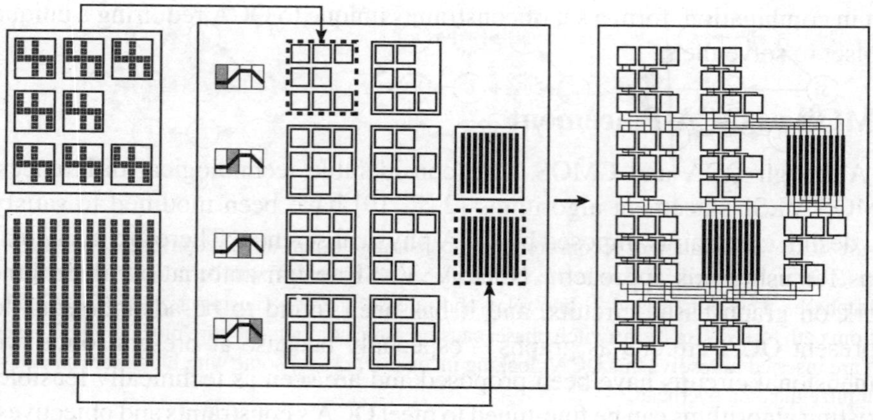

*Figure 10.1.* Overview of the QCA layout automation process. A logic and a wire block are shown. First, the input circuit is partitioned into logic and wire blocks (zone partitioning). Second, each block is placed onto 2D space while satisfying QCA timing constraints (zone placement). Third, QCA cells in each block is placed (QCA cell placement). Fourth, routing is performed to finish inter-block interconnect (global QCA routing) and intra-block interconnect (detailed QCA routing).

## 10.2 Problem Formulation

In this section, we provide an overview of the placement process for QCA circuits. We then present the formulation of three problems related to QCA placement–zone partitioning, zone placement, and cell placement problem. Our recent work on zone partitioning and zone placement work is available in [10] and cell placement in [16].

### Overview of the Approach

An overview of QCA physical design automation is shown in 10.1. QCA placement is divided into three steps: zone partitioning, zone placement, and cell placement. The purpose of zone partitioning is to decompose an input circuit such that a single potential modulates the inner-dot barriers in all of the QCA cells that are grouped within a clocking zone. Unless QCA cells are grouped into zones to provide zone-level clock signals, each individual QCA cell will need to be clocked. The wiring required to clock each cell individually would easily overwhelm the simplicity won by the inherent local interconnectivity of the QCA architecture. However, because the delay of the biggest partition also determines the overall clock period, the size of each partition must also be determined carefully. In addition, four-phase clocking imposes a strict constraint on how to perform partitioning. The zone placement step takes as input a set of zones–with each zone assigned a clocking label

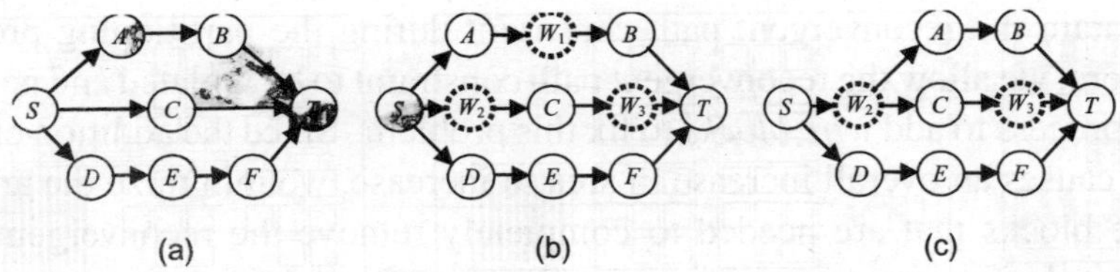


*Figure 10.2.* Illustration of reconverent path constraint. (a) all three reconvergent paths from $S$ to $T$ are unbalanced. If $S$ is in the switch phase, $A$, $B$, and $T$ will be in relax, release, and hold phase. This puts $C$ and $T$ into relax and release, thereby causing a conflict at $T$. The bottom path forces $T$ to be in switch phase, causing more conflict. (b) wire blocks $W_1$, $W_2$, and $W_3$ are inserted to resolve this QCA clocking inconsistency. (c) some wire blocks are shared to minimize the area overhead.

obtained from zone partitioning. The output of zone placement is the best possible layout for arranging the zones on a two dimensional chip area. Finally, cell placement visits each zone to determine the location of each individual logic QCA cell – (cells are used to build majority gates).

## Zone Partitioning Problem

A gate-level circuit is represented with a directed acyclic graph (DAG) $G(V, E)$. Let $P$ denote a partitioning of $V$ into $K$ non-overlapping and non-empty blocks. Let $G'(V', E')$ be a graph derived from $P$, where $V'$ is a set of logic blocks and $E'$ is a set of cut edges based on $P$. A directed edge $e(x, y)$ is *cut* if $x$ and $y$ belong to different blocks in $P$. Two paths $p$ and $q$ in $G'$ are *reconvergent* if they diverge from and reconverge to the same blocks as illustrated in Figure 10.2(a). If $l(p)$denotes the length of a reconvergent path $p$ in $G'$, then $l(p)$ is defined to be the number of cut edges along $p$. A formal definition of zone partitioning problem is as follows:

DEFINITION 10.1 *Zone partitioning: we seek a partitioning of logic gates in the given netlist into a set of zones so that cutsize (= total number of cut nets), wire block (= required during the subsequent zone placement) are minimized. The area of each partition needs to be bounded (area constraint), and there should not exist cyclic dependency among partitions (acyclic constraint). In addition, the length of all reconvergent paths should be balanced (clocking constraint).*

The reconvergent path constraint is illustrated in Figure 10.2. Cycles may exist among partitions as long as their lengths are multiples of four (i.e. because of an assumed 4-phase, QCA clock). However, it becomes difficult to enforce this constraint while handling other objectives and constraints. Therefore, we prevent any cycles from forming at the partition level. In addition, it is difficult to maintain the reconvergent path constraint during the partitioning process. Therefore, we allow the reconvergent path constraint to be violated and perform a post-process to add *wire blocks* to fix this problem. Since the addition of wire blocks causes an overall increase in area to increase, we minimize the amount of wire blocks that are needed to completely remove the reconvergent path problems during zone partitioning.

## Zone Placement Problem

Assuming that all partitions (= zone) have the same area, placement of zones becomes a geometric embedding of the partitioned network onto a $m \times n$ grid, where each logic/wire block is assigned to a unique location in the grid. In this case, a bipartite graph exists for every pair of neighboring clocking levels. We define the *k-layered bipartite graph* as follows:

DEFINITION 10.2 *K-layered bipartite graph: a directed graph $G(V, E)$ is k-layered bipartite graph iff (i) $V$ is divided into $k$ disjoint partitions, (ii) each partition $p$ is assigned a level, denoted $lev(p)$, and (iii) for every edge $e = (x, y)$, $lev(y) = lev(x) + 1$.*

Therefore, the zone placement problem is to embed a *zone-level k-layered bipartite graph* onto an $m \times n$ grid so that all blocks in the same layer are placed in the same row. All the I/O terminals are assumed to be located on the top and bottom boundary of each block, and we may insert routing channels between clocking levels for the subsequent routing. A formal definition of zone placement problem is as follows:

DEFINITION 10.3 *Zone placement: we seek to place the zones we obtain from zone partitioning onto a 2D space so that area, wire crossings and wire length are minimized. Each zone (= logic/wire block) is labeled with a clocking level (= longest path length from input zones), and all zones with the same clocking level should be placed in the same row (clocking constraint). In addition, all inter-zone wires need to connect two neighboring rows (neighboring constraint).*

## Cell Placement Problem

The input to the cell placement is the zone placement result, where all logic/wire blocks at the same clocking level are placed in the same row. The output of cell placement is an arrangement of QCA cells in each logic block. The reconvergent path problem does not exist in cell placement–it is perfectly fine to have unbalanced reconvergent path lengths among the logic gates in each logic block. The reason is that correct output values will eventually be available at the output terminals in each block if the clock period is longer than the maximum path delay in each block. We determine the clock period based on the maximum path delay among all logic/wire blocks. A formal definition of cell placement problem is as follows:

DEFINITION 10.4 *Cell placement: we seek a placement of individual logic gates in the logic block so that area, wire crossings and wire length are minimized. The following set of constraints exists during QCA cell placement: (1) the timing constraint: the signal propagation delay from the beginning of a zone to the end of a zone should be less than a clock period established from zone partitioning and the constraints of physical science (maximum zone delay) (i.e. we want to eliminate possible skew), (2) the terminal constraint: the I/O terminals are located on the top and bottom boundaries of each logic block, (3) the signal direction constraint: the signal flow among the logic QCA cells needs to be unidirectional-from the input to the output boundary for each zone.*

A signal's direction is dictated by QCA's clocking scheme, where an electric field $E$ created by underlying CMOS wire is propagating in uni-directionally within each block. Thus, cell placement needs to be done in such a way that the logic outputs will propagate in the same direction as $E$. In order to balance the length of intra-zone wires, we construct a *cell-level k-layered bipartite graph* for each zone and place it.

## 10.3 Zone Partitioning Algorithm

This section presents zone partitioning and wire block insertion algorithms. Zone partitioning algorithm is an iterative improvement based method, whereas wire block insertion is based on the longest path computation.

### Zone Partitioning

Let $lev(p)$ denote the longest path length from the input partitions (partitions with no incoming edges) to partition $p$, where the path length is the number of partitions along the path. Then $wire(e)$ denotes the total number of wire blocks to be inserted on an inter-partition edge $e$ to resolve the unbalanced reconvergent path problem (clocking constraint of the QCA zone partitioning problem). Simply, $wire(e) = lev(y) + lev(x) - 1$ for $e = (x, y)$, and the total number of wiring blocks required without resource sharing is $\sum wire(e)$. Thus, our heuristic approach is to minimize the $\sum wire(e)$ among all inter-zone edges while maintaining acyclicity. Then, during post-processing, any remaining clocking problems are fixed by inserting and sharing wire blocks. An illustration of zone partitioning and wire block insertion is shown in Figure 10.3.

First, the cells are topologically sorted and evenly divided into a number of partitions $(p_1, p_2, \cdots p_k)$. The partitions are then level numbered using a breadth-first search. Next, the acyclic FM partitioning algorithm [1] is per-

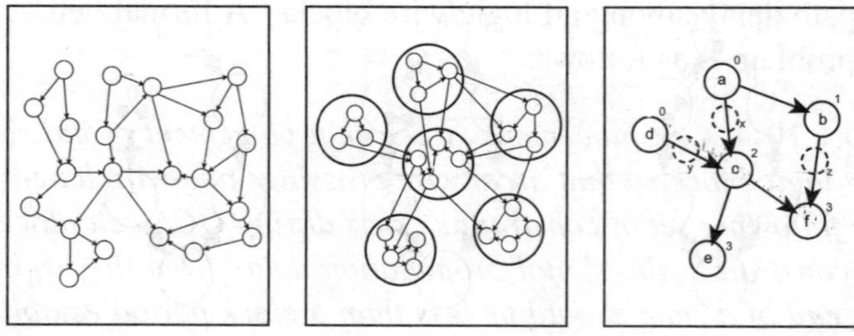


*Figure 10.3.* Illustration of zone partitioning and wire block insertion. (a) directed graph model of input circuit, (b) zone partitioning under acyclicity and reconvergent path constraint, (c) wire block insertion, where the numbers denote the longest path length. The dotted nodes indicate wire blocks.

formed on adjacent partitions $p_i$ and $p_{i+1}$. Constraints that must be met during any cell move include area and acyclicity. The cell gain has two components: cutsize gain and wire block gain. The former indicates the reduction in the number of inter-partition wires, whereas the latter indicates the reduction in the total number of wire blocks required. We then find the best partition based on a combined cost function for both cutsize and wire block gain. Multiple passes are performed on two partitions $p_i$ and $p_{i+1}$ until there is no more improvement on the cost. Then, this acyclic bipartitioning is performed on partitions $p_{i+1}$ and $p_{i+2}$, etc.

Movement of a single cell could change $lev(p)$, the level number of a partition $p$. Therefore every time a cell move is made, we check to see if this cell move affects the level number. Levels can change as a result of a newly introduced inter-zone edge or from completely removing an inter-zone edge. In Figure 10.4, cell $a$ in Figure 10.4(a) is moved from partition $A$ to $B$, thereby creating a new inter-partition edge in 10.4(b). This in turn changes the level of all downstream partitions. In Figure 10.4(c), cell $a$ in Figure 10.4(a) is moved from partition $A$ to $C$, thereby removing the inter-partition edge between $A$ and $C$ 10.4(c). This again changes the level of all downstream partitions. To update levels, we maintain a maxparent for each $p$ so that the level number of the parent of $p$ is $lev(p) - 1$. $lev(F)$ is defined as the level number of the "from block" of a cell $c$ and $lev(T)$ is defined as the level number of the "to block" of $c$. In the first case where a new inter-partition edge is created, $lev(T)$ is updated if $lev(F) \geq lev(T)$ after the cell move. In this case, $lev(T) = lev(F) + 1$. Then, we recursively update the maxparent and levels of all downstream partitions. The maxparent for partition $C$ was changed from $A$ to $B$ in Figure 10.4(b), and $lev(C)$ now becomes $lev(B) + 1 = 2$. This in turn requires the level number of all downstream nodes to change. In the second case where an existing

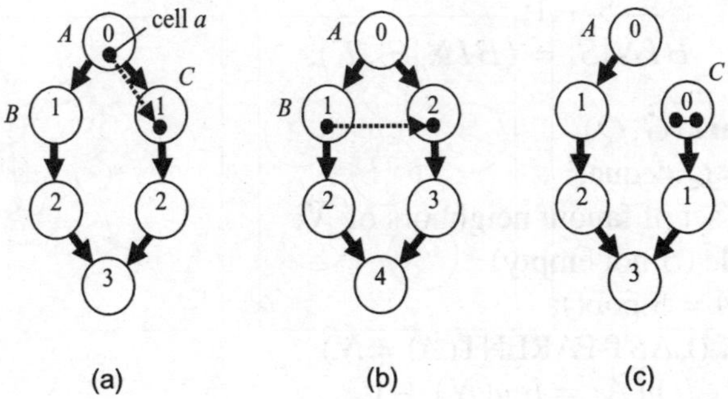


*Figure 10.4.* Illustration of clocking level update.

inter-partition edge is removed, the maxparent again needs to be update. The maxparent for partition $C$ was changed from $A$ to none in Figure 10.4(c), and $lev(C)$ now becomes $lev(C) = 0$.

## Wire Block Insertion

During post-processing, any remaining clocking problems are fixed by inserting and sharing wire blocks, while satisfying wire capacity constraints. The input to this algorithm is the set of partitions and inter-partition edges. First, a super-source node is inserted in the graph whose fan-out neighbors are the original sources in the graph. This is done to ensure that all sources are in the same clocking zone. Then the single-source longest path is computed for the graph with the super-source node as the source–and every partition is assigned a clocking level based on its position in the longest path from the source. For a graph with $E'$ inter-partition edges, this algorithm runs in exactly $O(E')$ iterations. In the algorithm's next stage, any edge connecting partitions that are separated by more than one clock phase is marked, and the edge is added to an array of bins at every index where a clocking level is missing in the edge. The following algorithms perform wire block insertion.

**wire_block_insertion**$(G(V, E))$
    $lev(SUPER) = -1$;
    $Q$.enque($SUPER$);
    BFS-mark($G, SUPER$);
    while ($E$ not empty)
        $N = E$.pop();
        $S = lev(N.source)$;
        $T = lev(N.sink)$;
        while ($S + 1 < T$)

```
        S = S + 1;
        BIN[S] = (BIN[S], E);

BFS-mark(G, Q)
    N = Q.deque;
    S = set of fanout neighbors of N;
    while (S not empty)
        A = S.pop();
        if (LAST-PARENT(A) = N)
            lev(A) = lev(N) + 1;
            Q.enque(A);
    BFS-mark(G, Q)
```

The number of wire blocks in each bin is calculated based on a predetermined capacity for the wire blocks. This capacity is calculated based on the width of each cell in the grid. Then the inter-partition edges are distributed amongst the wire block, filling one wire block to full capacity before filling the next. It might seem that a better solution would be to evenly distribute the edges to all the wire blocks in the current level. This is not true because the wire blocks with the most number of feed-throughs are placed closer to the logical blocks in the next stage. This minimizes wire length, and hence the number of wire crossings.

## 10.4 Zone Placement Algorithm

This section presents our zone placement algorithm. Our zone partitioning algorithm is an iterative improvement based method, where the initial placement of a zone-level $k$-level-bipartite-graph is refined via block swaps to minimize the total number of wire crossings and reduce wire length.

### Placement of k-Layered Bipartite Graph

The logical blocks (obtained from the partitioning stage) and the wire blocks (obtained from post-processing) are placed on an $m \times n$ grid with a given aspect ratio and skew. The individual zone dimensions and the column widths are kept constant to ensure scalability and manufacturability of this design as clocking lines would have to be laid underneath QCA circuits with great precision. The partitions are laid out on the grid, with the cells belonging to the first clocking zone occupying the leftmost cells of the first row of the grid, and the next level occupying the leftmost cell of the next row, etc., until row $r$. The next level of cells is placed again on row $r$ to the right of the rightmost placed cell amongst the $r$ placed rows. Then, the next level of cells is placed in row $r - 1$ and the rest of the cells are placed in a similar fashion until the first row is reached. This process is repeated until all cells are placed (thereby forming a "snake-shape").

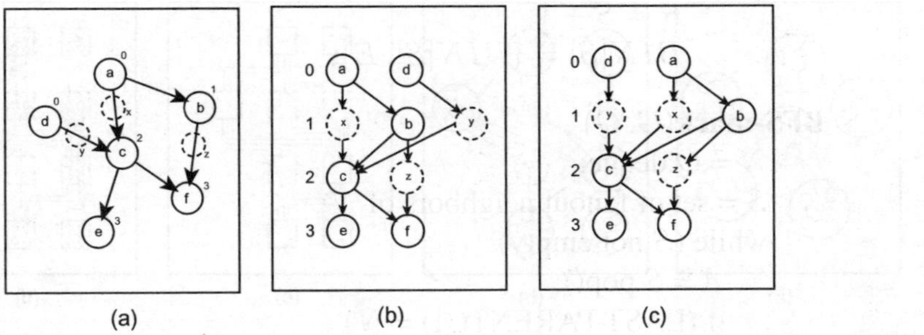


*Figure 10.5.* Illustration of zone placement and wire crossing minimization. (a) zone partitioning with wire block insertion, (b) zone placement, where a zone-level k-layered bipartite graph is embedded onto a 2D space, (c) wire crossing minimization via block re-ordering.

The white nodes are white space that is introduced because of variations in the number of wire and logic blocks among the various clocking levels. The maximum wire length between any two partitions in the grid determines the clock frequency for the entire grid as all partitions are clocked separately. For the first and last rows (where inter-partition edges are between partitions in two different columns), maximum wire length was given more priority as maximum wire length at these end zones can be twice as bad as the maximum wire length between partitions on the same column. An illustration of zone placement and wire crossing minimization is shown in Figure 10.5.

## Wire Crossing Minimization

During the next phase, blocks are reordered within each clocking level to minimize inter-partition wire length and wire crossings. Two classes of solutions were applied to minimize the above objectives: an analytical solution that uses a weighted barycenter method, and Simulated Annealing. The analytical method only considers wire crossings since as there is a strong correlation between wire length and the number of wire crossings.

Analytical Solution: A widely used method for minimizing wire crossings (introduced by Sugiyama et al. [18]) is to map the graph into $k$-layer bipartite graph. The vertices within a layer are then permuted to minimize wire crossings. This method maps well to this problem as we need to only consider the latter part of the problem (the clocking constraint provides the k-layer bipartite graph). Still, even in a two-layer graph, minimizing wire-crossings is NP-hard [18]. Among the many heuristics proposed, the *barycenter heuristic* [18] has been found to be the best heuristic in the general case for this class of problems. A modified version of the barycenter heuristic was used to accommodate edge weights. Edge weights represent the number of inter-partition edges that exist

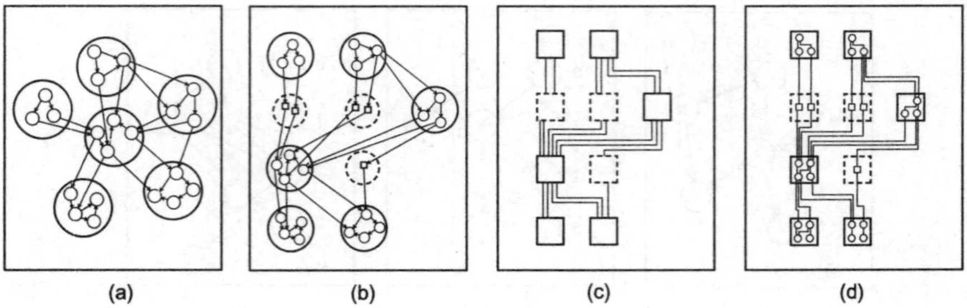

*Figure 10.6.* Illustration of cell placement, global routing and detailed routing. (a) zone placement result, (b) cell placement result, where cells in each partition (= zone) again forms a k-layered bipartite-graph, (c) global routing, where inter-zone connections are made, (d) detailed routing, where intra-zone connections are made.

between the same pair of partitions. The heuristic can be summarized as follows:

$$barycenter(v) = \frac{\sum_N [weight(n) \times position(n)]}{\sum_N weight(n)}$$

where $v$ is the vertex in the variable layer, $n$ is the neighbor in the fixed layer, and $N$ is the set of all neighbors in the fixed layer.

Simulated Annealing: A move is done by randomly choosing a level in the graph and then swapping two randomly chosen partitions $[p_1, p_2]$ in that level in order to minimize the total wire length and wire crossings. In our implementation, the initial calculation of the wire length is $O(n)$ and updating the number of wire crossings is $O(n^3)$ – where $n$ is the number of nodes in a layer of the bipartite graph. In our approach, we initially compute the wire length and wire crossing and incrementally update these values after each move so that the update can be done in $O(m)$ time where $m$ is the number of neighbors for $p_i$. This speedup allows us to explore a greater number of candidate solutions, and as a result, obtain better quality solutions.

## 10.5 Cell Placement Algorithm

This section presents our cell placement algorithm, which consists of feed-through insertion, row folding, and wire crossing and wire length optimization steps. Figure 10.6 shows an illustration of cell placement as well as QCA routing.

### Feed-Through Insertion

In order to satisfy the relative ordering and to satisfy the signal direction constraint, the original graph $G(V, E)$ is mapped into a k-layered bipartite graph $G'(V', E')$ which is obtained by insertion of feed-through gates, where $V'$ is

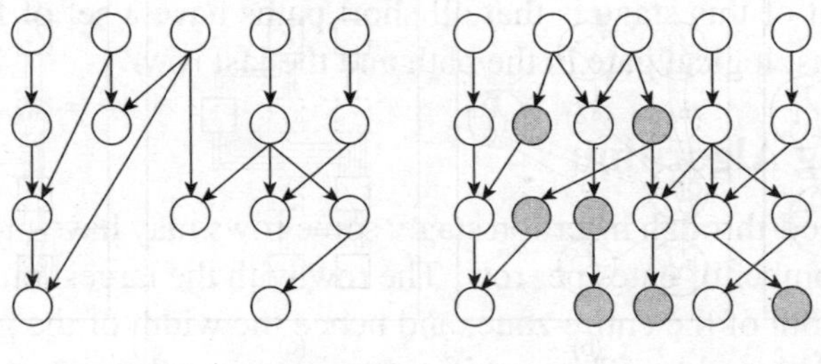

*Figure 10.7.* Illustration of feed-through insertion, where a cell-level k-layered bipartite-graph is formed via feed-through nodes.

the union of the original vertex set $V$ and the set of feed-through gates, and $E'$ is the corresponding edge set. The following algorithm performs feed-through insertion.

```
feed-through_insertion(G(V,E))
    if (V is empty)
        return;
    n = V.pop();
    if (n has no child with bigger level)
        return;
    g = new feed-through;
    lev(g) = lev(n) + 1;
    for (each child c of n)
        g = parent(c);
        c = child(g);
    n = parent(g);
    g = child(n);
    add g into G;
    feed-through_insertion(G(V,E))
```

In this algorithm, we traverse through every vertex in the vertex set of the graph. For a given vertex, if any of the outgoing edges terminate at a vertex with a topological order that is more than one level apart, a new feed-through vertex is added to the vertex set. The parent of the feed-through is set to the current vertex, and all children of the current vertex which have a topological order difference of more than one are set as the children of the feed-through. We do not need to specifically worry about the exact level difference between the feed-through and the child nodes, as this feed-through insertion is a recursive process. This algorithm runs in $O(k|V'|)$, where $k$ is the maximum degree of $V'$. Figure 10.7 shows the graph before and after feed-through insertion.

A trivial result of this stage is that all short paths have a set of feed-throughs between the last logical gate in the path and the last row.

## Row-folding Algorithm

After the feed-through insertion stage, some rows may have more gates than the average number of gates per row. The row with the largest number of gates defines the width of the entire zone, and hence the width of the global column that the zone belongs to. This significantly increases the circuit area. Hence, rows with a large number of cells are folded into two or more rows. This is accomplished by inserting feed-through gates in the place of logic gates, and moving the gates to the next row. Row-folding decreases the width of the row as a feed-through has a lower width than the gate it replaces. A gate $g$ is moved into the next existing row if it belongs to the row that needs to be folded, and all paths that $g$ belongs to contain at least one feed-through with a higher topological order than $g$. The reason for the feed-through condition is that $g$, along with all gates between $g$ and the feed-through can be pushed to a higher row, and the feed-through can be deleted without violating the topological ordering constraint. The following algorithm performs row folding.

```
row_folding(G, w)
    if (w is a feed-through)
        return(TRUE);
    if (w.level = G.max_level)
        return(FALSE);
    RETVAL = TRUE;
    k = w.out-degree;
    i = 0;
    while (RETVAL and i < k)
        RETVAL = row_folding(G,w.CHILD(i));
        i = i + 1;
    return(RETVAL);
```

This algorithm returns true if a node can be moved, and false if a new row has to be inserted. If this feed-through criterion is not met, and the row containing $g$ has to be folded, then a new row is inserted and $g$ is moved into that row.

The number of gates that need to be moved from a row that needs folding to a new row is given by the following trivial calculation. Let $n$ be the number of gates that need to be moved to the next row. Let $m$ be the original number of gates in the row, and let $M$ be the maximum number of gates allowed in a row. Further, let $a$ be the ratio of the width of a feed-through to the width of the gate. As the width of a gate is always greater than the width of a feed-through, ($a < 1$), for every gate that is moved to a new row, a feed-through has to be inserted in its original place. Hence, after moving $n$ gates to the next row, the width of the original row will now be $m-n+an$, so $n = (m-M)/(1-a)$. This calculation is repeated for the next row if $n$ is itself greater than the constraint $M$. The principal reason for increasing the height of a zone rather than increasing the width of the zone is that the width of global column that the zone belongs to is much smaller than the height of the column as the aspect ratio of the entire circuit layout is close to unity.

## Wire Length and Wire Crossing Minimization

During zone placement stage, a zone-level k-layered bipartite graph is formed via wire block insertion. This graph is then placed in such a way that all zones at the same clocking level are placed in the same row. The same graph transformation and placement is done during cell placement–a cell-level k-layered bipartite graph is formed via feed-through insertion, and this graph is placed in such a way that all cells of the same longest path length are placed in the same row. In both cases, iterative improvement is performed to reduce the wire crossing and wire length at the zone and cell level. We perform barycenter heuristic to build the initial solution and perform block/cell swaps to improve the solution quality.

To compute the net wire length in a circuit we traverse through every vertex and accumulate the difference between the column numbers of the vertex and all of its children. This runs in $O(N)$, where $N$ is the number of vertices. But, during the first calculation, we store the sum of all outgoing wire length in every vertex. This enables us to incrementally update if the position of only one node changes. A node cannot change its row number since at this stage the topological level is fixed. If a node changes its position within a level, then it is enough to calculate the difference in position with respect to its neighbors alone. Hence, subsequent wire length calculation is reduced to $O(K)$ where $K$ is the node's vertex degree.

Wire crossing computation can be done with either the adjacency list or matrix, depending on the sparseness of the graph. We used the adjacency matrix to compute the number of wire crossings in a graph. In a graph, there is a wire crossing between two layers $v$ and $u$ if $v_i$ talks to $u_j$ and $v_x$ talks to $u_y$, where $i$, $j$, $x$, and $y$ denote the relative positional ordering in the nodes, and either, $i < x < j < y$ or $i < x < y < j$ or $x < i < y < j$ or $x < i < j < y$ without loss of generality. In terms of an adjacency matrix, this can be regarded as if either the point $(i, j)$ is in the lower left sub-matrix of $(x, y)$ or vice versa, there is a crosstalk. Hence, our solution is to count the number of such occurrences. If this counting is done unintelligently, it can be in the order of $O(n^4)$. Our algorithm to compute the number of wire crossings runs in $O(n^2)$.

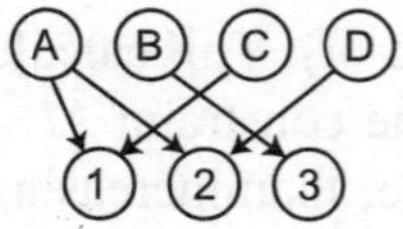


(a)

| | 1 | 2 | 3 |
|---|---|---|---|
| A | 1 | 1 | 0 |
| B | 0 | 0 | 1 |
| C | 1 | 0 | 0 |
| D | 0 | 1 | 0 |

(b)

| | 1 | 2 | 3 |
|---|---|---|---|
| A | 1 | 2 | 2 |
| B | 0 | 0 | 1 |
| C | 1 | 1 | 1 |
| D | 0 | 1 | 1 |

(c)

| | 1 | 2 | 3 |
|---|---|---|---|
| A | 2 | 4 | 5 |
| B | 1 | 2 | 3 |
| C | 1 | 2 | 2 |
| D | 0 | 1 | 1 |

(d)

*Figure 10.8.* Illustration of incremental wire crossing computation. (a) a bipartite graph with 3 wire crossings, (b) adjacency matrix of (a), (c) row-wise sum of (b) from left to right, (d) column-wise sum of (c) from bottom to top. Each entry in (d) now represent the total sum of entries in low-left sub-matrix. Using (b) and (d), wire crossing is $A2 \times B1 + B3 \times C2 = 3$, where the first entry is from (b) and second from (d).

Figure 10.8 shown and example of wire crossing computation. The graph in Figure 10.8(a) can be represented by the adjacency matrix shown in Figure 10.8(b). The number of crossings in Figure 10.8(a) is 3. This can be obtained from the matrix by adding the product of every matrix element and the sum of its left lower matrix elements (i.e. the number of crossings is $\sum(A_{ij} \times \sum\sum A_{xy})$, where $i+1 < x < n$ and $1 < y < j-1$). This formula gives a good intuition of the process but is computationally very expensive. We now illustrate our method to calculate wire crossing more efficiently. First we take the row-wise sum of all entries as in Figure 10.8(c). Then we use this to compute the column-wise sum as in 10.8(d). Finally, we multiply all the entries in the original matrix and the column-wise sum matrix to compute the total wire crossing–each entry $(r, c)$ in the original matrix is multiplied by the entry $(r+1, c-1)$ in the column-wise sum matrix as shown in 10.8(d). In the simulated annealing process, when we swap two nodes, it is identical to swapping the corresponding rows in the above matrices. Hence, it is enough if we just update the values of the rows in between the two rows that are being swapped. The pseudo-code for this incremental algorithm is as follows.

```
calc_wire_crossing(R1, R2, M)
    if (R2 < R1)
        return(calc_wire_crossing(R2, R1, M));
    sum = pos = neg = diff = j = 0;
    while (j < NumRows)
        tmp = diff;
        i = R2 - 1;
        while (i > R1)
            sum = sum + M[i][j] * (pos - neg);
```

$$
\begin{aligned}
&\quad\quad diff = diff + M[i][j];\\
&\quad\quad i = i + 1;\\
&\quad sum = sum - M[R_1][j] * (tmp + neg);\\
&\quad sum = sum + M[R_2][j] * (tmp + pos);\\
&\quad pos = pos + M[i][j];\\
&\quad neg = neg + M[R_2][j];\\
&\text{return(sum);}
\end{aligned}
$$

During cell placement, a move is done by randomly choosing a level in the graph and then swapping two randomly chosen gates $[g_1, g_2]$ in that level in order to minimize the total wire length and wire crossing. In our implementation, the initial calculation of the wire length takes $O(n)$ and updating the number of wire crossings takes $O(n^2)$ where $n$ is the number of nodes in a layer of the bipartite graph. In our approach, we initially compute the wire length and the number of wire crossings and incrementally update these values after each move so that the update can be done much faster as illustrated above. This speedup allows us to explore a greater number of candidate solutions, and as a result, obtain better quality solutions. We set the initial temperature such that roughly 50% of the bad moves were accepted. The final temperature was chosen such that less than 5% of the moves were accepted. We used three different cost functions. The first cost function only optimized based on the net wire length. The second cost function evaluated the number of wire crossings, while the last cost function looked at a weighted combination of both. The weights used were the ratio between the wire length and the number of wire crossings obtained in the analytical solution.

## 10.6 Experimental Results

Our algorithms were implemented in C++/STL, compiled with gcc v2.96 run on Pentium III 746 MHz machine. The benchmark set consists of seven biggest circuits from ISCAS89 and five biggest circuits from ITC99 suites due to the availability of signal flow information.

### Zone Partitioning Results

Table 10.2 shows the zone partition results for our QCA placement. The number of partitions is determined such there are $100 \pm 10$ majority gates per partition. We set the capacity of each wire block to 200 QCA cells. We compare acyclic FM [1] and QCA zone partitioning in terms of cutsize, white space, and wire blocks needed *after* zone placement. With QCA partition, we see a 20% improvement in cutsize at the cost of a 6% increase in runtime. A new algorithm was implemented to reduce the number of white space, by taking into account terminal propagation [2]. Our new algorithm for reducing the number of white nodes involves moving wire blocks to balance the variation in the number of

*Table 10.2.* QCA zone partitioning results.

| | Acyclic FM | | | Zone Partitioner | | |
|---|---|---|---|---|---|---|
| name | cut | white | wire | cut | white | wire |
| b14 | 2948 | 151 | 138 | 2566 | 168 | 127 |
| b15 | 4839 | 220 | 260 | 4119 | 144 | 256 |
| b17 | 16092 | 1565 | 1789 | 13869 | 1616 | 1710 |
| b20 | 6590 | 641 | 519 | 6033 | 642 | 518 |
| b21 | 6672 | 599 | 560 | 6141 | 622 | 557 |
| b22 | 9473 | 1146 | 1097 | 8518 | 1158 | 1098 |
| s13207 | 2708 | 143 | 138 | 1541 | 144 | 137 |
| s15850 | 3023 | 257 | 183 | 2029 | 254 | 181 |
| s35932 | 7371 | 875 | 1014 | 5361 | 734 | 1035 |
| s38417 | 9375 | 757 | 784 | 5868 | 775 | 773 |
| s38584 | 9940 | 1319 | 1155 | 7139 | 1307 | 1095 |
| s5378 | 1206 | 34 | 30 | 866 | 34 | 30 |
| s9234 | 1903 | 99 | 81 | 1419 | 104 | 76 |
| Ave | 6318 | 600 | 596 | 5036 | 592 | 584 |
| Ratio | 1 | 1 | 1 | 0.8 | 0.99 | 0.98 |
| time | | 14646 | | | 14509 | |

partitions per clocking level. Although our algorithm results in a 67% decrease in wire nodes and 66% decrease in white nodes, there is a tradeoff in a resulting increase in the number of wire crossings. Since wire crossings have been seen as a much more significant problem, we choose to sacrifice an increase in area for a decrease in the number of wire crossings.

## Zone Placement Results

Table 10.3 details our zone placement results, where we report placement area, wire length, and wire crossings for the benchmarked circuits. We compare the analytical solution to simulated annealing. Comparing simulated annealing to the analytical solution, we see an 87% decrease in wire length and slight increase in wire crossings.

## Cell Placement Results

Table 10.4 shows our cell placement results where we report net wire length and the number of wire crossings for the circuits using our analytical solution and all three flavors of our simulated annealing algorithm. We further tried simulated annealing from analytical start, and the results were identical to the analytical solution. We observe in general that the analytical solution is better than all three flavors of the Simulated Annealing methods, except in terms of wire length in the case of the weighted Simulated Annealing process. But, the

*Table 10.3.* QCA zone placement results.

| name | area | Analytical length | Analytical xing | SA-based length | SA-based xing |
|---|---|---|---|---|---|
| b14 | 20x17 | 81 | 67 | 23 | 67 |
| b15 | 20x24 | 59 | 90 | 34 | 90 |
| b17 | 69x52 | 3014 | 346 | 305 | 345 |
| b20 | 36x36 | 414 | 165 | 99 | 166 |
| b21 | 36x37 | 140 | 172 | 100 | 172 |
| b22 | 48x50 | 1091 | 230 | 188 | 230 |
| s13207 | 18x21 | 28 | 9 | 28 | 9 |
| s15850 | 24x23 | 81 | 16 | 11 | 14 |
| s35932 | 45x44 | 1313 | 64 | 78 | 68 |
| s38417 | 42x43 | 493 | 54 | 48 | 54 |
| s38584 | 55x48 | 1500 | 102 | 110 | 80 |
| s5378 | 10x10 | 3 | 10 | 2 | 9 |
| s9234 | 15x16 | 15 | 11 | 5 | 11 |
| | Ave | 633 | 103 | 79 | 101 |
| | Ratio | 1 | 1 | 0.13 | 0.98 |
| | time | 23 | | 661 | |

*Table 10.4.* QCA cell placement results.

| | Analytical wire | Analytical xing | SA+WL wire | SA+WL xing | SA+WC wire | SA+WC xing | SA+WL+WC wire | SA+WL+WC xing |
|---|---|---|---|---|---|---|---|---|
| b14 | 5586 | 1238 | 28680 | 23430 | 54510 | 3740 | 5113 | 4948 |
| b15 | 9571 | 1667 | 23580 | 40400 | 69030 | 7420 | 8017 | 8947 |
| s13207 | 3119 | 548 | 14060 | 15530 | 30610 | 1450 | 3250 | 1982 |
| s15850 | 3507 | 634 | 18610 | 22130 | 42700 | 2140 | 3919 | 2978 |
| s38417 | 9414 | 1195 | 45830 | 48400 | 80240 | 7320 | 9819 | 9929 |
| s38584 | 19582 | 4017 | 59220 | 75590 | 140130 | 9820 | 20101 | 33122 |
| s5378 | 1199 | 156 | 6280 | 6690 | 13600 | 730 | 1344 | 841 |
| s9234 | 2170 | 205 | 10720 | 11540 | 23290 | 980 | 1640 | 2159 |
| Ave | 4192 | 741 | 16980 | 19950 | 38950 | 2740 | 3880 | 6878 |
| Ratio | 1 | 1 | 4.05 | 26.9 | 9.29 | 3.69 | 0.92 | 9.27 |
| time | 180 | | 604 | | 11280 | | 12901 | |

tradeoff in wire crossings makes the analytical solution more viable, since wire crossings pose a bigger barrier than wire length in QCA architecture.

One interesting note is that when comparing amongst the three flavors of simulated annealing, we find that simulated annealing with wire crossing minimization alone has the best wire crossing number. However, surprisingly, in terms of wire length the simulated annealing procedure with wire length alone as the cost function is not as good as the simulated annealing procedure which optimizes both wire length and wire crossing. We speculate that this behavior is because a lower number of wire crossings has a strong influence on wire length, but smaller wire length does not necessarily dictate a lower number of crossings in our circuits.

## 10.7 Conclusions and Ongoing Work

In this article, we proposed a QCA partitioning and placement problem and present an algorithm that will help to automate the process of design within the constraints imposed by physical scientists. Work to address QCA routing and node duplication for wire crossing minimization are underway. Our ongoing work for zone placement includes a 2D placement solution, where the partitions are placed anywhere in the grid with the help of properly clocked routing channels. The outputs from this work and the work discussed here will be used to generate computationally interesting and optimized designs for experiments by QCA physical scientists. Finally, this work is an example of how systems-level research can positively affect physical device development - and why we should integrate both veins of research. Lastly, during this work it became apparent that a better picture of the QCA circuit design could be painted if we could compare the results from QCA placement to the placement of a CMOS circuit with the same functionality, and our ongoing work focuses on this issue.

## References


[1] J. Cong and S. K. Lim. Performance driven multiway partitioning. In *Proc. Asia and South Pacific Design Automation Conf.*, pages 441–446, 2000.

[2] A. Dunlop and B. Kernighan. A procedure for placement of standard-cell VLSI circuits. *IEEE Trans. on Computer-Aided Design*, pages 92–98, 1985.

[3] M. R. Garey and D. S. Johnson. *Computers and Intractability: A Guide To the Theory of NP-Completeness*, pages 209–210. Freeman, San Francisco, 1979.

[4] S. Hamilton. Taking moore's law into the next century. *IEEE Computer*, pages 43–48, 1999.

[5] K. Hennessy and C. Lent. Clocking of molecular quantum-dot cellular automata. *Journal of Vacuum Science and Technology*, pages 1752–1755, 2001.

[6] R. Ho, K. Mai, and M. Horowitz. The future of wires. *Proceedings of the IEEE*, pages 490–504, 2001.

[7] J. M. Kleinhans, G. Sigl, F. M. Johannes, and K. J. Antreich. GORDIAN : VLSI placement by quadratic programming and slicing optimization. *IEEE Trans. on Computer-Aided Design*, pages 356–365, 1991.

[8] M. Lieberman, S. Chellamma, B. Varughese, Y. Wang, C. Lent, G. Bernstein, G. Snider, and F. Peiris. Quantum-dot cellular automata at a molecular scale. *Annals of the New York Academy of Science*, pages 225–239, 2002.

[9] Carver Mead and Lynn Conway. *Introduction to VLSI Systems*. Addison-Wesley Publishing Company, 1980.

[10] J. Nguyen, R. Ravichandran, S. K. Lim, and M. Niemier. Global placement for quantum-dot cellular automata based circuits. Technical Report GIT-CERCS-03-20, Georgia Institute of Technology, 2003.

[11] M. Niemier and P. Kogge. Exploring and exploiting wire-level pipelining in emerging technologies. In *Proc. Great Lakes Symposum on VLSI*, page Int. Conf. on Computer Architecture, 2001.

[12] M. Niemier and P. Kogge. The 4-diamond circuit: A minimally complex nano-scale computational building block in qca. In *IEEE Sym. on VLSI*, pages 3–10, 2004.

[13] Mike Niemier. The effects of a new technology on the design, organization, and architectures of computing systems. PhD Dissertation, Univ. of Notre Dame, 2003.

[14] P. Packan. Pushing the limits. *Science*, pages 2079–2081, 1999.

[15] J. M. Rabaey. *Digital Integrated Circuits: A Design Perspective*. Prentice Hall Electronics, 1996.

[16] R. Ravichandran, N. Ladiwala, J. Nguyen, M. Niemier, and S. K. Lim. Automatic cell placement for quantum-dot cellular automata. In *Proc. Great Lakes Symposum on VLSI*, 2004.

[17] P. Rutten. Is moore's law infinite? the economics of moore's law. *Kellog Tech Venture*, pages 1–28, 2001.

[18] K. Sugiyama, S. Tagawa, and M. Toda. Methods for visual understanding of hierarchical system structures. *IEEE Trans. Syst. Man,. Cybern*, pages 109–125, 1981.

[19] W. J. Sun and C. Sechen. Efficient and effective placement for very large circuits. *IEEE Trans. on Computer-Aided Design*, pages 349–359, 1995.

IV

# VALIDATION OF NANO-SCALE ARCHITECTURES

# Preface

The tremendous density that we expect from scaling down device sizes leads us to another very important problem. We are already quite far behind in our validation technologies today when total device count is on the order of several millions, and each single device can be assumed to function more or less correctly. The so called "validation gap" in the semiconductor industry translate to about 70% of any design cycle being devoted to validating the correctness of a design.

Some of us believe not only that the tools for complete formal verification are inadequate, but also that the methodologies are not completely developed in order to exploit the full power of formal verification. For example, exploiting modularity, hierarchy, and composability in design may enable us to formally verify larger designs.

The authors of the single chapter in this section address some of these issues, and speculate how to apply various formal verification techniques applicable to large scale systems, to future large-scale nanotechnology based systems.



*S.K. Shukla and R.I. Bahar (eds.), Nano, Quantum and Molecular Computing*, 321.

Chapter 11

# VERIFICATION OF LARGE SCALE NANO SYSTEMS WITH UNRELIABLE NANO DEVICES

Michael S. Hsiao
*Department of Electrical and Computer Engineering*
*Virginia Tech*
hsiao@vt.edu

Shuo Sheng
*Mentor Graphics Corporation*
*Wilsonville, OR*
shuo_sheng@mentor.com

Rajat Arora
*Department of Electrical and Computer Engineering*
*Virginia Tech*
raarora@vt.edu

Ankur Jain
*Haas School of Business*
*University of California at Berkeley*
ankurjain@mba.berkeley.edu

Vamsi Boppana
*Zenasis Technologies*
vamsi@zenasis.com

**Abstract** Any nano-system that designers build must guarantee functional correctness. The sheer scale factor and the added layers of uncertainty in nano-systems demand revolutionary breakthroughs in system design tools and algorithms. Formal



*S.K. Shukla and R.I. Bahar (eds.), Nano, Quantum and Molecular Computing,* 323-351

verification of nano systems, then, must be able to deal with large state spaces, together with the presence of unknowns and uncertainties. The methods described in this chapter present a suite of algorithms that can offer potential in reducing the problem complexity in verification of nano-systems.

**Keywords:** Verification, model checking, unknowns, uncertainties

## 11.1 Introduction

Nanotechnology has brought us the promise of building multi-billion device systems. However, with such minuscule devices, where the logic and architecture are built on *defect-prone* nanostructured devices, verification and validation of such large-scale nano systems will be a nightmare for designers and system integrators.

Conceptually, any nano-system that designers build must guarantee functional correctness as if it was built using classical components composed of CMOS gates. As illustrated in Figure 11.1, let the classical system be the virtual golden model (with multi-billion devices), then the system built using nanotechnology must produce the correct output for every input pattern without being affected by the inherent uncertainties in the system.

While these challenges are not difficult to comprehend, the sheer scale factor and the added layers of uncertainty in nano-systems demand revolutionary breakthroughs in verification tools and algorithms. These technologies must

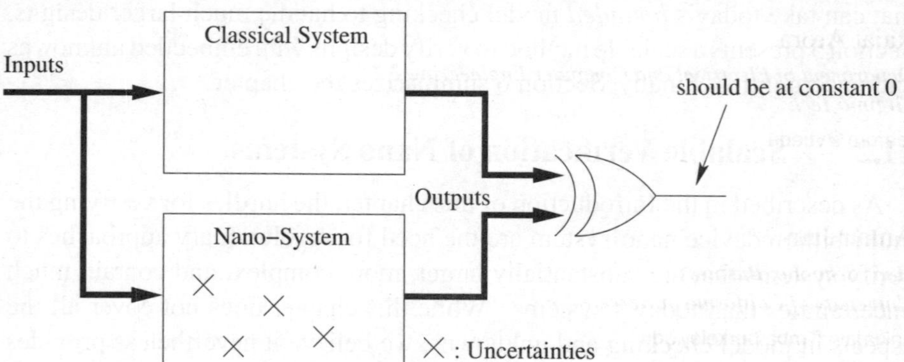


*Figure 11.1.* A robust nano-system should be as reliable as a classical system.

consider and efficiently deal with (1) large state spaces and (2) presence of unknowns and uncertainties within the design. These two characteristics of nano-systems make up two of the fundamental challenges that enabling verification technologies should possess for nano-scale systems.

This chapter presents a suite of recent advances in verification that can significantly reduce the problem complexity as well as handling unknowns/uncertainties within the design. In particular, an automatic test pattern generation (ATPG) based approach has been used to successfully compute the preimage, a key step in unbounded model checking. In addition, this ATPG not only is memory efficient, it can also be computationally superior than traditional BDD-based approaches. The proposed ATPG uses a novel success-driven learning to uncover and avoid previously visited search spaces. Next, a global sequential learning engine that can be very scalable for bounded model checking (BMC) is presented. As the bound in BMC increases, the search space increases as well. While satisfiability (SAT) based BMC solvers have shown success in recent years, the increase in problem size generally translates to exponential increase in computational complexity. Therefore, global learning can play an important role in reducing the SAT complexity. This chapter describes a global learning technique that spans multiple time-frames without the need to explicitly unroll the sequential design. And the learned global constraint clauses can be quickly replicated throughout the BMC instance, making the learning very attractive and effective. Finally, an approach that can verify the design in the presence of unknowns is described. The proposed method offers an efficient and scalable approach to handle uncertainties within the design, offering an exciting and promising solution to nano-system verification.

The rest of the chapter is organized as follows. First, Section 2 gives a brief overview of the verification using model checking (both bounded and unbounded). Next, Section 3 describes a novel method to scale *unbounded* model checking using ATPG-based method. Section 4 discusses approaches that can take today's *bounded* model checking to handle much larger designs. Section 5 presents a scalable method to verify designs with embedded unknowns and uncertainties. Finally, Section 6 summarizes the chapter.

## 11.2 Scalable Verification of Nano Systems

As described in the Introduction of this chapter, the hurdles for verifying the multi-billion device nano-system are the need for revolutionary approaches to verify systems that are substantially larger, more complex, and contain much uncertainties than today's systems. While this chapter does not cover all the aspects in model checking and unknowns, we believe it nevertheless provides an initial attack on the problem of nano-scale verification.

### Brief Overview of Model Checking

Model Checking is a widely accepted formal verification method. It involves the exploration of the design space to see if desired properties are upheld in the design model. If a property is violated, a counterexample can be generated. Symbolic Model Checking [1, 2], based on Reduced Ordered-Binary Decision Diagrams (ROBDDs), has shown to hold promise. However, BDDs are known to suffer from the memory explosion problem, and hence are not scalable for bigger circuits with large number of state variables.

There are two broadly categorized approaches to solving this problem. One category of solutions is by restricting the model checker to an unbounded circuit model, termed unbounded model checking (UMC), in which the property to be checked is not restricted to a given number of time-frames. The other category is the bounded circuit model, termed Bounded Model Checking (BMC), where the verification is performed only within the specified bound on the number of time-frames. We will first discuss the UMC formulation that can significantly achieve orders of magnitude speedup, followed by scalable techniques for BMC, and finally the framework for verification with embedded uncertainties and unknowns in the system.

## 11.3 Scalable Unbounded Model Checking

The advantage of UMC over BMC is that it is complete – it can falsify the property as well as prove that the property holds because of its fixed-point check capability. The disadvantage with conventional UMC is that ROBDD is very sensitive to the variable-ordering. The BDD size can blow up if a bad variable-ordering is chosen. In some cases (e.g. a multiplier unit), no variable-ordering exists that can yield a compact ROBDD representation of the circuit function. In addition, for many problem instances, even the ROBDD for transition relation can be constructed, memory can still easily blow up during the quantification operation. Current research in this area continues in improving BDD algorithms to reduce memory explosion [25] and using abstraction and reduction techniques to reduce model sizes [26].

A complement to BDDs is to use SAT solvers in model checking [27]. The transition relation of a system is unrolled $K$ time frames, allowing any counterexample of length up to $k$ to be found by a SAT solver. The advantage is that SAT solvers are less sensitive to the problem size and do not suffer from space explosion. Recent advances in SAT solver has made it capable of solving very large industrial problems in seconds [36]. However, the major disadvantage is that this methodology can only verify a property within a bounded number ($K$) of transitions. If no counterexample is found in $K$ time frames, nothing can be concluded for $K + 1$ and longer cases. We need to unroll the transition $K + 1$ time frames and do the SAT check again. This procedure could continue infinitely. Therefore, it is called Bounded Model Checking (BMC). BMC is incomplete as compared to BDD-based UMC. It can only find counterexamples but can not prove the property hold unless a bound on the maximum length of counterexample is known.

Since both BDD and SAT solvers have their pros and cons, researchers have been trying to make a perfect combination of the two [28–30]. In [28], the transition relation is represented in a data structure called "Reduced Boolean Circuits". Some simplification method is applied to reuse and manipulate the subformulas. SAT solver is used to do fixed-point checking. However, quantification is still expensive. In [29], a method of combining SAT-solver and BDD for image computation is proposed. The transition relation is represented by a CNF formula. SAT-solver performs a high-level decomposition of the search space and BDD is used to compute all solutions below the intermediate points in SAT decision tree, which is referred to as "BDDs at SAT leaves". As an extended work, a decision heuristic based on separator-set induced partitioning for SAT-solver was proposed in [30], which yielded simpler BDD subproblems. However, these two work still fall into the framework of partitioned BDDs: SAT-solver is just used to compute disjunctive decomposition of the problem and the decomposed problems are handled by BDDs.

On the other hand, a couple of recent works have been focusing on using a pure SAT solver to do unbounded model checking [31, 32]. This nails down an initial attempt to the key problem of how to find an efficient SAT procedure to perform quantifier elimination and thus compute images and preimages. The basic idea in [31, 32] is to modify the SAT procedure such that the solver continues to search for the next solution when a satisfying assignment (called a "solution" in the rest of this paper) is found, as if a conflict occurs. The solutions that have already been found, named "block clauses" in [31] and "excluding clauses" in [32], serve for two purposes: they prevent SAT solver from being trapped into the same solution again and store the information for constructing the final results – the quantified formula. Essentially, the SAT quantification problem is equivalent to the problem of enumerating all solutions to a given SAT problem.

However, while the single-solution SAT problem is NP-complete, the all-solution SAT problem is known to be #P-complete [33]. #P-complete problems are generally harder than NP-complete problems. In fact, the BDD-construction problem is #P-complete. Therefore, by switching from BDD to SAT one does not actually get away from the intrinsic complexity of the problem. The two methods are just a trade-off between space and time. If the hardness of the problem manifest itself as space explosion in BDDs, the corresponding SAT "symptom" is likely to be time explosion. In fact, the case could be worse for SAT. The reason is: most decision procedures employed in SAT-solver are based on branch-and-bound algorithms (e.g., Davis-Putnam, PODEM, etc.); these algorithms are optimized to derive one solution at a time, as opposed to BDD which captures all solutions simultaneously. To use SAT to enumerate all solutions, backtracks are enforced so that the algorithm can continue searching for the next solution when one is found. If the solution set is large, e.g. it contains billions of solutions, then enumerating and storing them one at a time is obviously impossible due to both time and memory limitation. SAT methods, in this case, will suffer from both time and space explosion problems! This phenomenon is termed "solution explosion". This is also the reason the authors of [29] avoid having the SAT-solver run to a leaf node of the decision tree, and instead they let BDDs finish off the decision tree starting at intermediate nodes. The solution explosion problem have not been well addressed in either [31] or [32].

As a result, learning plays an important and critical role in non-BDD-based unbounded model checking for large designs. In particular, recent Automatic Test Pattern Generator (ATPG)-based and SAT-based Unbounded Model Checking [21–23, 13–15, 24], can offer potential memory savings over BDD-based methods. In particular, we will discuss a new technique that centers around learning from "successes".

## The Basic Idea

We now describe a novel ATPG approach that has been developed to address the "solution explosion" problem in computing preimages for UMC. The prototype ATPG algorithm we use is based on PODEM [39], which is a Davis-Putnam-like decision procedure but employs circuit structural information in making decisions. As mentioned before, a naive way of using such a decision procedure to enumerate all solutions is to enforce a backtrack whenever a solution is found so that the algorithm could continue to search for the next solution until the entire search space is exhausted. Various search-space pruning techniques have been proposed [34–37] to improve search efficiency. However, these methods, e.g. conflict-driven learning, dependency-directed non-chronological backtracking, and conflict clauses, target pruning *conflict spaces*, which contains no solution. In other words, they *learn from mistakes*. In the all-solution SAT problem scenario, this is far from being sufficient because a lot of subspaces contain solutions and they can overlap heavily. They do not cause any conflicts with one another. This is explained in Figure 11.2, where multiple solutions are found by PODEM for a preimage computation SAT problem. In the figure, a circle with a label denotes a decision node, which is either a state element or a primary input in a sequential circuit; the label is marked using its gate ID; the left branch corresponds to the decision of 0 and the right branch corresponds to 1; a triangle denotes a conflict subspace (no solution exists); a rectangle marks the terminal node of a solution. Each solution is defined as a cube characterized by the decision path from the root node to a terminal node. There are three solutions under the left branch of decision node 107, marked as Soln#1, Soln#2 and Soln#3. Actually they are marked in the order they are found. The Soln#2 and Soln#3 were found by simply enforcing a backtrack and making ATPG continue after Soln#1 has been found.

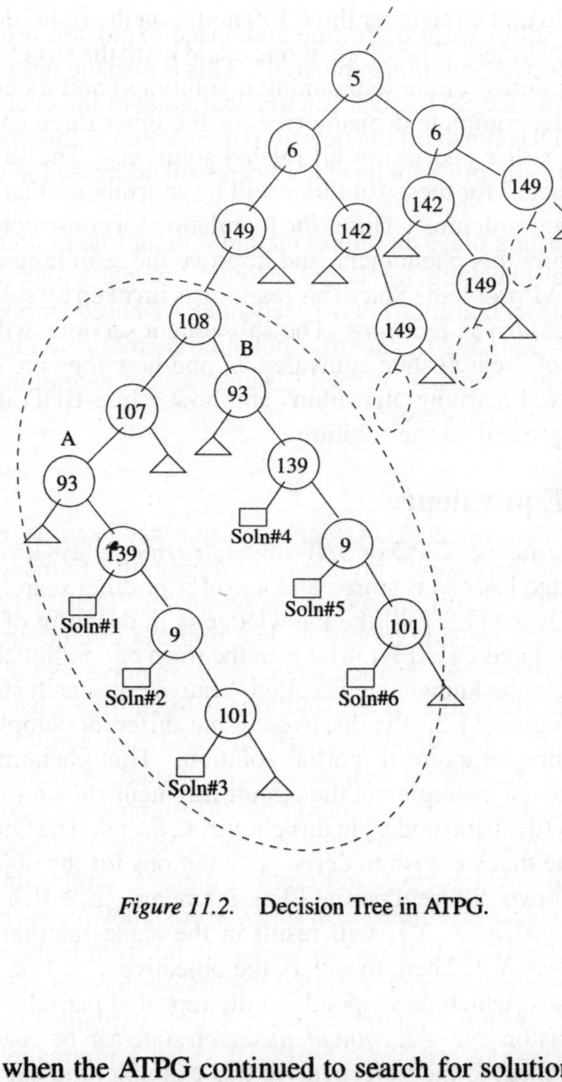


*Figure 11.2.* Decision Tree in ATPG.

However, when the ATPG continued to search for solutions and found the fourth, fifth and the sixth solutions (the three lying under the right branch of node 108, marked as Soln#4, Soln#5 and Soln#6), an interesting phenomenon can be observed: the same *partial* assignments for decision nodes $\{93, 139, 9, 101\}$ are repeated. This implies that the subspace immediately before the two 93 nodes (marked as $A$ & $B$ in the figure) are somehow "equivalent". In other words, earlier node assignments $\{5 = 0, 6 = 0, 149 = 0, 108 = 0, 107 = 0\}$ and $\{5 = 0, 6 = 0, 149 = 0, 108 = 1\}$ resulted the same circuit state. Therefore, if one could learn from the first three solutions, the subspace $B$ can be skipped and the search can directly return solutions #4, #5, and #6. When the search is advanced further, such phenomenon was observed again. It is found that the entire subtree under the left-most decision node 149 (within the big dotted circle) was repeated again under the other three 149 nodes on the right (denoted by the three small dotted circles). Therefore, if one could learn the structure in the area inside the largest dotted circle (containing 6 solutions) and its corresponding "search space", the complete decision tree for the other three can be skipped by backtracking earlier and return all related solutions. The savings for the "enforced backtracks" for these solutions will be enormous when a lot of them have such "repeated structure". This is the foundation for constructing a learning algorithm to capture this phenomena and improve the searching efficiency for the all-solution SAT problem. Since this learning is invoked by solutions found, it is called *success-driven learning*. The subsequent sections will explain the proposed notion of "search state equivalence" and how they are implemented as a "success-driven learning algorithm" and how a free-BDD is constructed accordingly to represent all the solutions.

## Search State Equivalence

In a high-performance ATPG or SAT-solver, *learning* plays a very important role: the knowledge learned is stored and used for pruning search space in the future, e.g. in SOCRATES [38] the knowledge is in the form of implications and in GRASP [35] and CHAFF [36] it is in the form of conflict clauses. In the proposed approach, the knowledge is called "equivalent search state".

As shown in Figure 11.2, it is discovered that different complete solutions may share the same structure of partial solution. This phenomenon can be explained again by an example via the circuit fragment shown in Figure 11.3. There are four PIs (decision nodes) in this circuit: $a, b, c, d$. The OR-gate $z$ is the PO. Let us assume that we wish to derive all solutions for the objective $z = 1$. It is observed that two different partial PI assignments, $\{a = 0, b = X, c = 1\}$ and $\{a = 0, b = 0, c = X\}$, will result in the same internal circuit state $\{g = 0, f = 0, h = X\}$. Then, to satisfy the objective $z = 1$, $d = 1$ needs to be set in both cases, which corresponds to the repeated partial solution at the bottom of the decision tree. An *equivalent* search state can be characterized by its cut set with respect to the objective. In this example, the cut set is simply $\{g = 0, f = 1, d = X\}$, which is a unique *internal* circuit state.

The cut set $\{g = 0, f = 1, d = X\}$ consists of two parts: internal specified gates $\{g = 0, f = 1\}$ and unspecified PI gates $\{d = X\}$. They are obtained by the following procedures: 1. beginning from the objective site $z$ (currently unsatisfied), backtrace to PIs along all the X-paths in its fanin cone; 2. record every specified input to all unspecified gate-output encountered in this depth-first search (in this example, there is only one X-path $z - h - d$, $f = 1$ is the specified input of the unspecified gate $h$ and $g = 0$ is the one for unspecified gate $z$.); 3. record the unspecified PIs at the end of each X-Path ($d = X$ in the example). After performing this depth-first-search all the marked gates

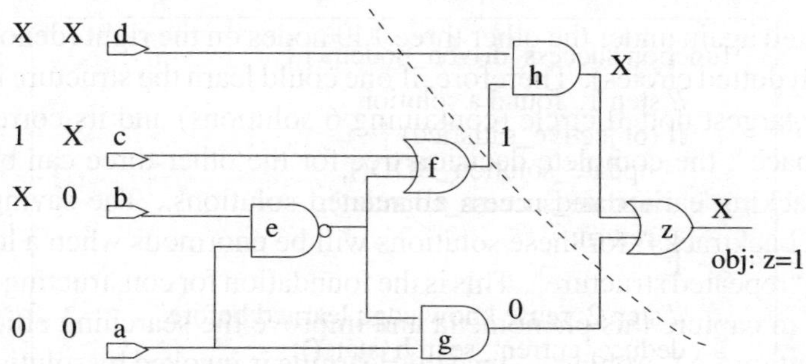


*Figure 11.3.* Search State Equivalence.

(specified gates and unspecified PIs) and their values (1,0,X) define a cut set of the circuit state (shown in dashed line in Figure 11.3). Notice that *this cut set is rather with respect to the objective than to the entire circuit*. Cut sets are stored in a hash table which is managed as the knowledge database. The algorithm uses this database to determine if an equivalent search space is encountered, and if so, it skips this search space.

## Success-Driven Learning

The success-driven learning algorithm controls when to invoke the search state equivalence learning and when to reuse it. It is built into a basic recursive PODEM algorithm that enumerates the entire search space, shown in Figure 11.4. Note that unlike conventional PODEM which returns either SUCCESS or FAIL, this function does not have a return value because a backtrack is always enforced when a solution is found so that it can continue to search for the next solution. However, the number of solutions found under each node in the decision tree must be recorded. A *success-counter* (SC) pair is set up for each decision node in the decision stack. This pair, $SC^0(i)$ and $SC^1(i)$, counts the total number of solutions found for the 0 and 1 branch of node $i$ in the decision tree. For example, in Figure 11.2, for the left most decision node 93 (marked 'A' in figure), $SC^0(93) = 0$ and $SC^1(93) = 3$. The subroutine $update\_success\_counter()$ at step 1 in Figure 11.4 is called every time a solution is found. It increments all the success counters in the current decision stack. Using the same example in Figure 11.2, when the solution at the 0 (left)-branch of the left-most decision node 139 (below point 'A') is found, $update\_success\_counter()$ will increment the following success counters: $SC^0(139), SC^1(93), SC^0(107), SC^0(108), SC^0(149), SC^0(6), SC^0(5)$ and all such counters above decision node 5 till the top of the decision tree. Next, when the solution at the left branch of succeeding decision node 9 is found, $update\_success\_counter()$ will perform the same update again, except that

```
function success_driven_podem() {
    // step 1.  found a solution
    if (objective_satisfied) {
        update_solution_BDD();
        update_success_counter();
        return;
    }

    // step 2. reuse knowledge learned before
    deduce_current_search_state();
    if (lookup_search_state_dateBase()== HIT) {
        update_solution_BDD();
        update_success_counter();
        return;
    }

    // step 3. get a new decision node
    next_decision_node = get_obj();
    if (next_decision_node == NULL) return;

    // step 4. try left branch of the decision node
    next_decision_node = 0;
    if (imply()== CONFLICT)
        nBackTrack++;
    else
        success_driven_podem();
    // step 5. try right branch of the decision node
    next_decision_node = 1;
    if (imply()== CONFLICT)
        nBackTrack++;
    else
        success_driven_podem();
    // step 6. pop decision node out of decision stack
    if (check_success_counter() > 0) {
        update_search_state_datebase();
    } }
```

*Figure 11.4.* Success-Driven Learning Alg.

$SC^1(139)$ rather than $SC^0(139)$ is updated, since it is on the right branch of node 139. In addition, $SC^0(9)$ is initialized to 1. Note that through this mechanism, all the success counters are synchronized as they are dynamically maintained. When a decision node is popped out of the decision stack, it indicates that the subspace below it has been *fully* explored. Therefore, the two success counters for this node should have the *final* numbers of all solutions in the subspace below it. If any of them is greater than 0 (indicating that there exist at least one solution), the search state equivalence learning is performed by computing the corresponding cut set and delete the success counters in the subtree. This operation is performed by subroutine $check\_success\_counter()$ and $update\_search\_state\_datebase()$ at step 6. Because only the equivalent search space is computed when there is at least one success (solution), this algorithm is called "success-driven learning". Note that since success counters are allocated only for decision nodes that are active in the current decision stack and since only PIs and FFs can be decisions in PODEM, the maximum number of success counters managed is $2 \times |PIs + FFs|$.

At step 2 in the algorithm, the function computes the current search state and look it up in the knowledge database. If it detects a $Hit$, then it immediately returns and the entire subspace below is pruned, with solution BDD updated. At steps 4 and 5, the subroutine $imply()$ performs logic simulation and implication when a new decision variable is assigned a specified value. If it detects a conflict, it increments the counter $nBackTrack$ and skips the exploration of the subspace below. At step 6, $update\_search\_state\_datebase()$ creates a new entry in the hash table if a new search state cut set is found. The subroutine $update\_solution\_BDD()$ at steps 1 and 2 constructs the BDD to record every partial solution found by the function. This BDD grows from *partial* to *complete* when the entire search space has been explored and the last decision node in decision stack is popped out.

## Constructing a BDD from the ATPG Decision Tree

To avoid solution explosion, a graph representation of the preimage set is desirable. The main idea comes from the observation that the decision tree in Figure 11.2 resembles a free-BDD, although this BDD may not be canonical.

The BDD node data structure is shown in Figure 11.5. There are six members in the structure. The first is the decision node label, which identifies the PI. The second is the address of the left child. The third is the number of total solutions under the left branch (value passed from the success counter $SC\_0$ during success-driven learning). The next two fields are for the right branch. The last member is the index to the hash table that identifies the search state cut set.

Based on the success-driven learning algorithm, the BDD is built in a *bottom-up* fashion. Whenever a new solution is found, a BDD node is created for the leaf decision node and its associated search state. Its parent nodes are created when the ATPG backtracks to higher levels in the decision tree and detects there are solutions beneath by success counters. Figure 11.6 shows how the solution BDD is constructed in chronological order for the solutions found in the decision tree in Figure 11.2. The BDD grows gradually from a

```
typedef struct _BDD_NODE {
    int decision_node_id;
    int left_branch_successor;
    int num_solutions_in_left_branch;
    int right_branch_successor;
    int num_solutions_in_right_branch;
    int search_state_hash_table_index;
} BDD_NODE;
```

*Figure 11.5.* BDD Node Data Structure.

single node (Figure 11.6(a)) to a complete graph (Figure 11.6(f)), in which each path corresponds to a solution. When a search state equivalence is detected and this knowledge is reused, instead of creating a brand new BDD node, subroutine $update\_solution\_BDD()$ will link the current decision node to an existing BDD node, as shown by those dashed edges in Figure 11.6. Through this reduction, significant space is saved for storing the overlapping portion of solutions.

An interesting question would be what decides the variable ordering for this free-BDD. As a matter of fact, the variable ordering is implicitly decided by the ATPG decision procedure when it picks a new decision variable. It is dynamically generated by the subroutine $get\_obj()$ at step 3 in the function $success\_driven\_podem()$. The function $get\_obj()$ is a standard function in ATPG that uses controllability and observability as heuristics to find next decision PIs through backtracing along X-paths [39]. Notice that unlike SAT-solvers in which the variable that directly satisfies the largest number of clauses is chosen [35], ATPG's decision variable selection is more guided by the structural information of the circuit and the objective. It has been shown in [40] that such testability-based heuristics often yield a good variable ordering for constructing compact shared ROBDDs. Note that although our resulting solution is in the form of a free-BDD, it can be converted to an ordered BDD later if desired.

## Experimental Results

The success-driven learning algorithm together with a basic PODEM were implemented in 5,000 lines of C code. ATPG experiments were conducted on a Pentium 4 1.7 GHz machine with 512MB RAM and Mandrake Linux OS 8.0. The experiments were designed to evaluate the effectiveness of the proposed method and compare its performance to a CUDD-based BDD package, BINGO

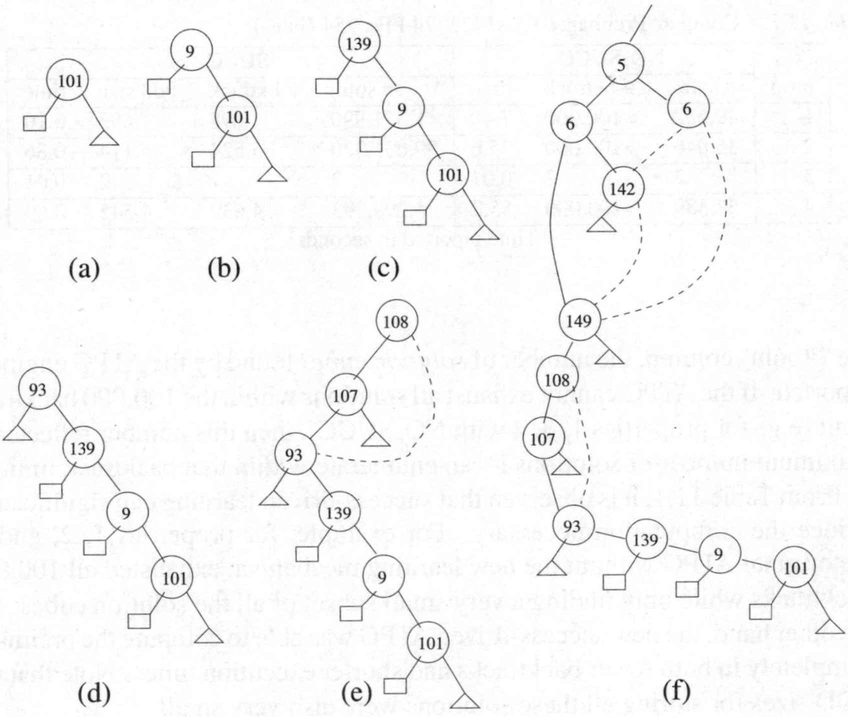


*Figure 11.6.* BDD constructed chronologically.

[41]. The BINGO experiments were conducted on a SUN Ultra-1 200MHz machine with 512MB RAM.

The first set of experiments conducted was for computing 1-cycle preimages for some "properties" of ISCAS benchmark circuits *s1423* which has 74 FFs and 754 gates. The properties used are random conjectures of 10 state variables. Therefore, each property actually specifies a set of target states. The proposed success-driven learning algorithm was applied to compute all states which not only can reach the target states in a single transition but also contain the property themselves. This is a typical step for checking *liveness (EG) properties* in model checking, where the property needs to be included in every state in the path.

Table 11.1 reports the results for *s1423* over four properties. First, the results obtained from the original ATPG without success-driven learning are shown under column **NO_SUCC**; the number of solutions, execution time, and the number of backtracks occurred are first reported. Then, the results for the new ATPG with success-driven learning are shown under column **SUCC**. Here, the resulting BDD size that contains the complete preimage is also reported. Under

*Table 11.1.* Compute Preimages for s1423 (74 FFs, 754 Gates)

| | NO_SUCC | | | SUCC | | | |
|---|---|---|---|---|---|---|---|
| prop | # soln. | # bktrack | time | # soln. | # bktrack | bdd size | time |
| 1 | 49,682 | >100,000 | 7.10 | 1,371,990 | 698 | 693 | 0.10 |
| 2 | 36,046 | >100,000 | 35.6 | 99,631,320 | 5,524 | 5,114 | 0.86 |
| 3 | 2 | 7 | 0.01 | 2 | 7 | 2 | 0.01 |
| 4 | 57,389 | >100,000 | 55.2 | 4,298,293 | 4,639 | 4,545 | 0.69 |

Time reported in seconds

the "# soln" column, the number of *solution cubes* found by the ATPG engine is reported. If the ATPG cannot exhaust all solutions within the 100,000 backtrack limit (e.g. for properties 1, 3, 4 with NO_SUCC), then this number reflects the maximum number of solutions it can enumerate within that backtrack limit.

From Table 11.1, it is observed that success-driven learning can significantly reduce the computation necessary. For example, for properties 1, 2, and 4, the original ATPG without the new learning mechanism exhausted all 100,000 backtracks while only finding a very small subset of all the solution cubes. On the other hand, the new success-driven ATPG was able to compute the preimage completely in both fewer backtracks and shorter execution times. Note that the BDD sizes for storing all these solutions were also very small.

Next, preimages for a larger benchmark circuit, *s5378*, with 3043 gates and 179 FFs are reported in Tables 11.2 and 11.3. Three methods were compared: BINGO [41], ATPG without success-driven learning (NO_SUCC), and ATPG with success-driven learning (SUCC). Note that the results for BINGO are reported in a separate Table (Table 11.2). Again, a backtrack limit of 100,000 is imposed. Since the success-driven learning algorithm constructs a final BDD, the "bdd-size" (number of BDD nodes) and the "mem" (peak memory) columns are also reported for ATPG with SUCCess-driven learning so as to compare with BINGO results.

From Tables 11.2 and 11.3 one can see that success-driven learning significantly reduced the execution time (about 2 to 3 orders of magnitude speedup) for finding all solutions when compared to ATPG without success-driven learning, while being memory efficient than BINGO. For example, for property #2, both NO_SUCC and SUCC found all 7,967 solutions; however, NO_SUCC consumed more than eighteen thousand backtracks and 22.6 sec, while SUCC took only 200 backtracks and 0.42 sec in time and 10Mb in memory to finish the job. BINGO took 14.7 sec and 27Mb memory to do the job. The bdd size for BINGO is 230 nodes while general BDD obtained by the proposed method only contained 139 nodes. For property #6, NO_SUCC could not exhaust all solutions; it only found 22,266 solutions within the 100,000 backtrack limit and took 60.49 sec; SUCC can enumerate all 67,379,200 solutions in 1.03 sec,

*Table 11.2.* Compute Preimages using BDDs for s5378 (179 FFs, 3043 gates)

| | BINGO [41] | | |
|---|---|---|---|
| prop. | bdd size | time(s) | mem |
| 1 | 229 | 14.7 | 27M |
| 2 | 230 | 14.2 | 27M |
| 3 | 219 | 14.7 | 27M |
| 4 | 1689 | 14.8 | 27M |
| 5 | 1075 | 14.7 | 27M |
| 6 | 3064 | 14.8 | 27M |
| 7 | 965 | 14.7 | 27M |
| 8 | 2796 | 14.7 | 27M |
| 9 | 3893 | 14.8 | 27M |
| 10 | 506 | 14.7 | 27M |

*Table 11.3.* Compute Preimages using ATPG for s5378 (179 FFs, 3043 gates)

| | NO_SUCC | | | SUCC | | | | |
|---|---|---|---|---|---|---|---|---|
| prop. | soln. | bktrack | time(s) | soln. | bktrack | bdd size | time(s) | mem |
| 1 | 2,990 | 12,034 | 20.1 | 2,990 | 319 | 250 | 0.62 | 10M |
| 2 | 7,967 | 18,117 | 22.67 | 7,967 | 200 | 139 | 0.42 | 10M |
| 3 | 5 | 67 | 0.08 | 5 | 29 | 23 | 0.07 | 10M |
| 4 | 14,165 | >100,000 | 115.34 | 250,880 | 509 | 283 | 0.51 | 10M |
| 5 | 1,024 | 8,218 | 4.83 | 1,024 | 77 | 76 | 0.1 | 10M |
| 6 | 22,266 | >100,000 | 60.49 | 67,379,200 | 806 | 739 | 1.03 | 10M |
| 7 | 14,630 | >100,000 | 44.18 | 31,928 | 611 | 606 | 1.1 | 10M |
| 8 | 16,331 | >100,000 | 61.54 | 8,630,272 | 3,555 | 3551 | 1.97 | 10M |
| 9 | 4,008 | 19,826 | 30.7 | 4,008 | 517 | 395 | 0.98 | 10M |
| 10 | 20,626 | >100,000 | 77.04 | 22,750 | 1,149 | 494 | 1.71 | 10M |

using only 806 backtracks and 739 nodes to represent these solutions. Note that a separate graph traversal is not needed to obtain those solution numbers - the sum of the two solution counters in the root node in the BDD because they have been synchronized to reflect the number of total solutions in the subtrees. Also, even though ATPG trades off time for space, the ATPG's performance (less than 2 sec. on a 1.7 GHz Pentium machine) was on the same order in execution time with BINGO (about 14 sec on a 200MHz Sun UltraSparc). For all ten properties, the proposed ATPG engine completed the preimage computation using only 10MB for each property, while BINGO required 27MB for each property.

Finally, another similar experiment was conducted for a property of circuit *s38417*, which contains 1636 FFs and 23950 gates. The results are shown in Table 11.4. For this circuit, BINGO failed to construct the transition function

*Table 11.4.* Compute Preimages for s38417 (1636 FFs, 23950 gates)

| | NO_SUCC | SUCC | BINGO |
|---|---|---|---|
| # solutions | 24,938 | 129,171,456 | abort |
| # backtracks | >100,000 | 440 | NA |
| Time (s) | 517 | 0.77 | abort |
| BDD size | NA | 438 | abort |
| Memory (MB) | NA | 187 | > 512 |

due to the 512Mb memory limitation while the proposed method successfully finishing enumerating all 129,171,456 solutions within 0.77 sec, using only 187Mb memory.

## 11.4 Scalable Bounded Model Checking

While BMC can be solved using various methods, including BDDs, ATPG, and SAT, we will focus only on SAT-based BMC, as advances in SAT in recent years have made it extremely effective for such applications.

In SAT-based BMC, given a safety/liveness property, the SAT-solver tries to determine its satisfiability (or unsatisfiability) in the bounded length $k$. The sequential circuit is first unrolled into $k$ time-frames. While unrolling, the D flip-flops in the first time frame are treated as Pseudo Primary Inputs (PPIs); for subsequent time-frames they are converted into buffers and fed to the combinational portion of the sequential circuit. Finally, for the last time-frame these flip-flops are treated as Pseudo Primary Outputs (PPOs). Next, a BMC Circuitry called Monitor Circuit is constructed for the unrolled circuit corresponding to the property to be verified. A CNF database is built for this transformed circuit and the SAT solver is asked to satisfy the Monitor Circuit output to logic 1. For example, consider that the sequential circuit under verification has 6 flop-flops ($S_1S_2S_3S_4S_5S_6$). Suppose the starting/initial state is (101010) and the safety property EF(0X0X10) needs to be verified. A monitor circuit is constructed such that it evaluates to logic 1 if the target state (0X0X10) can be reached in any of the $k$ time frames. The starting state (101010) is also added to the existing CNF as a constraint.

Previously, efforts have been made to compute relations across the circuit to improve the SAT-based BMC. In [3], the authors perform BDD-based approximate reachability analysis and convert this reachability information to clauses added to the original CNF. These clauses in turn restricts the search space of the SAT-solver. In [4], the authors induce signal correlation into the original CNF by locally building up BDDs around the seed node (which is selected statically or dynamically). Every path leading to 0 in such a BDD denotes a conflict, and is added as multi-literal clause to the existing CNF. However, in general, the locally built BDDs are not helpful in extracting global relations among signals across time frames. In [5], equivalence and implication relations are learned within a set of observable variable list and converted to clauses. In [6], incremental learn-from conflict strategy was introduced for SAT-based combinational equivalence checking. All these previous techniques only target relationships among *local* nets; no global learning was performed and the performance improvements would subsequently be limited. Finally, in a recent deduction engine Hypre [7], hyper resolution and equality reduction was performed in a preprocessing step to learn additional clauses; these deduced clauses comprise only a subset of learned clauses compared to the proposed method. In addition, the proposed method explores global sequential relationships that span multiple time frames to achieve orders of magnitude performance gains over other methods. We note that the global sequential learning *never* needs to explicitly unroll the sequential circuit, making our method very attractive and scalable to large designs.

## The formulation to enhance global learning

In [8], non-trivial relations among signals across the sequential circuit are efficiently identified, especially those crossing time-frame boundaries. This is unlike the previous works, where only relationships among signals in the combinational portion of the circuit are learned. Moreover, the sequential relations are learned *without* unrolling the circuit; thus, the global learning is extremely fast. The preprocessing phase involves building the sequential implication graph for the circuit under verification, converting the nontrivial implications into two-literal clauses, replicating these clauses over the entire unrolled circuit, and finally appending the new clauses to the existing CNF database. Figure 11.7 shows the concept of sequential implications in a 5 time-frame unrolled circuit. These implications not only help us to identify relations within the combinational portion of the sequential circuit (of type $a_i \rightarrow b_i$ in the figure), but also the relations spanning multiple time frames (of type $w_i \rightarrow x_j$ and $y_i \rightarrow z_j$) which can play a very significant role. The clauses corresponding to sequential implications crossing time-frame boundaries will involve variables from different time frames. The two-literal clauses resulting from these non-trivial implications can be quickly replicated throughout the k-frame unrolled sequential circuit. For instance, the implication, $w_0 \rightarrow x_1$ in time-frame 1 is replicated as $w1 \rightarrow x2$, $w2 \rightarrow x3$, and so on. Note that the sequential implications crossing time-frames only need to be computed *once* in the proposed approach, and the subsequent replication is applied automatically. This is different from combinational learning on the unrolled circuit, where each relation crossing time-frame boundary (each of $w_i \rightarrow x_j$ and $y_i \rightarrow z_j$) is regarded as a distinct relation and must be learned individually. Finally, these added clauses constrain

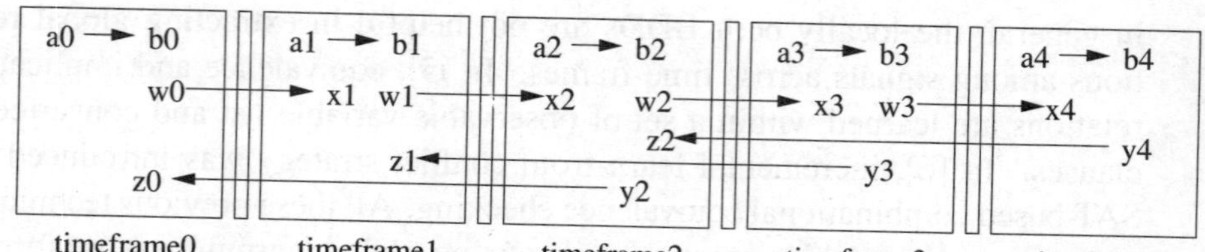


*Figure 11.7.* Global implications in a sequential design

the overall search space of the SAT-solver and provide correlation among the different variables, which enhances the Boolean Constraint Propagation (BCP).

Before going further, a brief overview of how these implications are computed is described. We will use the notation $[a, v]$ to indicate logic value $v$ assigned to node $a$ in the current time-frame 0; the '0' is implicit and is omitted in the notation. $[b, u, t]$ indicates value $u$ assigned to node $b$ in time-frame $t$ (for sequential circuits), and $impl[a, v]$ to denote the implication set of assigning logic value $v$ to node $a$ in the current time-frame of the circuit.

The implications are stored in a directed graph (called implication graph), where each node corresponds to a circuit node assignment $[node, value]$, and each directed edge denotes an implication. Since sequential circuits are considered, some edge may have a non-zero integer weight associated with it, indicating the time frame to which the implication spans; this is similar to the implication graph described in [9–11]. Within the combinational portion of the circuit, all implication edges will have zero edge weights. The non-zero edge weights come in at the flip-flop boundaries only.

The global sequential implication relations that we compute are composed of direct implications and indirect implications. Direct implications of a gate G consist of implications associated with the gates directly connected to G. Such implications are easily computed by traversing through the immediate fanins and fanouts of the gate. When an implication propagates across a D flip-flop, the time frame is incremented or decremented accordingly. Figure 11.8 shows how implications can go beyond the current time frame. Part (a) of this figure illustrates an iterative logic array expansion of a sequential circuit, where four nodes have implication relations as shown (e.g., $[a, 0] \rightarrow [b, 1, 1]$, which is node $b = 1$ in the next time frame). Figure 11.8(b) shows the corresponding implication graph for the four nodes. The transitive law is also reflected by the implication graph. For instance, since $[a, 0] \rightarrow [b, 1, 1]$, and $[b, 1] \rightarrow [c, 0, 2]$, by the transitive law, $[a, 0] \rightarrow [c, 0, 3]$ is obtained. Note that during transitive propagation, the time-frame value is summed across the directed

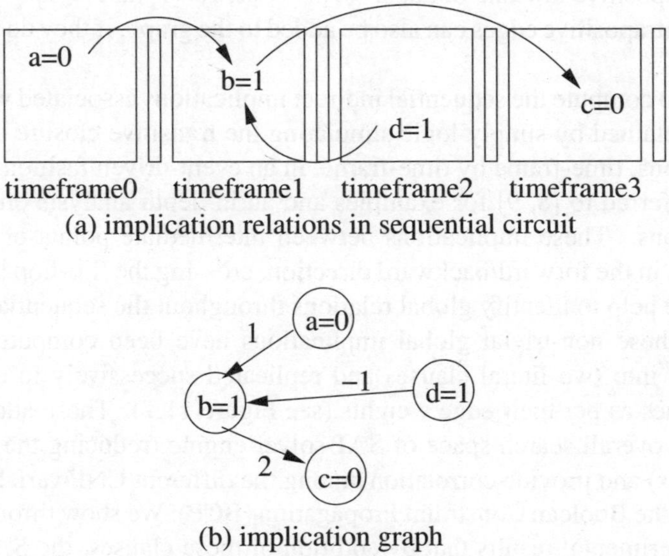


*Figure 11.8.* Implication graph example

edges. In general, the transitive law helps to deduce implication relations with edge weights ranging from $-n$ to $+n$ ($n$ being a whole number). However, in case of a loop, $n$ can be infinity. Hence, we restrict this $n$ in our implementation and make it user-specified. We call this $n$ as *Maximum Edge Weight*.

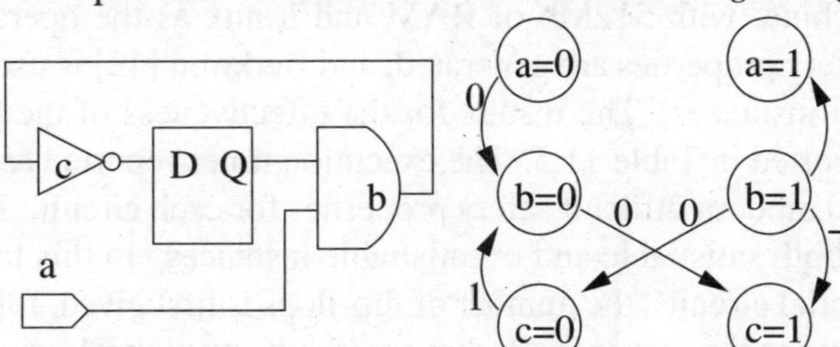


*Figure 11.9.* Sequential implication graph

To illustrate where the non-zero edge weights arise around flip-flops, a simple sequential circuit with its implication graph is used (Figure 11.9). In this example, one can see from the implication graph that $[b, 1] \rightarrow [c, 1, -1]$ directly, and that $[b, 1]$ indirectly implies $[c, 1, 1]$ via the nodes $c = 0$ and $b = 0$.

The contrapositive law states that if $[a, u] \rightarrow [b, v, t]$, then $[b, \bar{v}] \rightarrow [a, \bar{u}, -t]$. Using the example in Figure 11.8, since $impl[a, 0] = \{[b, 1, 1], [c, 0, 2]\}$, using the contrapositive law one obtains: $[b, 0] \rightarrow [a, 1, -1]$ and $[c, 1] \rightarrow [a, 1, -2]$. These contrapositive edges can also be added to the graph, if they do not already exist.

We also compute the sequential indirect implications associated with a node. This is obtained by simply logic simulating the transitive closure of its direct implications, time-frame by time-frame, in an event-driven fashion. The readers are referred to [8, 9] for examples and an in-depth analysis on sequential implications. These implications between intermediate points of the circuit propagate in the forward/backward direction, crossing the flip-flop boundaries, and hence help to identify global relations throughout the sequential circuit.

Once these non-trivial global implications have been computed they are converted into two literal clauses and replicated successively in each of the time-frames as per their edge weights (see Figure 11.7). These added clauses prune the overall search space of SAT-solver engine (reducing the number of backtracks) and provide correlation among the different CNF variables, which enhances the Boolean Constraint Propagation (BCP). We show through preliminary experimental results that by addition of these clauses, the SAT instance complexity is reduced significantly, thereby resulting in orders of magnitude speedup over the conventional approach.

## Preliminary results for bounded model checking with global learning

The proposed concept was implemented in an algorithm called SIMP2C (Sequential Implications to Clauses). All experiments were run on Pentium-4, 1.8GHz machine, with 512Mb of RAM and Linux as the operating system. Arbitrary safety properties are generated, and Berkmin [12] is used as the SAT solver for all instances. The results for the effectiveness of the proposed approach is reported in Table 11.5. The execution times reported are the average on a set of 10 random difficult safety properties for each circuit. These properties include both satisfiable and unsatisfiable instances. In this table, for each of the sequential circuits, the number of flip-flops is first given, followed by the average execution time taken by Berkmin without any global learning. Next, the average execution time taken by the proposed approach including global constraints is reported. In the final column, the speedup attained with the proposed approach is shown.

According to Table 11.5, the proposed method achieved speedups ranging from 7.7X (7700%) for s1423 to 149.0X (14900%) for circuit s9234.1 when applied on the state-of-the-art SAT solver Berkmin. The vast range in speedup is due to the fact that the execution time is both circuit and property dependent. Some properties can be quickly solved by Berkmin alone, whereas some are computationally expensive. For instance, some random properties generated

*Table 11.5.* Bounded model checking with added global constraints

| Ckt. | # FFs | Berkmin [12] | Proposed | Speedup |
|---|---|---|---|---|
| s382 | 21 | 49.51 s | 2.05 s | 24.2 |
| s400 | 21 | 38.30 s | 0.89 s | 43.0 |
| s444 | 21 | 34.25 s | 1.04 s | 32.9 |
| s820 | 5 | 414.60 s | 16.60 s | 25.0 |
| s1423 | 74 | 62.35 s | 8.15 s | 7.7 |
| s1488 | 6 | 714.50 s | 55.72 s | 12.8 |
| s1512 | 57 | 440.30 s | 7.70 s | 57.2 |
| s9234.1 | 211 | 1445.20 s | 9.70 s | 149.0 |
| s38417 | 1636 | 585.12 s | 97.31 s | 16.3 |

Average speedup of 41X over state-of-the-art solver Berkmin [12]

for circuit s1423 were solved very quickly with Berkmin (alone). After the non-trivial global implication clauses were added with the proposed approach, the average time taken by Berkmin would be reduced at a less dramatic effect. Nevertheless, 7700% speedup is very significant considering that other proposed methods generally achieved speedups of less than two-fold. On the other hand, for circuit s9234.1, the average execution time to solve a set of 10 safety properties was reduced from 1445.2 seconds to only 9.7 seconds, thereby achieving a speedup of over one-hundred fold.

Note that the time taken by the proposed pre-processing engine SIMP2C is very low, ranging from 0.14 seconds to 25.12 seconds, thus making this method very attractive - little effort is sufficient to significantly reduce SAT complexity.

The proposed method can also be applied to the combinational equivalence checking problem. Table 11.6 shows the results when Berkmin [12] is used to solve the equivalence checking with and without global learning on a few difficult ISCAS and ITC benchmark instances. Since we are dealing with combinational circuits there is no notion of time frames or sequential implications. Here we compute implication relationships only in the combinational portion of the circuit and in addition to direct and indirect implications we also compute highly non-trivial extended backward implications [19, 9]. In Table 11.6 for each combinational equivalence-checking instance (each original benchmark circuit mitered with its optimized version), the original Berkmin results are reported, followed by the combined SIMP2C and Berkmin, as well as the speedup. Note that the times taken by SIMP2C is included under the third column. According to the experiments, more than 10,000 times speedup can be achieved. For the instance (c6288_eqv), global learning takes a fraction of a second, and those learned clauses significantly affect the way SAT solvers perform the search. Before the added clauses, Berkmin could not finish even after 7200 seconds. But after the global constraints were added, it took only a fraction of a second to complete the search.

*Table 11.6.* Equivalence checking with added global constraints on hard instances

| Instance | Berkmin [12] | SIMP2C+Berkmin | Speedup |
|---|---|---|---|
| c6288_eqv | >7,200.00 s | 0.34 + 0.01 s | 13,478.60 |
| c7552_eqv | 102.31 s | 1.71 + 0.29 s | 51.15 |
| b14_eqv | 417.13 s | 26.05 + 4.37 s | 13.71 |
| b18_eqv | >150,000.00 s | 2132.50 + 527.20 s | 56.39 |

Note: At least one order of magnitude improvement is achieved for each instance

## 11.5 Verification in the Presence of Unknowns and Uncertainties

Given a nano-system composed of billions of devices, many of which may behave in an uncertain manner due to potentially many defects at the nano-scale, functional verification of the design must be able to handle such uncertainties. In other words, in the presence of a defect, one would like to verify if the target system still works to its desired functionality. Allowing "unknowns" in the design verification procedure would be critically useful because of the natural manner in which it supports abstraction of such uncertainties. A side benefit to handling unknowns is the capability to handle black-box verification [17, 18], in which implementations with unfinished blocks can be verified against the specification.

### Modeling unknowns and uncertainties

Satisfiability (or SAT for short), has gained much prominence in the verification community as much stride in SAT solvers has been made in recent years [35, 36, 12]. A method to model unspecified assignments in a Boolean satisfiability framework in order to find minimum-sized prime implicants has been presented [42]. Here, a brief description is given for the proposed technique used to model unknown constraints on arbitrary nodes in a circuit for a Boolean satisfiability solver which determines if a conjunctive normal form (CNF) formula from the propositional logic is satisfiable.

Given a CNF formula, $F = c_1 \wedge c_2 \wedge \ldots \wedge c_m$, where $c_i$ for $1 \leq i \leq m$ is a clause from propositional logic over a finite set $V$ of variables, such that $1 \leq | c_i | \leq 3$ for $1 \leq i \leq m$, an assignment $S : V \rightarrow \{1, 0, X\}$ is said to *satisfy* $F$ if and only if under the assignment, $F$ evaluates to a Boolean 1. If there is no satisfying assignment, $F$ is said to be *unsatisfiable*. Given a circuit, a CNF formula can be readily extracted, where each node in the circuit maps to a unique variable in the formula. The formula evaluates to a Boolean 1 for any valid combination of Boolean values at the inputs and outputs of the nodes in the circuit. For example, the CNF formula of the NAND gate 1 in Figure 11.10

is as follows: $F = (\overline{d} + \overline{a} + \overline{b})(d + a)(d + b)$. The CNF formula for the entire circuit can be expressed as follows:

$$F = (\overline{d} + \overline{a} + \overline{b})(d + a)(d + b)(\overline{e} + \overline{b} + \overline{c})(e + b)(e + c)$$
$$(\overline{f} + \overline{d} + \overline{e})(f + d)(f + e)$$

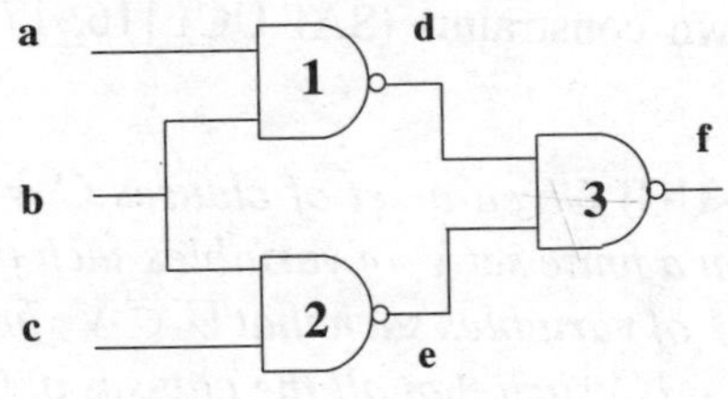


*Figure 11.10.* Example circuit

In the preceding discussion, each node in a circuit is represented by a single variable in the CNF formula for the circuit. But, a single variable is able to encode only two different assignments (Boolean 0 and Boolean 1) to its corresponding node. Therefore, in order to encode an unknown constraint on a node (in addition to a Boolean 0 and a Boolean 1), a new variable encoding scheme in the CNF formula is necessary for that node. Two variables $n^0$ and $n^1$ are used to represent the value assignments to each node $n$ in the propositional logic [16, 17]. Table 11.7 shows one possible encoding scheme for variables representing node $n$.

*Table 11.7.* Encoding scheme for assignments to node $n$

| $n$ | $n^0$ | $n^1$ |
|---|---|---|
| 0 | 1 | 0 |
| 1 | 0 | 1 |
| X | 0 | 0 |

Note that the combination $n^0 = 1$ and $n^1 = 1$ is *illegal* in the encoding scheme given in Table 11.7. Unknown and uncertainty constraints on a node $n$ can be encoded by adding two unary clauses, $\overline{n^0}$ and $\overline{n^1}$, to the CNF formula. Also, since the combination $n^0 = 1$, $n^1 = 1$ is *illegal*, the clause $(\overline{n^0} + \overline{n^1})$ is added to the CNF formula. Consider the NAND gate 1 in Figure 11.10 again. If the node $a$ can have unknown constraints in addition to assignments of Boolean 0 and 1, it has to be represented using two variables in the CNF formula for the gate. To be able to propagate the unknown constraints across gate 1, node $d$ needs two variable representation as well. The CNF formula for the gate can be derived from the truth assignments in the encoded truth table. Further to prevent the illegal value assignment on $a^0$, $a^1$, $d^0$ and $d^1$, clauses $(\overline{a^0} + \overline{a^1})$ and $(\overline{d^0} + \overline{d^1})$ are added to the CNF formula.

## Complexity of SAT with uncertainties

The SAT under unknown constraints (SAT-UC) [16, 17] can be formally defined as:

DEFINITION 11.1 **(SAT-UC)** *Given a set of clauses $C = \{c_1, c_2, \ldots, c_m\}$ from propositional logic on a finite set $V$ of variables such that $1 \leq |\, c_i \,| \leq 3$ for $1 \leq i \leq m$, and a set $U$ of variables such that $U \subset V$, find a set of Boolean assignments for $V' \subseteq (V - U)$ such that all the clauses in $C$ are satisfied.*

THEOREM 11.2 *SAT-UC is NP-complete.*

*Proof:* **(SAT-UC is in NP)** Given an instance of SAT-UC, a non-deterministic algorithm can guess a set of Boolean assignments to the variables in $V'$ and check in polynomial time if the assignments satisfy the clauses in $C$.

*Proof by restriction* is used to show that SAT-UC is NP-complete. If the set $U$ is empty, the SAT-UC problem reduces to the general SATISFIABILITY problem. Hence, under *proof by restriction*, SATISFIABILITY is a special case of SAT-UC. Therefore, SAT-UC is NP-complete. $\diamond$

## Preliminary results on verification with uncertainties

Preliminary results were collected to demonstrate the constrained Boolean satisfiability solutions developed [16, 17]. The setup is to miter each of the ISCAS85 benchmark circuits with itself; however, a copy is injected with a number of **uncertainties**, similar to the setup illustrated in Figure 11.1. Uncertainty constraints were placed at nodes that are at one level (fanout) away from PIs. Such nodes were chosen because these represent the **most difficult cases** of circuit modification for the problem. Results from these experiments are reported in Table 11.8. The constraints were increased starting from zero nodes (no constraints) and increased up to 4 nodes. The columns in Table 11.8 represent the number of constraints, the time required to create a structural modification of the circuit (to encode the unknown constraints), the time required for Boolean comparison and the maximum memory used in the comparison process, respectively.

Although the benchmark instances are small designs, Table 11.8 clearly demonstrates that it is indeed possible to solve constrained versions of the Boolean comparison problem *rapidly* and without extensive memory requirements. It is also evident that the time required for constrained Boolean comparison is larger than the ordinary Boolean comparison without constraints. This

*Table 11.8.* Verifying classical system against nano-system with increasing uncertainty constraints

| Ckt. | # Constr | ModTime (sec.) | BCTime (sec.) | Mem (MB) |
|---|---|---|---|---|
| c880 | 0 | 0.00 | 0.40 | 3.12 |
| | 1 | 0.05 | 0.47 | 3.28 |
| | 2 | 0.08 | 0.50 | 3.36 |
| | 3 | 0.10 | 0.43 | 3.37 |
| | 4 | 0.28 | 5.06 | 5.37 |
| c1355 | 0 | 0.00 | 3.88 | 3.89 |
| | 1 | 0.32 | 5.11 | 5.35 |
| | 2 | 0.37 | 12.33 | 7.21 |
| | 3 | 0.33 | 11.31 | 6.56 |
| | 4 | 0.28 | 8.86 | 6.66 |
| c1908 | 0 | 0.00 | 1.21 | 3.87 |
| | 1 | 0.50 | 1.05 | 6.42 |
| | 2 | 0.22 | 1.61 | 4.64 |
| | 3 | 0.23 | 2.19 | 4.69 |
| | 4 | 0.57 | 0.96 | 5.94 |
| c2670 | 0 | 0.00 | 0.87 | 4.92 |
| | 1 | 0.45 | 1.07 | 5.33 |
| | 2 | 0.38 | 1.08 | 5.36 |
| | 3 | 0.33 | 1.09 | 5.39 |
| | 4 | 0.33 | 1.11 | 5.45 |
| c5315 | 0 | 0.00 | 1.62 | 6.62 |
| | 1 | 0.82 | 1.91 | 7.34 |
| | 2 | 0.83 | 2.06 | 7.37 |
| | 3 | 0.72 | 2.19 | 7.41 |
| | 4 | 0.92 | 2.29 | 7.45 |
| c7552 | 0 | 0.00 | 3.60 | 8.70 |
| | 1 | 0.88 | 12.53 | 23.65 |
| | 2 | 0.80 | 5.96 | 23.95 |
| | 3 | 1.03 | 4.27 | 9.87 |
| | 4 | 1.18 | 4.42 | 9.90 |

occurs due to the fact that the tool used in the experiments uses internal correspondences to verify the required equivalence relation between the two Boolean networks. Nevertheless, it illustrates that the formulation is *very scalable*, in which *the complexity in verification does not increase with increasing number of uncertainties.*

## 11.6 Summary

In this chapter, some of the challenges faced in verifying nano systems have been addressed; in particular, countering the sheer scale factor and the handling of uncertainties. Techniques to extend model checking (both bounded and unbounded) using SAT and ATPG were described, whose results reported one to two orders of magnitude speedup. In addition, a method for verification in the presence of uncertainties was described in which the verification complexity scales well with the number of added unknowns. As we move into the nano era, we believe the methods described in this chapter can offer potential in reducing the both the complexity and cost of nano-system verification.

## 11.7 Acknowledgments

This research has been funded in part by the National Science Foundation under Grants CCR-0196470 and CCR-0305881, and a grant from Fujitsu Labs of America.

## References

[1] J. R. Burch, E. M. Clarke, K. L. McMillan, D. L. Dill, and L. J. Hwang. Symbolic model checking: $10^{20}$ states and beyond. In *Proc. Logic in Computer Science*, pages 428–439, 1990.

[2] K. L. McMillan. *Symbolic Model Checking: An Approach to State Explosion Problem*. Kluwer Academic publishers, 1993.

[3] G. Cabodi, S. Nocco, and S. Quer. Improving sat-based bounded model checking by means of BDD-based approximated traversals. In *Proc. Design Automation and Test in Europe Conf.*, pages 898–903. IEEE/ACM, 2003.

[4] A. Gupta, M. Ganai, C. Wang, Z. Yang, and P. Ashar. Learning from 03 in sat-based bounded model checking. In *Proc. Design Automation Conf.*, pages 824–829. IEEE/ACM, 2003.

[5] Y. Novikov. Local search for boolean relations on the basis of unit propagation. In *Proc. Design Automation and Test in Europe Conf.*, pages 810–815. IEEE/ACM, 2003.

[6] F. Lu, L.-C. Wang, K. Cheng, and R. C.-Y. Huang. A circuit sat solver with signal correlation guided learning. In *Proc. Design Automation and Test in Europe Conf.*, pages 892–897. IEEE/ACM, 2003.

[7] F. Bacchus and J. Winter. Effective preprocessing with hyper-resolution and equality reduction. In *Proc. International Conf. Theory and App. Satisfiability Testing*. IEEE, 2003.

[8] R. Arora and M. S. Hsiao. Enhancing sat-based bounded model checking using sequential logic implications. In *Proc. VLSI Design Conf.*, pages 784-787. IEEE, 2004.

[9] J. Zhao, J. A. Newquist, , and J. Patel. A graph traversal based framework for sequential logic implication with an application to c-cycle redundancy identification. In *Proc. VLSI Design Conf.*, pages 163–169. IEEE, 2001.

[10] M. S. Hsiao. Maximizing impossibilities for untestable fault identification. In *Proc. Design Automation and Test in Europe Conf.*, pages 949–953. IEEE, 2002.

[11] M. Syal and M. S. Hsiao. A novel, low-cost algorithm for sequentially untestable fault identification. In *Proc. Design Automation and Test in Europe Conf.*, pages 316–321. IEEE, 2003.

[12] E. Goldberg and Y. Novikov. Berkmin: a fast and robust sat-solver. In *Proc. Design Aut. and Test in Europe Conference*, pages 142–149. IEEE, 2002.

[13] S. Sheng and M. S. Hsiao. Efficient preimage computation using a novel success-driven ATPG. In *Proc. Design Automation and Test in Europe Conf.*, pages 822–827. IEEE, 2003.

[14] S. Sheng, K. Takayama, and M. S. Hsiao. Effective safety property checking based on simulation-based ATPG. In *Proc. Design Automation Conf.*, pages 813–818. IEEE, 2002.

[15] M. S. Hsiao and J. Jain. Practical use of sequential ATPG for model checking: going the extra mile does pay off. In *Proc. Int'l Workshop High Level Design Validation and Test.*, pages 39–44. IEEE, 2001.

[16] A. Jain. On arbitrary defects: modeling and applications. In *M.S. Thesis*, Department of Electrical and Computer Engineering, Rutgers University, June 1999.

[17] A. Jain, V. Boppana, R. Mukherjee, J. Jain, M. S. Hsiao, and M. Fujita. Testing, verification, and diagnosis in the presence of unknowns. In *Proc. VLSI Test Symp.*, pages 263–269. IEEE, 2000.

[18] C. Scholl and B. Becker. Checking equivalence for partial implementations. In *Proc. Design Automation Conf.*, pages 238-243, 2001.

[19] R. Arora and M. S. Hsiao. Enhancing sat-based equivalence checking with static logic implications. In *Proc. IEEE High-Level Design Validation and Test Workshop*, pages 63–68. IEEE, 2003.

[20] Q. Wu and M. S. Hsiao. Efficient ATPG for design validation based on partitioned state exploration histories. In *to appear Proc. IEEE VLSI Test Symp.*. IEEE, 2004.

[21] B. Li, M. S. Hsiao, and S. Sheng, A novel SAT all-solutions solver for efficient preimage computation In *Proceedings of the IEEE/ACM Design Automation and Test in Europe (DATE) Conference*, pages 272-277. February, 2004.

[22] K. Chandrasekar and M. S. Hsiao, ATPG-based preImage computation: efficient search space pruning With ZBDD In *Proceedings of the IEEE High-Level Design Validation and Test Workshop*, November 2003, pp. 117-122.

[23] V. Boppana, S. P. Rajan, K. Takayama, and M. Fujita, Model checking based on sequential ATPG In *Proc. Computer Aided Verification,* 1999, pp. 418-429.

[24] A. Biere, A. Cimatti, E. M. Clarke, and Y. Zhu, Symbolic model checking without BDDs In *Tools and Algorithms for Analysis and Construction of Systems,* 1999, pp. 193-207.

[25] N. Narayan, J. Jain, M. Fujita, A. Sangiovanni-Vincentelli, Partitioned ROBDDs: A compact, canonical and efficiently manipulable representation for Boolean functions In *Proc. Intl Conf. Computer Aided Design*, 1996, pp. 547-554.

[26] D. Wang, P.-H. Ho, J. Long, J. Kukula, Y. Zhu, T. Ma, and R. Damiano, Formal Property Verification by Abstraction Refinement with Formal, Simulation and Hybrid Engines In *Proc. of Design Automation Conference*, 2001.

[27] A.Biere, A.Cimatti, E.M.Clarke, M.Fujita, and Y.Zhu, "Symbolic Model Checking Using SAT Procedures Instead of BDDs," *Proc. DAC*, 1999, pp.317-20.

[28] P. A. Abdulla, P. Bjesse, and N. Een., Symbolic Reachability Analysis Based on SAT-solvers In *Proc. of TACAS*, 2000.

[29] A. Gupta, Z. Yang, P. Ashar and A. Gupta, SAT-based Image Computation with Application in Reachability Analysis In *Proc. FMCAD*, 2000.

[30] A. Gupta, Z. Yang, P. Ashar, L. Zhang, and S. Malik, Partition-Based Decision Heuristics for Image Computation using SAT and BDDs In *Proc. ICCAD*, 2001, pp. 286-292.

[31] K. L. McMillan, Applying SAT methods in Unbounded Symbolic Model Checking In *Proc. of CAV*, 2002.

[32] H. Kang and I. Park, SAT-based Unbounded Symbolic Model Checking In *Proc. of DAC*, 2003.

[33] D. Roth, On the Hardness of Approximate Reasoning In *Journal of Artificial Intelligence*, 1996.

[34] J. P. Marques and K. A. Sakallah, Dynamic Search-Space Pruning Techniques in Path Sensitization *Proc. DAC*, 1994.

[35] J. P. Marques-Silva and K. A. Sakallah, GRASP: A Search Algorithm for Propositional Satisfiability In *IEEE Trans. Computers*, vol. 48, no. 5, pp. 506-521, May, 1999.

[36] L. Zhang, C. F. Madigan, M. H. Moskewicz and S. Malik, Efficient Conflict Driven Learning in a Boolean Satisfiability Solver *Proc. ICCAD*, 2001, pp. 279-285.

[37] L. Zhang and S. Malik, Towards Symmetric Treatment of Conflicts And Satisfaction in Quantified Boolean Satisfiability Solver In *Proc. 8th Intl. Conf. on Principles and Practice of Constraint Programming*(CP2002), 2002.

[38] M. H. Schulz, E. Trischler and T. M. Sarfert, SOCRATES: A Highly Efficient Automatic Test Pattern Generation System In *IEEE Trans. CAD*, vol.7, no.1, pp. 126-137, 1988.

[39] M. Abramovici, M. A. Breuer and A. D. Friedman, *Digital Systems Testing and Testable Design*, IEEE Press, 1990.

[40] P. Chung, I. N. Hajj and J. H. Patel, Efficient Variable Ordering Heuristics for Shared ROBDD *Proc. ISCAS*, 1993, pp. 1690-1693.

[41] H. Iwashita and T. Nakata, Forward Model Checking Techniques Oriented to Buggy Designs In *Proc. ICCAD*, pp.400-404, Nov. 1997.

[42] J. P. M. Silva, "On Computing Minimum Size Prime Implicants," *Proc. International Workshop on Logic Synthesis*, 1997.

## Biographies

**Rajat Arora** obtained his Bachelor's Degree in Electrical Engineering from Punjab Engineering College, India in Spring 2002. He came to Virginia Tech, USA in Fall 2002 to pursue graduate studies in Computer Engineering. Currently, he is working on his Master's thesis with Dr. Michael Hsiao. His research interests include VLSI Testing and Formal Verification Methods especially Equivalence Checking and Bounded Model Checking.

**R. Iris Bahar** is an Associate Professor in the Division of Engineering at Brown University. She received her Ph.D. in Electrical and Computer Engineering from the University of Colorado, Boulder in 1996. Between 1987 and 1992, Iris worked for Digital Equipment Corporation in their microprocessor design center. She earned her Masters and Bachelors degrees from the University of Illinois, Urbana. Dr. Bahar's research focuses on CAD tools for VLSI, power-aware computer architectures, and architectures and methodologies compatible with future nanoscale technologies. Dr. Bahar is a recipient of the National Science Foundation CAREER Award.

**Debayan Bhaduri** is a Ph.D. student in the Electrical and Computer Engineering Department of Virginia Polytechnic Institute and State University. He received his Bachelor's degree in Computer Science and Engineering from Manipal Institute of Technology, Manipal, India in 2001, and his Masters degree in Computer Engineering in 2004 from Polytechnic Institute and State University. His research interests include defect tolerant computing for nano-architectures, formal verification of probabilistic systems, embedded systems design, modeling and validation of network protocols, software design and high Level Specification of complex control systems.

**Vamsi Boppana** received the B.Tech (Hons) in computer science and engineering from the Indian Institute of Technology, Kharagpur, India, in 1993 and the M.S. and Ph.D. degrees from the department of electrical and computer engineering at the University of Illinois at Urbana-Champaign in 1995 and 1997, respectively. He is a Co-founder and VP, Engineering of Zenasis Technologies, a VLSI design company creating leading-edge automated transistor-level optimization technologies. He has authored or co-authored over 35 technical papers and has seven granted/pending patents. He has served on the program committee and as a session chair for several computer-aided design and test conferences, including the International Conference on Computer-Aided Design. His current research interests include all aspects of VLSI design, test and verification. Dr. Boppana received the Indian Institute of Technology TCS Best Project Award in 1993, the University of Illinois Van Valkenburg Fellowship for demonstrate excellence in research in 1995, the Best Paper Award at the VLSI Test Symposium in 1997, and a Fujitsu Laboratories of America Intellectual Property Contribution Award in 1999.

**Jie Chen** is an Assistant Professor in the Division of Engineering at Brown University. He received his Ph.D. in Electrical and Computer Engineering from University of Maryland, College Park. Between 1999 and 2002, Jie worked for Bell Labs. Later he co-founded Ibiquity Digital Corporation (www.ibiquity.com) and Flarion Wireless Corporation (www.flarion.com). Dr. Chen's multidisciplinary research focuses on wireless multimedia communications (Ultra-wideband communication and its cross-layer design), probabilistic-based design for nanoscale computation, and genomic signal processing. Dr. Chen is a recipient of the IEEE distinguished lecturer award.

**Fred Chong** is an Associate Professor and Chancellor's Fellow in the Department of Computer Science at the University of California at Davis. He received his Ph.D. in 1996 from MIT and is a recipient of the National Science Foundation's CAREER award. His current research focuses on architectures and applications for novel computing technologies.

**André DeHon** received S.B., S.M., and Ph.D. degrees in Electrical Engineering and Computer Science from the Massachusetts Institute of Technology in 1990, 1993, and 1996 respectively. From 1996 to 1999, André co-ran the BRASS group in the Computer Science department at the University of California at Berkeley. Since 1999, he has been an Assistant Professor of Computer Science at the California Institute of Technology. He is broadly interested in how we physically implement computations from substrates, including VLSI and molecular electronics, up through architecture, CAD, and programming models. He places special emphasis on spatial programmable architectures (*e.g.* FPGAs) and interconnect design and optimization.

**Lisa Durbeck** conducts research in next generation computers and is interested in all aspects of them: the design of their hardware, software, operating systems, programming, I/O, and user interfaces; their applicability to new types of problems; their integration into products other than computers; their environmental impact; and their role in transforming society. Durbeck and Nicholas J. Macias founded Cell Matrix Corporation in 1999 to conduct long range research in processing and control systems design, application, and manufacture, in collaboration with researchers at universities and research institutions. Durbeck is the recipient of grants for research in fault tolerant systems and very large scale distributed computing, and an NSF Traineeship in graduate school. She has an M.S. degree in Computer Science from the University of Utah for work in robotics and scientific computing, and an undergraduate education in Systems Engineering from the University of Pennsylvania, Biology from Bryn Mawr College, and Computer Science from U.C. Berkeley. She has co-authored sixteen publications and one patent. Durbeck is also an Adjunct Professor involved in advising graduate student research at Utah State University.

**Diana Franklin** is an Assistant Professor in the Department of Computer Science at California Polytechnic State Institute, San Luis Obispo. She is supported by a Forbes Endowment. She received her Ph.D. in 2002 from UC Davis. Her current research focuses on parallel architectures in traditional and emerging technologies.

**Maya Gokhale** is a Project Leader at Los Alamos National Laboratory leading multiple projects, including the Scalable Reconfigurable Computing project funded by the Lab to explore the use of nano-scale technologies to develop high-performance embeddable processors and scalable approaches for programming them. Gokhale's career experiences span computer companies, academia, and government-sponsored research institutes and she has received multiple awards and patents for her work. Her contributions to the high-performance computing community have been in developing programming models, languages, and software systems for novel high-performance embeddable architectures. Most recently, Gokhale's team for the Deployable Adaptive Processing Systems project has developed a language and compiler targeting embedded systems composed of a heterogeneous collection of reconfigurable-computing and conventional processors. Gokhale has a PhD from the University of Pennsylvania, Philadelphia, and serves as an associate editor for the Engineering of Reconfigurable Systems and Algorithms (ERSA) Conference in addition to serving on the program committees for the ERSA, Field-Programmable Custom Computing Machines (FCCM), and High-Performance Embedded Computing (HPEC) conferences.

**Seth Copen Goldstein** is an associate professor in the Department of Computer Science at Carnegie Mellon University. His research focuses on computing systems and nanotechnology. Currently, he is working on architectures, compilers, and tools for computer systems built with electronic nanotechnology. He believes that the fundamental challenge for computer science in the twenty-first century is how to effectively harness systems which contain billions of potentially faulty components. In pursuit of meeting this challenge he is working on novel circuit techniques, defect and fault tolerance, reconfigurable architectures, scalable optimizing compilers for spatial computing, and self-organizing systems. Dr. Goldstein joined the faculty at Carnegie Mellon University in 1997. He received his Masters and Ph.D. in Computer Science at the University of California at Berkeley. Before attending UC Berkeley, Seth was CEO and founder of Complete Computer Corporation. His undergraduate work was undertaken at Princeton University.

**Paul Graham** is a Technical Staff Member at Los Alamos National Laboratory, where his current research interests include the use of nano-scale technologies to build reliable computing systems and the use of FPGA-based reconfigurable computing in harsh environments. His contributions to the reconfigurable computing community have been in the development of tools for simulating and mitigating the effects of radiation-induced single-event upsets in SRAM FPGAs, the development of hardware debugging strategies and tools for reconfigurable computing applications, and the development of several early reconfigurable computing applications illustrating the benefits of the technology. He is currently a co-principal investigator for the NASA AIST project called Reconfigurable Hardware in Orbit (RHinO) and is involved in the Lab's Scalable Reconfigurable Computing project. Graham received a PhD in Electrical Engineering from Brigham Young University in 2001.

**Michael S. Hsiao** received the B.S. in computer engineering with highest honors from the University of Illinois at Urbana-Champaign in 1992, and M.S. and Ph.D in electrical engineering in 1993 and 1997, respectively, from the same university. Between 1997 and 2001, Michael was an Assistant Professor in the Department of Electrical and Computer Engineering at Rutgers University. Since 2001, he has been an Associate Professor in the Bradley Department of Electrical and Compüter Engineering at Virginia Tech. Michael is a recipient of the National Science Foundation CAREER Award, and he has published more than 90 refereed journal and conference papers in the areas of verification, testing, and power management. He currently serves on the program committee for several of the related conferences and workshops.

**Ankur Jain** has six years of experience in semiconductor testing and CAD tool development. He spent four years at Intel where, as a CAD engineer, he helped resolve critical issues in test and diagnosis of the PentiumT4 microprocessor and its successors. Ankur has a M.S. in Electrical and Computer Engineering from Rutgers University and is currently an MBA student at the Haas School of Business at UC Berkeley.

**Peter Kogge** was with IBM, Federal Systems Division, from 1968 until 1994, and was appointed an IBM Fellow in 1993. In August, 1994 he joined the University of Notre Dame as first holder of the endowed McCourtney Chair in Computer Science and Engineering (CSE). Starting in the summer of 1997, he has been a Distinguished Visiting Scientist at the Center for Integrated Space

Microsystems at JPL. He is also the Research Thrust Leader for Architecture in Notre Dame's Center for Nano Science and Technology. In August, 2001 he became the Associate Dean for Research, College of Engineering. Starting in the fall of 2003, he also is a Concurrent Professor of Electrical Engineering. His current research areas include massively parallel processing architectures, advanced VLSI technology and architectures, non von Neumann models of programming and execution, parallel algorithms and applications, and their impact on computer architecture.

**Sung Kyu Lim** received B.S. (1994), M.S. (1997), and Ph.D. (2000) degrees all from the Computer Science Department of University of California at Los Angeles. During 2000-2001, He was a post-doctoral scholar at UCLA, and a senior engineer at Aplus Design Technologies, Inc. He joined the School of Electrical and Computer Engineering at Georgia Institute of Technology as an assistant professor in August 2001 and the College of Computing as an adjunct assistant professor in September 2002. He is currently the director of the Georgia Tech Computer Aided Design Laboratory. Dr. Lim's primary research focus is on physical design automation for high performance, low-power, and reliable computing systems targeting 3D chip/packages, quantum circuits, field programmable gate/analog array, and microarchitecture design space exploration. He has been on the advisory board of ACM/SIGDA since 2003. He is currently serving the technical program committee of IEEE International Symposium on Circuits and Systems, ACM Great Lakes Symposium on VLSI, and IEEE International Conference on Computer Design. He has been awarded the ACM/SIGDA DAC Graduate Scholarship in June 2003.

**Nicholas Macias** is a Director and co-founder of Cell Matrix Corporation. He holds degrees in Electrical Engineering and Computer Science from George Washington University, in Mathematics from Duke University, and is an Adjunct Professor at Utah State University. He has 25 years experience working in the government and private sectors, and has been working on the Cell Matrix architecture since 1986. His other interests include philosophy, music and bicycling.

**Mahim Mishra** is a graduate student in the Computer Science Department at Carnegie Mellon University. His research focuses on defect tolerance in electronic nanotechnology and end-of-the-roadmap CMOS. Specifically, he is working on test strategies for large reconfigurable fabrics made of future-generation technologies, as well as layout algorithms that are scalable and defect-aware. Mahim obtained an undergraduate degree in Computer Science and Engineering from the Indian Institute of Technology, Kanpur in 2001.

**Joseph L. Mundy** is a Professor of Engineering (research) in the Division of Engineering at Brown University. He received his PhD in Electrical Engineering from Rensselaer Polytechnic Institute in 1969. He joined Brown in 2002 after a career of nearly 40 years at General Electric's Research Center. While at GE he pursued research in a number of fields including, microwave theory, solid state physics, large scale integrated circuit design, and computer vision. He is currently pursing research in object recognition in video, medical image analysis, and nano-scale architecture.

**Michael Niemier** received his Ph.D. from the University of Notre Dame in 2003. He is currently an assistant professor at the Georgia Institute of Technology. His research interests lie in developing circuit designs and architectures for nano-scale devices by working in close conjunction with physical scientists.

**Arijit Raychowdhury** received his B.E. degree in electronics and telecommunication engineering from Jadavpur University, Calcutta in 2001. He has been an analog circuit designer in Texas Instruments India. He is presently pursuing his PhD degree in electrical and computer engineering in Purdue University, IN. His research interests include device/ circuit design for scaled silicon and non-silicon devices. He has received academic excellence awards in 1997,2000 and 2001 and Messner Fellowship from Purdue University in 2002. He has been awarded the 'Best Student Paper" award in the IEEE Nanotechnology Conference, 2003.

**Kaushik Roy** received the B.Tech. degree in electronics and electrical communications engineering from the Indian Institute of Technology, Kharagpur, India, and the Ph.D. degree in electrical and computer engineering from the University of Illinois at Urbana-Champaign in 1990. He was with the Semiconductor Process and Design Center of Texas Instruments, Dallas, where he worked on FPGA architecture development and low power circuit design. In 1993, he joined the electrical and computer engineering faculty at Purdue University, West Lafayette, IN, where he is currently a University Faculty Scholar and Professor. His research interests include Nano-scale device and circuit design, computer-aided design (CAD) with particular emphasis in low-power electronics for portable computing and wireless communications, VLSI testing and verification, and reconfigurable computing. He has published more than 300 papers in refereed journals and conferences, holds five patents, and is a coauthor of a Low-Power CMOS VLSI Circuit Design (New York: Wiley, 2000). Dr. Roy received the National Science Foundation Career Development Award in 1995, the IBM Faculty Partnership Award, the ATT/Lucent Foundation Award, and the Best Paper Awards at the 1997 International Test Conference and 2000 International Symposium on Quality of IC Design, and

Latin American Test Workshop, 2003. He is a Fellow of IEEE.

**Shuo Sheng** received his B.E. in major of automatic control and minor of applied mathematics from Huazhong University of Science and Technology, Wuhan, P.R.China, in 1995. He received his M.E. degree from department of Automation, Tsinghua University, Beijing, P.R.China in 1998 and Ph.D. in computer engineering from Department of Electrical and Computer Engineering, Rutgers University in 2003. Since July 2003, he has been with the Design for Test Division, Mentor Graphics Corporation, Wilsonville, Oregon as a research and development staff engineer. Dr. Sheng's research interest includes automatic test pattern generation and formal verification.

**Sandeep K. Shukla** is currently an assistant professor of Electrical and Computer Engineering at the Virginia Polytechnic and State University. Sandeep earned his Bachelors in Computer Science and Engineering in 1991 from Jadavpur University, Kolkata, India. He received his M.S and Ph.D degrees in Computer Science from the State University of New York at Albany, in 1995 and 1997 respectively. In 2002 Sandeep received the NSF CAREER award for his work on using stochastic specification and verification frameworks to evaluate trade-offs in system design. Sandeep has published almost 50 technical papers in journals and conferences, and he has built his research lab FERMAT. Sandeep received faculty visiting fellowships from University of Birmingham, UK, and INRIA, France. Sandeep has co-edited special issues of ACM Transactions on Embedded Systems, Journal of Circuits and Systems, and Formal Methods in System Design. Sandeep co-chaired the program committee of ACM/IEEE Formal Models and Methods of Co-Design conference in 2003, and is one of the general chairs of the same conference in 2004. He also served on program committee of Design Automation and Test in Europe, IEEE Workshop on High Level Design, Validation and Test in 2003. He also co-chaired workshops such as Formal Methods for GALS Design at the International FME Conference in 2003, , and Formal Approaches to Component Based Embedded Software Design at ETAS 2004.